Energy Crisis

Volume 2

1974-75

Energy Crisis

Volume 2

1974-75

Edited by Lester A. Sobel

Contributing editors: Hal Kosut, Joseph Fickes, Mary Elizabeth Clifford, Chris Hunt, Steve Orlofsky, Maurie Sommer, Gerry Satterwhite

FACTS ON FILE, INC. NEW YORK, N.Y.

Energy Crisis

Volume 2

1974-75

Published by Facts on File, Inc.,
119 West 57th Street, New York, N.Y. 10019.
Library of Congress Catalog Card No. 74-75154
ISBN 0-87196-279-9

9 8 7 6 5 4 3 2 1
PRINTED IN
THE UNITED STATES OF AMERICA

Contents

Foreword

MANY SCIENTISTS, ECONOMISTS and world leaders had been aware for more than 40 years that the world must eventually face the problem of dwindling energy sources. Nevertheless, the situation was largely disregarded until the beginning of the 1970s. And it was not until October 1973, when Arab petroleum producers used their oil as a political weapon, that most people of the developed world seemed to realize the seriousness of the "energy crisis" that had been so quietly developing.

The emergence of this world problem and the steps taken to solve it during the years 1969 through 1973 are recorded in the FACTS ON FILE book *Energy Crisis, Volume 1, 1969–73*. The events of the following year and a half are detailed in this current book, which brings the narrative to about the midpoint of 1975.

The dangers and significance of the situation were discussed by U.S. Secretary of State Henry Kissinger in early 1975 before the National Press Club in a speech that could serve as a brief summary of and comment on the problem presented in these two volumes. "The international and national dimensions of the energy crisis are crucially linked," Kissinger noted. "What happens with respect to international energy policy will have a fundamental effect on the economic health of this nation. And the international economic and energy crisis cannot be solved without purposeful action and leadership by the United States. Domestic and international programs are inextricably linked." Kissinger continued:

> The energy crisis burst upon our consciousness because of sudden, unsuspected events. But its elements have been developing gradually for the better part of two decades. In 1950, the United States was virtually self-sufficient in oil. In 1960, our

reliance on foreign oil had grown to 16% of our requirements. In 1973, it had reached 35%. If this trend is allowed to continue, the 1980s will see us dependent on imported oil for fully half of our needs....

This slow but inexorable march toward dependency was suddenly intensified in 1973 by an oil embargo and price increases of 400% in less than a single year. These actions—largely the result of political decisions—created an immediate economic crisis, both in this country and around the world. A reduction of only 10% of the imported oil, lasting less than half a year, cost Americans half a million jobs and over 1% of national output; it added at least five percentage points to the price index, contributing to our worst inflation since World War II; it set the stage for a serious recession; and it expanded the oil income of the OPEC [Organization of Petroleum Exporting Countries] nations from $23 billion in 1973 to a current annual rate of $110 billion, thereby effecting one of the greatest and most sudden transfers of wealth in history.

The impact on other countries much more dependent on oil imports has been correspondingly greater. In all industrial countries economic and political difficulties that had already reached the margin of the ability of governments to manage have threatened to get out of control.

... If we permit our oil consumption to grow without restraint, the vulnerability of our economy to external disruptions will be grossly magnified.... Unless strong, corrective steps are taken, a future embargo would have a devastating impact on American jobs and production. More than 10% of national employment and output, as well as a central element of the price structure of the American economy, would be subject to external decisions over which our national policy can have little influence....

In a sense, we in America are fortunate that political decisions brought the energy problem to a head before economic trends had made our vulnerability irreversible. Had we continued to drift, we would eventually have found ourselves swept up by forces much more awesome than those we face today. As it is, the energy crisis is still soluble. Of all nations, the United States is most affected by the sudden shift from near self-sufficiency to severe dependence on imported energy. But it is also in the best position to meet the challenge. A major effort now—of conservation, of technological innovation, of international collaboration—can shape a different future for us and for the other countries of the world. A demonstration of American resolve now will have a decisive effect in leading other industrial nations to work together to reverse present trends toward dependency. Today's apparently pervasive crisis can in retrospect prove to have been the beginning of a new period of creativity and cooperation....

The material in this book is taken largely from the printed record compiled by Facts on File in its weekly coverage of world events. Much of the subject matter is highly controversial, but a conscientious effort was made to record all developments without bias and to produce, as far as possible, a balanced and reliable reference work.

Oil & International Tensions

Cooperation Attempts

The Arab oil-producing nations had curtailed petroleum output during the October 1973 Arab-Israeli war and had halted oil shipments to the U.S. and the Netherlands because these two countries had helped the Israelis. Some industrialized nations, their economies suffering from the resulting oil shortage, offered political and economic inducements to the oil producers in hopes of making bilateral deals to assure themselves of adequate oil supplies. And negotiations were started early in 1974 in efforts to bring about cooperation between producers and consumers of petroleum. The Arabs soon restored their oil production to former levels and ended their embargo in 1974 after it became clear that both embargoed nations had found it possible to import sufficient quantities of oil despite the embargo.

U.S. seeks foreign ministers' meeting. Secretary of State Henry A. Kissinger Jan. 10 called on oil-producing and oil-consuming nations to seek a long-term multinational agreement to deal with the energy shortage.

Kissinger's appeal followed an announcement by the White House Jan. 9 that President Richard M. Nixon had asked foreign ministers of eight oil-consuming nations meet in Washington Feb. 11 to discuss world energy problems. Invitations were sent to the heads of government of Britain, Canada, France, Italy, Japan, the Netherlands, Norway and West Germany. Nixon also had sent messages to the 13 states belonging to the Organization of Petroleum Exporting Countries (OPEC), inviting them to join in the discussions with the consumer nations at a later date.

According to the text of the letter to the eight nations made public by the White House Jan. 10, Nixon warned that "the energy situation threatens to unleash political and economic forces that could cause severe and irreparable damage to the prosperity and stability of the world." The way was open either to "progressive division" and "increasing political conflict" or "enlightened unity and cooperation for the benefit of mankind," Nixon said.

The President said the purpose of the foreign ministers' meeting would be to analyze the situation and then "establish a task force" to "formulate a consumer action program." The program, he said, would "deal with the explosive growth of global energy demand" and would "accelerate the coordinated development of new energy sources." According to Nixon, the oil-consuming nations would seek to "meet the legitimate interests of oil-producing countries while assuring the consumer nations adequate supplies at fair and reasonable prices."

Kissinger's remarks on the fuel crisis

were made at a joint news conference with William E. Simon, head of the Federal Energy Office. The secretary said the goal of multilateral agreements lay behind President Nixon's proposal for the Feb. 11 energy conference. Kissinger advised the oil-consuming nations not to seek individual agreements with oil-producers to protect their supplies because such "unrestricted bilateral competition will be ruinous for all countries concerned."

Kissinger also cited the problems of the developing nations, most of whom, he noted, could not pay for the increased price of Arab oil.

EEC members to attend talks—The nine foreign ministers of the European Economic Community (EEC) Jan. 15 accepted President Nixon's invitation to participate in a conference of energy consuming nations.

The ministers, meeting in Brussels Jan. 14–15 as the Council of Ministers, appointed its current president, West German Foreign Minister Walter Scheel, and EEC Commission President Francois-Xavier Ortoli to represent the community. Each of the nine members would also send its own representative to speak for itself. Nixon had sent invitations only to France, Britain, the Netherlands, Italy and West Germany, although the U.S. later said that the four uninvited nations could also attend. Eight of the member nations indicated they would send representatives, while France, which feared that the Arab oil producers would interpret the conference as a consumers' cabal, remained silent on its intention.

France proposes world energy meeting—French Foreign Minister Michel Jobert, in a letter to United Nations Secretary General Kurt Waldheim dated Jan. 18, proposed the urgent convening of a world energy conference under U.N. auspices. Jobert wrote that the conference should aim at determining the general principles of future cooperation between the energy producers and consumers and to devise practical steps likely to achieve such cooperation. The planning would seek to forestall difficulties between nations or groups of nations, and would be of particular interest to developing countries hurt by energy shortages.

U.S. & EEC positions differ—The U.S. and the EEC adopted differing positions for the oil consumers' talks.

In a note sent Jan. 30 to the countries invited to the conference (published by the New York Times Feb. 1), the U.S. called for a "comprehensive cooperative action program" to meet the energy crisis. The proposals, which included restraints on public energy demand and development of additional conventional fuels, branched out to cover intensified economic and monetary cooperation to deal with the consequences of the present situation.

The note proposed that finance ministers, energy officials and science and technology officials attend the conference, along with foreign ministers, and hold their own discussions in separate working groups. The U.S. also suggested establishment of an "international task force of senior officials to prepare for a follow-up foreign ministers conference" that would arrange ways to involve oil producing and underdeveloped countries in the program.

The EEC, setting sharply narrower conference goals in its joint position adopted in Brussels Feb. 5, called for a general review of the energy crisis and of possible areas of cooperation. Yielding mainly to French pressure against the U.S. approach, the EEC said the conference "cannot, above all in its present composition," be transformed into a permanent organization.

The declaration said the conference should "avoid all confrontation" between consumer and producer countries, adding that a dialogue with the oil producers should be begun by April 1.

France announced Feb. 6 that it would participate in the Washington conference, although the government spokesman said France would continue to press for a meeting between EEC and Middle East oil producing nations and a separate United Nations conference on energy and raw materials.

U.S. warns Arabs on embargo. U.S. Defense Secretary James R. Schlesinger said in a TV interview Jan. 7, 1974 that if the Arabs continued their oil embargo against the U.S., the American public might be provoked into demanding force be taken to end the ban. But Schlesinger said he believed the oil producers recognized the problem and would not push too far.

Another warning on the oil boycott was voiced Jan. 8 by Vice President Gerald R. Ford, who said the economic disorder caused by the cutoff might result in a reduction of U.S. food shipments to the Middle East and North Africa. Ford did not threaten a deliberate American move to end food shipments to those regions but emphasized the "circular flow" of the world economy "that requires the cooperation of all to keep things moving."

Kuwait and Saudi Arabia, regarding Schlesinger's remarks as a threat of military intervention, were reported Jan. 9 to have made plans to blow up their oil wells in the event of an American attempt to occupy them. Kuwaiti newspapers quoted Foreign Minister Sheik Ahmed Sabah al-Jaber as saying that mines had been planted near the fields and could be detonated at a moment's notice.

Arab press reports from Riyadh said Saudi Arabia also had wired its oil fields with explosives that would be set off in the event of a U.S. attack.

The chief Arab oil spokesman, Saudi Arabian Petroleum Minister Ahmed Zaki al-Yamani, warned major oil-consuming nations Jan. 12 that any counteraction they might take against the Arab oil embargo would result in international economic "disaster" and a possible head-on "confrontation." Yamani's statement was in reference to the conference of eight oil-consuming nations called by President Nixon Jan. 9.

Speaking at a news conference in Rome on a tour of European capitals, Yamani rejected a suggestion by a Kuwait newspaper that Arab oil producers blacklist all nations that attended the U.S. energy conference.

Yamani said Saudi Arabia was not supplying "a drop of oil" to the U.S. and was boycotting "any refinery that supplies petroleum products" to the U.S.

Yamani reversed himself in Bonn Jan. 17 when he admitted the boycott against the U.S. and the Netherlands was ineffective. Yamani, accompanied by Algerian Industry and Energy Minister Belaid Abdelsalam, made the statement at the conclusion of a three-day visit to the West German capital.

The Organization of Arab Petroleum Exporting Countries (OAPEC) had warned Libya of "serious consequences" of refusing to comply with the organization's oil boycott against the U.S., a Beirut newspaper reported Jan. 5.

The newspaper Al Hayat said that at its meeting in Kuwait, Dec. 25, 1973, the OAPEC had called Libya's attention to the "large quantities" of its oil that was reaching the U.S. despite the embargo. The Libyan delegation had been told that "between 60% and 90% of the oil exported to the United States originated from Libya," the report said.

Libya and Iraq were later reported to oppose proposals by Egyptian President Anwar Sadat that the embargo be lifted.

Libyan Premier Abdel Salam Jalloud was quoted by the government radio Jan. 23 as saying that, contrary to Sadat's views, the U.S. had not changed its policies in the Middle East enough to warrant an end to the ban. If the Arab states abandoned the embargo, Jalloud added, "they would be condemning the oil weapon as a failure."

The Baghdad press reported Jan. 24 that Iraqi President Ahmed Hassan al-Bakr had refused to receive a special envoy sent by Sadat Jan. 22 to explain Egypt's troop disengagement agreement with Israel. The Iraqi government also announced that Bakr had rejected an invitation by President Nixon to attend a world energy conference in Washington with other oil exporters.

Kuwaiti Foreign Minister Sheikh Sabah al-Ahmad al-Jaber Jan. 21 had dismissed as "premature" speculation that the Arabs might ease their oil embargo against the U.S. because of an Israeli-Egyptian pullback agreement. "Lifting the oil measures is still linked to Israeli withdrawal from occupied Arab territories and the restoration of the rights of the Palestinian people," he said.

A U.S. government official in Washington said Jan. 29 that Sadat's appeal to other Arab leaders to lift the oil embargo was in response to a pledge to Secretary of State Henry A. Kissinger that he would urge such a move in appreciation for the secretary's efforts in helping achieve the Israeli-Egyptian troop disengagement agreement. Sadat had informed Arab leaders during his eight-nation tour Jan. 18–23 that Kissinger's mediation was a reflection of change in

American policy in the Middle East that warranted a shift in Arab oil policy, the U.S. official said.

Kissinger said Feb. 6 that the U.S. had been led to expect an end to the embargo in view of recent American peace efforts in the Middle East. The continuance of this ban "under these conditions must be construed as a form of blackmail and would be considered highly inappropriate by the United States and cannot but affect the attitude with which we will have to pursue our diplomacy," he said.

This was the first time Kissinger had used the word blackmail to describe the embargo.

President Nixon had said in his State of the Union message Jan. 30 that he had been assured by friendly Arab leaders that a meeting would be held to discuss the lifting of the embargo.

Kissinger's remarks followed subsequent reports from the Middle East that the Arabs had no intentions at that time to resume the flow of oil to the U.S.

The secretary also sought to assure oil producers and some European countries that were apprehensive about the forthcoming international energy conference to be held in Washington Feb. 11–12. Kissinger discounted as untrue reports that the U.S. would try to use the parley "to organize a consumers' cooperative" or "to produce a confrontation with the producers." He also said the conference did not reflect U.S. desire to "neglect the less developed countries."

Kissinger reiterated his criticism of bilateral oil deals, saying "they tend to support a price which is ruinous to nations making them and generally to the world economy." This, he said, pointed up the fact that "all nations have a common problem of how to insure their supply and how to relate their crisis to the structure of the world economy."

Zaki al-Yamani had said Feb. 5 that he had "no knowledge of any Arab country promising to lift the embargo against the United States." Yamani's statement was in reply to a question about President Nixon's Jan. 30 announcement that an early meeting of Arab oil producers would end the boycott. As for Saudi Arabia's intentions of lifting the ban against the U.S., Yamani said this would have to be decided jointly "at the meeting of Arab oil ministers." This was a reference to a conference to be held in Tripoli, Libya Feb. 14.

Yamani spoke in Lebanon where he was on a tour to explain Arab oil policies.

Beirut newspapers reported Feb. 4 that Saudi Arabia and Kuwait had assured Syrian President Hafez al-Assad that they would continue the oil stoppage against the U.S. until Israel and Syria agreed to disengagement of their forces on Syrian terms. Assad was given the pledge in meetings with Saudi Arabian King Faisal in Riyadh Feb. 2 and in talks with officials in Kuwait Feb. 3.

OPEC delays price changes. The Organization of Petroleum Exporting Countries (OPEC) announced Jan. 9 after a three-day meeting in Geneva that there would be no "increase or decrease in the [basic] price of crude oil until April 1."

Iranian Finance Minister Jamshid Amouzegar, who issued that statement, said the OPEC would raise its prices after April 1 if import costs continued to rise. According to Amouzegar, a four-point agreement reached at the OPEC meeting provided for continued pricing studies, a decrease in the fixed relationship between posted prices and actual market prices, possible creation of a development bank to assist underdeveloped countries to pay their higher oil bills and a renewed plea to consumer nations to combat inflation.

Algeria Jan. 17 increased posted prices of its oil by 75%, effective Jan. 1. The boost raised the price of a barrel from $9.25 to $16.21.

Shah says U.S. continues to get oil. Shah Mohammed Riza Pahlevi of Iran said Feb. 24 that the U.S. was importing at least as much oil as it did before the Arabs imposed their embargo during the October 1973 war in the Middle East and indicated that the American oil industry may have deliberately manipulated gasoline shortages in the U.S. to increase their profits. The charge was disputed by U.S. officials.

The shah's statement was made to interviewer Mike Wallace and broadcast on WCBS-TV. Suggesting that the U.S. may be getting oil surreptitiously, the Iranian leader told of tankers changing their destinations "two or three times" in mid-ocean and of oil being sold for one destination and being delivered somewhere else.

The shah predicted that oil-producing countries would raise their prices further but the extent depended on whether the industrial consumers adopted "a hostile or a cooperative attitude." He said the degree of the price increase would also depend on whether the industrial nations would control their inflation. "The bills [for their goods] that they are presenting us every day is something ridiculous," the shah said.

In another interview broadcast by NBC Feb. 24, the monarch warned that the world's oil resources would be depleted in 30 years unless other sources of energy were found to heat homes and operate trains.

Shah offers $1 billion for poor lands—The shah proposed Feb. 21 the creation of a new development fund to ease the problems of poorer countries resulting from high oil prices. He pledged more than $1 billion to the fund.

The shah made the offer at a meeting in Teheran with World Bank President Robert McNamara and Hendrikus Johannes Witteveen, managing director of the International Monetary Fund. The two financial bodies would work closely with the proposed development fund, which would receive annual contributions of $2 billion–$3 billion from oil-exporting countries and the major industrial nations. Iran's proposal was to be submitted to the next meeting of the Organization of Petroleum Exporting Countries.

Saudi Arabian Petroleum Minister Ahmed Zaki al-Yamani said Jan. 27 that King Faisal was preparing to "take very important steps" aimed at reducing the price of crude oil to avoid harming the world economy.

Yamani made the statement in Tokyo where he had arrived Jan. 26 on a five-day mission to explain his government's oil policy to Japanese officials. He told a news conference that although Saudi Arabia regarded current prices as "fair and reasonable," it was concerned that the "present prices of oil will create some serious problems in the balance of payments of so many nations, whether they are developing nations or industrialized nations."

Algerian Industry and Energy Minister Belaid Abdelsalam, who accompanied Yamani, expressed disagreement and said he was opposed to any cut in petroleum prices. He said the current high rates were necessary in view of Algeria's balance of payments deficit and its heavy indebtedness to foreign countries.

The two Arab officials held discussions Jan. 27 with Japanese leaders, including Deputy Premier Takeo Miki and Finance Minister Takeo Fukuda.

Yamani said Jan. 28 that any cut in oil prices must be taken jointly by the oil producers. "If we can convince the others, we will reduce our prices," Yamani said. He reiterated his warning against participation by other countries in the forthcoming U.S.-sponsored world energy conference to be held in Washington Feb. 11. Yamani said Japan and other oil-consuming nations "cannot afford any sort of confrontation."

Before leaving Tokyo Jan. 30, Yamani disclosed that as a result of his talks with Japanese leaders, both countries were nearing agreement on a bilateral treaty of economic and technical cooperation.

Arabs vs. users' talks. The scheduled oil users' conference was criticized Feb. 10 by the major Arab oil-producing states of Libya, Kuwait and Saudi Arabia.

Tripoli radio called the meeting "an aggressive act against the oil-producing states, particularly the Arab states." It said the parley was "an American trap to sanction an American tutelage in Europe and internationalize oil resources by means of force."

Kuwaiti Petroleum Minister Abdel Rahman Atiki said his government approved of the conference in principle, but opposed "the idea of allowing opposing blocs to discuss such an important topic."

Prince Saud of Saudi Arabia said "if the aim of the Washington conference is to mount pressure on the producing states, then we do not think it will be fruitful." He questioned whether the meeting would produce a unified position of the oil consumers in view of "the discrepancy in the attitudes of the participants."

Algerian President Houari Boumedienne had said Feb. 4 that the conference was "directed toward creation of an imperialist protectorate over energy resources." He said it would bring no results "because the oil-producing states are opposed to it."

The Soviet Communist Party newspaper Pravda described the conference Feb. 11 as a U.S. attempt "to create a coalition of industrial states against the oil-exporting developing countries."

Arabs cancel oil conference—The Arab oil-producing conference scheduled to open in Tripoli, Libya Feb. 14 was suddenly canceled indefinitely.

The official Libyan press agency reported Feb. 13 that the decision was taken "at the request of Saudi Arabia and Egypt." No official clarification for the action was given, but diplomatic sources in Cairo said Egypt and Saudi Arabia needed more time to persuade other Arab leaders that the oil embargo against the U.S. must be lifted soon. The oil ban and whether to restore the 15% cut in oil production were to have been discussed at the Tripoli meeting.

Oil Users' Conference

Joint action planned. A U.S. proposal for international cooperation to combat the world's energy crisis was endorsed in a communique adopted Feb. 13, 1974 at the end of a three-day conference in Washington of 13 major oil-consuming nations. France, one of the participants, objected strenuously to several key points of a 17-point proposal.

Attending the U.S.-sponsored meeting were the foreign ministers of Belgium, Britain, Canada, Denmark, France, West Germany, Ireland, Italy, Japan, Luxembourg, the Netherlands, Norway and the U.S. The finance ministers of all those countries, except France, also were present.

The proposals opposed by France:

- "A comprehensive action program to deal with all facets of the world energy situation." This included several U.S. proposals that called for joint measures in conserving energy, in allocating oil supplies in times of emergency and in speeding research for new energy sources.
- Establishment of a coordinating group to prepare for a conference of oil-producing and oil-consuming countries "at the earliest possible opportunity."
- Formation of a coordinating group of senior officials "to coordinate the development of the actions" recommended by the conference.
- Adoption of financial and monetary measures to avoid "competitive depreciation and the escalation of restrictions on trade and payments or disruptive actions in external borrowing."

French Foreign Minister Michel Jobert criticized the conference, asserting that energy matters were used by the U.S. as "a pretext." The parley's real purpose, he said, was a "political" desire by the U.S. to dominate the relationships of Western Europe and Japan.

Despite France's objections to the key elements of the communique, U.S. Secretary of State Henry A. Kissinger called the meeting a success and possibly a major step toward "dealing with world problems cooperatively."

The conference had opened Feb. 11 and was scheduled to end the following day. Its conclusion, however, was delayed a day because of a dispute between France and its EEC allies on how to respond to the U.S. proposal for joint efforts to cope with the energy crisis. Delegates spent Feb. 12 behind closed doors in an unsuccessful effort to work out a compromise that would be entirely acceptable to France. Among the principal stumbling blocks was formation of the coordinating group that would prepare for further conferences. It was incorporated into the final communique over Jobert's objections.

Jobert had said in a speech at the opening session Feb. 11 that the conference should not take specific action, that it should be confined to merely an exchange of views. He criticized West German Finance Minister Helmut Schmidt for going too far in supporting the U.S. position at the meeting. Jobert also was said to have rebuked the West German and British delegates for not taking a unified stand, as members of the EEC, on how to approach the energy conference.

The final communique adopted was based largely on a seven-point program Kissinger had submitted in an address to the Feb. 11 session. Calling his plan "Project Interdependence," the secretary said the U.S. was prepared to share its technological expertise with others in the development of new energy sources and in arranging for improved cooperation between oil producers and consumers.

The proposals set forth by Kissinger:

- The U.S. would join other consumer nations in studying conservation methods to cut down on the use of energy.
- In order to cushion the impact on major industrial nations, an international program would be set up to handle any interruption in, or manipulation of fuel supplies.
- The U.S. would make "a major contribution" toward international research and development of new technology in energy programs.
- The U.S. would be willing to share its energy with other consumers in times of shortages if those nations would offer to do the same.
- International cooperation, including "new mechanisms," must be carried out to deal with distribution of capital from oil revenues.
- New foreign aid programs were required to help less wealthy consumers that suffered even more from high oil prices.
- Consumers and producers should discuss what constituted a "just price" for oil and how to insure long-term investments.

Kissinger said his proposals were aimed at making certain of future "abundant energy at reasonable cost to meet the entire world's requirements for economic growth and human needs."

The secretary took note of Foreign Minister Jobert's opposing position toward the conference. He said: "Will we consume ourselves in nationalistic rivalry which the realities of interdependence make suicidal? Or will we acknowledge our interdependence and shape cooperative solutions."

At a dinner in honor of the foreign ministers later Feb. 11, President Nixon emphasized that bilateral deals caused the price of oil to increase. He said that while such transactions might be "good short-term politics," they were "bad long-term statesmanship."

*Text of Feb. 13 Communique of Energy Conference**

Summary Statement

1. Foreign Ministers of Belgium, Canada, Denmark, France, the Federal Republic of Germany, Ireland, Italy, Japan, Luxembourg, The Netherlands, Norway, the United Kingdom, the United States met in Washington from Feb. 11 to 13, 1974. The European Community was represented as such by the president of the council and the president of the commission. Finance ministers, ministers with responsibility for energy affairs, economic affairs and science and technology affairs also took part in the meeting. The secretary general of the OECD also participated in the meeting. The ministers examined the international energy situation and its implications and charted a course of action to meet this challenge which requires constructive and comprehensive solutions. To this end they agreed on specific steps to provide for effective international cooperation. The ministers affirmed that solutions to the world's energy problem should be sought in consultation with producer countries and other consumers.

Analysis of Situation

2. They noted that during the past three decades progress in improving productivity and standards of living was greatly facilitated by the ready availability of increasing supplies of energy at fairly stable prices. They recognized that the problem of meeting growing demand existed before the current situation and that the needs of the world economy for increased energy supplies require positive long-term solutions.

3. They concluded that the current energy situation results from an intensification of these underlying factors and from political developments.

4. They reviewed the problems created by the large rise in oil prices and agreed with the serious concern expressed by the International Monetary Fund's Committee of Twenty at its recent Rome meeting over the abrupt and significant changes in prospect for the world balance of payments structure.

5. They agreed that present petroleum prices presented the structure of world trade and finance with an unprecedented situation. They recognized that none of the consuming countries could hope to insulate itself from these developments, or expect to deal with the payments impact of oil prices by the adoption of monetary or trade measures alone. In their view, the present situation, if continued, could lead to a serious deterioration in income and employment, intensify inflationary pressures, and endanger the welfare of nations. They believed that financial measures by themselves will not be able to deal with the strains of the current situation.

6. They expressed their particular concern about the consequences of the situation for the developing countries and recognized the need for efforts by the entire international community to resolve this problem. At current oil prices the additional energy costs for developing countries will cause a serious setback to the prospect for economic development of these countries.

General Conclusions

7. They affirmed, that, in the pursuit of national policies, whether in the trade, monetary or energy fields, efforts should be made to harmonize the interests of each country on the one hand and the maintenance of the world economic system on the other.

Concerted international cooperation between all the countries concerned including oil producing countries could help to accelerate an improvement in the supply and demand situation, ameliorate the adverse economic consequences of the existing situation and lay the groundwork for a more equitable and stable international energy relationship.

8. They felt that these considerations taken as a whole made it essential that there should be a substantial increase of international cooperation in all fields. Each participant in the conference stated its firm intention to do its utmost to contribute to such an aim, in close cooperation both with the other consumer countries and with the producer countries.

9. They concurred in the need for a comprehensive action program to deal with all facets of the world energy situation by cooperative measures. In so doing they will build on the work of the OECD. They recognized that they may wish to invite, as appropriate, other countries to join with them in these efforts. Such an action program of international cooperation would include, as appropriate, the sharing of means and efforts, while concerting national policies, in such areas as:

- *The conservation of energy and restraint of demand.*
- *A system of allocating oil supplies in times of emergency and severe shortages.*
- *The acceleration of development of additional energy sources so as to diversify energy supplies.*
- *The acceleration of energy research and development programs through international cooperative efforts.*

10. With respect to monetary and economic questions, they decided to intensify their cooperation and to give impetus to the work being undertaken in the IMF, the World Bank and the OECD on the economic and monetary consequences of the current energy situation, in particular to deal with balance of payments disequilibria. They agreed that:

- *In dealing with the balance of payments impact of oil prices they stressed the importance of avoiding competitive depreciation and the escalation of restrictions on trade and payments or disruptive actions in external borrowing.*
- *While financial cooperation can only partially alleviate the problems which have recently arisen for the international economic system, they will intensify work on short-term financial measures and possible longer-term mechanisms to reinforce existing official and market credit facilities.*
- *They will pursue domestic economic policies which will reduce as much as possible the difficulties resulting from the current energy cost levels.*
- They will make strenuous efforts to maintain and enlarge the flow of development aid bilaterally and through multilateral institutions, on the basis of international solidarity embracing all countries with appropriate resources.

11. Further, they have agreed to accelerate wherever practicable their own national programs of new energy sources and technology which will help the overall worldwide supply and demand situation.

12. They agreed to examine in detail the role of international oil companies.

13. They stressed the continued importance of maintaining and improving the natural environment as part of developing energy sources and agreed to make this an important goal of their activity.

14. They further agreed that there was need to develop a cooperative multilateral relationship with producing countries, and consuming countries that takes into account the long-term interests of all. They are ready to exchange technical information with these countries on the problem of stabilizing energy supplies with regard to quantity and prices.

15. They welcomed the initiatives in the U.N. to deal with the larger issues of energy and primary products at a worldwide level and in particular for a special session of the U.N. General Assembly.

Coordinating Group

16. They agreed to establish a coordinating group headed by senior officials to direct and to coordinate the development of the actions referred to above. The coordinating group shall decide how best to organize its work. It should:

- *Monitor and give focus to the tasks that might be addressed in existing organizations.*
- *Establish such ad hoc working groups as may be necessary to undertake tasks for which there are presently no suitable bodies.*
- *Direct preparations of a conference of consumer and producer countries which will be held at the earliest possible opportunity and which, if necessary, will be preceded by a fourth meeting of consumer countries.*

17. They agreed that the preparations for such meetings should involve consultations with developing countries and other consumer and producer countries.

*Italicized matter represents sections rejected by France.

Financial aspects of oil price rise—The U.S. Feb. 14 released the text of Treasury Secretary George P. Shultz's address to the Washington energy conference.

In his speech on the world financial ramifications of soaring oil prices, Shultz warned that current price levels "spell misery and even starvation" to developing countries unable to pay for needed oil. Since industrial nations already faced severe payments crises because of higher oil prices, and "can't be expected to pay for the cost of increased oil" for developing nations, Shultz said, "that responsibility must fall primarily on the oil producers."

Any basic solution to the financial dilemma confronted by industrial and developing nations alike, Shultz said, involved a reduction in oil prices.

"Cooperation is essential . . . [but it] is not a substitute for changing the problem so that it is more manageable. There is no way to print up money and use it to 'paper over' a real problem," Shultz declared. To deal with the special problems of developing countries, Shultz said industrial nations must maintain their "historical levels of aid," but that the newly affluent oil-producing countries should "take immediate steps to greatly expand their aid programs." He also proposed establishment of "a kind of multinational joint venture," a mutual fund in which Arabs could safely invest their surplus wealth and also facilitate the redistribution of their profits for the benefit of poorer nations hardest hit by oil price increases. Investor and recipient nations would share in the management of the organization, according to Shultz's plan.

Other U.S. suggestions for dealing with money problems arising from the oil price crisis included increased use of reciprocal currency arrangements or swaps among central banks, increased Arab involvement in international financial organizations, a "rechanneling" of World Bank money from oil-producing states to poorer countries and early repayment of existing bank loans by petroleum-exporting nations.

Western energy group meets. The energy group established by the 13-nation international energy conference in Washington Feb. 13 held its first meeting there Feb. 25. This first meeting of the Energy Coordination Group (ECG) was limited to the single day and to minor matters only.

The meeting dealt only with procedural matters. It transferred some problems to the energy committee of the Paris-based Organization for Economic Cooperation and Development and decided to establish committees to study a code of conduct for transactions between oil-consuming and oil-producing states and for settling criteria for oil prices.

The U.S. had sought to have the session deal immediately with the world energy crisis, but deferred to a European request that the meeting be confined to procedure and that it last only one day.

Participaing nations were Belgium, Britain, Canada, Denmark, Ireland, Italy, Japan, Luxembourg, the Netherlands, Norway, the U.S. and West Germany. France did not send a representative. U.S. Undersecretary of State William H. Donaldson presided.

At its second meeting, held in Brussels March 13–14, the energy group decided to call for a high-level conference with oil producers and other consumer nations to discuss oil prices, supply stability, guarantees and related energy problems.

OPEC official accuses oil consumers—OPEC Secretary General Abderraham Khene of Algeria charged March 19 that major oil-consuming nations were "conspiring" to force down the market price of oil. He asserted that the alleged conspiracy was a result of the meeting of the major oil-consuming nations in Washington in February

Khene said: "We had doubts about the meeting in Washington, and now we are concerned about some of the results . . . we have the facts that big pressure has already been put on the oil companies by all consumer nations to resist the market price asked for."

Arabs offered cooperation plan. The European Economic Community (EEC) offered March 4 to explore long-term economic, technical and cultural cooperation with 20 Arab nations. The joint offer, approved at a meeting of EEC foreign ministers in Brussels, provoked a controversy with U.S. Secretary of State Henry A. Kissinger.

The pro-Arab move omitted mention of cooperation on political and oil questions. According to reports, France had wanted to include those issues.

Three initial steps were envisaged as part of the plan: diplomatic contacts in Bonn between the Arabs and West German Foreign Minister Walter Scheel, current president of the EEC Council of Ministers; establishment of joint working groups to discuss possible areas of cooperation; and eventually an EEC-Arab ministerial meeting.

Scheel said later that the EEC would seek to avoid conflicts between its initiative and energy moves in other forums, a reference to the U.S.-sponsored conference of major oil-consuming nations that met in Washington in February.

(Scheel had agreed with French Foreign Minister Michel Jobert in Bonn March 1 to try to repair EEC disunity over the energy crisis at the March 4 ministerial meeting.)

After the foreign ministers' meeting, Scheel presented the plan to U.S. Secretary of State Kissinger, who was in Brussels briefing representatives of the North Atlantic Treaty Organization (NATO) on his recent peace efforts in the Middle East.

Islamic conference discusses crisis. A summit meeting of heads of state and government of 38 Moslem nations was held in Lahore, Pakistan Feb. 22–24 to discuss Israel's occupation of Arab territories, the future status of Jerusalem and the oil crisis.*

A "Declaration of Lahore" established an eight-nation committee to study ways of easing the pressure of high oil prices on developing Moslem nations in Africa and Asia. However, the delegates rejected a proposal that would have provided a specific aid program by the Arab states to those nations. The committee members were Algeria, Egypt, Kuwait, Libya, Pakistan, Saudi Arabia, Senegal and the United Arab Emirates.

*Participants in the conference: Afghanistan, Algeria, Bahrain, Bangla Desh, Chad, Egypt, Gabon, Gambia, Guinea, Guinea Bissau, India, Indonesia, Iran, Iraq, Jordan, Kuwait, Lebanon, Libya, Malaysia, Mali, Mauritania, Morocco, Niger, Nigeria, Oman, Pakistan, Palestine Liberation Organization, Qatar, Saudi Arabia, Senegal, Somalia, South Yemen, Sudan, Syria, Tunisia, Turkey, Uganda, United Arab Emirates and Yemen.

Among the heads of state attending were King Faisal of Saudi Arabia, and Presidents Anwar Sadat of Egypt, Muammar el-Qaddafi of Libya, Houari Boumedienne of Algeria and Hafez al-Assad of Syria.

Oil Embargo Ended

Further meetings of the Arab oil producers finally resulted in the piecemeal lifting of the oil embargo.

Arabs meet on lifting ban. Petroleum ministers of nine Arab states met in Tripoli, Libya March 13 to consider lifting the oil embargo against the U.S. and restoring the production cutbacks imposed after the 1973 Arab-Israeli war.

According to an unidentified Libyan official March 13, the ministers agreed to lift the embargo, but wouldn't announce the step in Libya, possibly out of deference to the host country's known opposition to lifting the ban.

Attending the Tripoli meeting were the nations that joined in the embargo and production curtailment: Abu Dhabi, Algeria, Bahrain, Egypt, Kuwait, Libya, Qatar, Saudi Arabia and Syria. A terse communique did not mention action taken on the embargo or the production curbs. It merely said the meeting had discussed the world tours taken by Sheik Ahmed Zaki al-Yamani and Belaid Abdelsalam, Saudi and Algerian petroleum ministers, to apprise other nations of the Arab oil stand. The ministers re-examined earlier decisions "in light of developments in the Middle East situation," the statement said. The communique concluded by announcing the time and place of the OPEC meeting.

The oil conference originally had been scheduled in Cairo March 10 at President Anwar Sadat's invitation. However, it was called off that day after representatives of only six of the nine nations appeared. The absent states were Algeria, Libya and Syria, which were opposed to removing the oil restrictions until there was evidence that progress was being made in forcing Israel to withdraw from the Golan Heights.

Soviets back continued oil ban—The Soviet Union called for the continued oil embargo against the U.S. and criticized those Arab states that favored lifting the ban.

One Moscow broadcast monitored in London March 12 said if "some Arab leaders are ready to surrender in the face of American pressure and lift the oil ban before those demands [for Israeli withdrawal] are fulfilled, they are taking a chance by challenging the whole Arab world and the progressive forces of the world, which insist on the continued use of the oil weapon."

Another Soviet broadcast March 5 had said "United States imperialism has hidden behind the mask of a friend of the Arabs in order to break up Arab unity."

Nixon stresses peace over oil—President Nixon, during a TV speech in Chicago March 15, stressed that permanent peace in the Mideast was a goal that the U.S. would pursue whether the embargo was lifted or not.

The U.S., he said, "is not going to be pressured by our friends in the Mideast or others who might be our opponents to doing something before we are able to do it," and any action on the embargo that "has any implications of pressure" on the U.S. would have "a countereffect on our effort to go forward on the peace front, the negotiation front. . . . It would simply slow down . . . our very real and earnest effort to get the disengagement on the Syrian front and also to move toward a permanent settlement."

7 Arab nations lift ban. Seven of nine Arab petroleum-producing countries agreed at a meeting in Vienna March 18 to lift the oil embargo they had imposed against the U.S.

One of the seven nations, Algeria, said it was removing the ban provisionally until June 1. The Arab producers were to meet again on that date in Cairo to review their decision. The embargo was to remain in effect against the Netherlands, as was a curtailment of oil shipments to Denmark. The delegates placed Italy and West Germany on the list of "friendly nations," assuring them of larger supplies.

The Arab action, taken at a meeting of the Organization of Petroleum Exporting Countries (OPEC), was approved by Algeria, Saudi Arabia, Kuwait, Qatar, Bahrain, Egypt and Abu Dhabi. Libya and Syria refused to join the majority. Iraq boycotted the talks.

The OPEC meeting had begun March 17. The Vienna delegates had agreed at the March 17 meeting that oil prices would not be rolled back, despite protests and appeals from consumer nations. A communique said the oil ministers would convene a new meeting if any of the countries asked for one before July "with a view to revising the posted prices" for oil. The posted price for the next three months was to remain at $11.65 a barrel for Arabian light crude oil, in effect since Jan. 1.

Saudi Arabia was the only country which pressed for a lower price for oil. Other countries, led by Indonesia, Algeria, Nigeria and Iran, sought an increase above the $11.65 posted price.

With the announcement of the agreement to end the oil embargo, Saudi Arabia pledged March 18 an immediate production increase of a million barrels a day for the U.S. market.

A formal statement on the Arab decision did not mention the restoration of production cutbacks. The communique explained that a shift in American policy away from Israel had prompted the producers to terminate the embargo. The new U.S. "dimension, if maintained, will lead America to assume a position which is more compatible with the principle of what is right and just toward the Arab-occupied territories and the legitimate rights of the Palestinian people."

Algerian Petroleum Minister Belaid Abdelsalam said March 18 his country believed the U.S. had shown enough "goodwill" in using its influence to get Israel to carry out military disengagement with Egypt and to agree to contacts with Syria to negotiate a similar pullback to warrant lifting the embargo. Libya and Syria, however, were not convinced of this and decided not to go along with the majority opinion, Abdelsalam said.

Syrian resistance to abandon the "oil weapon" because of its belief that the U.S. had not done enough to insure Israeli withdrawal had delayed until March 18 the formal announcement of the end of the embargo, despite the majority decision taken at the Tripoli meeting March 13.

Saudi Arabia Petroleum Minister Sheik Ahmed Zaki al-Yamani said March 18 the decision against resuming the flow of Arab oil to the Netherlands and Denmark was taken because these two countries "have not made clear their position on asking for a full [Israeli] withdrawal from occupied territories."

On leaving Vienna March 20, Kuwaiti Oil Minister Abdel Rahman Atiki said his country regarded the Arab production cutback ordered in December 1973 to still be in effect.

Libyan Oil Minister Ezzedin Mabrouk said March 20 that his country would not only continue the embargo against the U.S., but also would retain its current production level of 1,850,000 barrels a day, compared with 2,300,000 barrels in September 1973.

Mabrouk said March 21 that the Arab producers had been prepared to continue the embargo against the U.S. but a "threat" from Washington prompted them to lift the ban. Mabrouk said the Arab oil ministers had decided in principle at a meeting in Tripoli March 13 to maintain the embargo for two months and that it was to be renewed automatically if Israel did not withdraw from all occupied Arab territories.

Mabrouk added: "But, faced with the American threat, the Arab oil ministers changed their mind and took an unqualified decision to lift the ban." Mabrouk did not disclose the nature of the alleged threat.

Yamani had warned Western countries March 20 that if no progress were made in resolving the Middle East dispute, the Arabs might reimpose the oil embargo and also mobilize other developing nations to take similar action by withholding their natural resources.

In Houston, Tex. March 19, President Nixon had said that ending the embargo was in the interest of the countries that imposed it as well as in the U.S. interest. "Inevitably, what happens in one area affects the other, and I am confident that the progress we are going to continue to make on the peace front in the Mideast will be very helpful in seeing to it that an oil embargo is not reimposed."

Saudis renew oil shipments to U.S. Saudi Arabia had resumed shipment of oil to the United States, it was reported March 25.

Resumption of the oil shipments was coupled with Saudi Arabia's formal notification to the Arabian-American Oil Co. (Aramco) March 21 to increase production by one million barrels a day to approximately the pre-October 1973 level of 8.3 million barrels a day, a company spokesman said March 24.

U.S. got oil during embargo. The U.S. Federal Energy Office announced April 9 that the first major post-embargo shipment of Arab oil had reached the U.S. However, another government report, released April 8, indicated that Saudi Arabia had continued to ship small amounts of crude oil to the U.S. from November 1973 through February 1974 despite the boycott.

The import data, which was published by the Commerce Department on a country of origin basis, revealed that although other Arab countries shipped significant amounts of crude to the U.S. during the first two months of the embargo, the shipments virtually dried up during the remainder of the embargo period. (Officials said that leakage during the early period may not have been intentional, because the oil could have been loaded prior to the ban and been in transit for 30–60 days.)

However, the U.S. did receive crude oil shipments during the period from several European countries which previously had produced little or no crude, prompting speculation that the oil had originated from Arab states. Leakage also occurred in another form, officials said, that was not included in the government report. Arab crude was shipped to refinery points and these secondary petroleum products were then transported to the U.S.

The Federal Energy Office had ordered the information kept secret during the period of the embargo on national security grounds, but with the end of the ban, the material was released after the Wall Street Journal invoked the Freedom of Information Act and sought access to the statistics.

Libya, which was widely believed to have been a major source of leaks, shipped 4.8 million barrels to the U.S. in November 1973, 1.2 million barrels in December 1973, and then halted shipments for the remainder of the boycott. (Libya's total embargo shipments were 6,046,226 barrels.)

During the same period, Saudi Arabia shipped 25,898,200 barrels to the U.S.: 18 million in November 1973; 7.1 million in December 1973; 257,187 in January; and 552,212 in February. Shipments from other Arab countries: 515,299 barrels (none after November 1973) from Iraq; 3,163,311 (none after December) from Kuwait; 962,316 (none after November) from Qatar; 3,067,855 (none after December) from the United Arab Emirates; 3,768,957 (none after

(December) from Algeria; and 1,010,435 (spread throughout the embargo period) from Tunisia.

Canada was the major source of oil for the U.S. during the four-month period, shipping 112,396,628 barrels. Other major exporters were Venezuela (64,101,137 barrels); Nigeria (55,399,880 barrels); Iran (46,289,453 barrels); and Indonesia (29,972,781 barrels). Their exports were increased during the embargo period.

Small amounts of oil were received from five European countries—the Netherlands, Belgium, Italy, France and West Germany. Eight Latin American nations and the Netherlands Antilles also exported oil to the U.S. Trailing Venezuela in size of shipments were Ecuador (10 million barrels) and Trinidad & Tobago (9.2 million barrels).

A total of 386,406,678 barrels of crude oil, valued at $2.3 billion were exported to the U.S. from 30 countries during November 1973–February 1974.

Leakage statistics disputed—The published report prompted denials from government officials and major oil companies. Federal Energy Office Director William Simon said April 12 that Saudi Arabian "oil arriving in the U.S. in November and December [1973] was probably loaded prior to the mid-October embargo and therefore did not constitute a leak." Simon had no comment on possible leaks from other Arab countries.

The Arabian American Oil Co. (Aramco) also contradicted the government's data on Saudi Arabia, charging April 11 that its export records showed no oil was shipped from its terminals to the U.S. "either directly or indirectly" after the boycott was announced. Any oil reaching the U.S. during the embargo period was loaded prior to imposition of the ban, company spokesmen said.

Each of the U.S. partners in Aramco—Texaco, Exxon, Mobil, and Standard (California)—also disputed the accuracy of the Commerce Department report, citing transshipment delays in receipt of the oil. The Saudi Arabian government, which was the fifth and major member of Aramco, was reported April 15 to have challenged the report.

Another analysis of the embargo was prepared by Harvard Business School Professor Robert B. Stobaugh for the Senate multinationals subcommittee. The report was released July 25.

The study concluded that it was a misconception to believe that the import curb was widely violated—the embargo was nearly 100% effective in cutting off the supply of oil imported to the U.S. Crude oil imports were reduced from 1.2 million barrels a day in September 1973 to 19,000 barrels a day in January and February.

The study also refuted charges that major oil companies contrived the winter fuel shortage in order to raise prices. Oil available for consumption in the U.S. dropped 6.1% during the worst four-month period of the embargo, compared with a 3.4% worldwide drop in supplies, the report said.

Contrary to public belief, Japan, which was 80% dependent on foreign oil supplies, did not feel the embargo's sharpest impact, the study added. West Germany, Britain and France were hardest hit, followed by the U.S. and Japan.

The U.S. received about 2% less non-Arab oil imports than usual, which the report attributed to cutbacks in exports of Canadian and Venezuelan oil to the U.S.

The report concluded that despite the global cutback in oil supplies, most of the rest of the world received proportionally as much oil as it would have obtained under normal conditions. That figure included the Netherlands, which was a special target of the Arab weapon. The general equality of distribution was achieved through a "diversion" of world oil exports from normal destinations to compensate for extraordinary shortages elsewhere, the report stated.

The report was based on national statistics compiled mainly by the oil industry.

U.S. military got oil during embargo—During the height of the oil embargo, the U.S. military was able to purchase 100,000 barrels a day for use by overseas forces, the Washington Post reported March 21. During January and February, the Post said, Arab oil producers began "to look the other" way and let their crude oil be sold to U.S. overseas military forces by way of refineries in non-embargoed countries.

Soviet oil shipped to U.S.—The U.S.S.R. continued to export Soviet petroleum products to the U.S. during the months that Moscow was publicly encouraging Arab nations to maintain their oil embargo against Washington, according to an Associated Press report April 1.

U.S. Customs records showed that two Greek and two U.S. flagships had brought 845,128 barrels of Soviet gasoline products (gasoline, kerosene, benzine and heating oil) to New York and New Jersey ports between Jan. 11 and March 18. The four shipments had originated in the Black Sea port of Tuapse, the site of major oil refineries.

(According to the Commerce Department's report April 8, there had been no crude Soviet petroleum imports since December 1973; the records showed that 171,090 barrels had been imported in November 1973, after the Arab oil embargo had begun.)

OAPEC split on embargo. All but one of the 10 members of the Organization of Arab Petroleum Exporting Countries (OAPEC) agreed June 2 to maintain the total embargo of oil shipments to the Netherlands and to continue the reduced deliveries to Denmark following a two-day meeting in Cairo. Algeria expressed opposition to the ban, and its petroleum minister, Belaid Abdelsalam, announced that "the embargo against the Netherlands has been lifted by Algeria from today."

The OAPEC delegates also decided to establish a fund to provide 20-year interest-free loans to poor Arab countries that did not produce oil and were adversely affected by increasing oil prices.

Saudi Petroleum Minister Sheik Ahmed Zaki al-Yamani, who attended the OAPEC meeting, proposed in Cairo June 3 a limited but high-level meeting of major oil-producing and oil-consuming nations for a "practical dialogue" on the problems relating to energy, raw materials and technology. He suggested Iran, Algeria, Venezuela and Saudi Arabia as among the oil-producing participants in such a conference, the U.S., Japan and the European Economic Community, among the industrial consumers, and India, Brazil, and Zaire, among the raw material exporters that were also oil consumers.

Arabs end oil ban against Holland. The Organization of Arab Petroleum Exporting Countries (OAPEC) decided at a meeting in Cairo July 10 to lift its nine-month old oil embargo against the Netherlands.

The ban had been imposed in October 1973 against the Netherlands and the U.S. for their support for Israel during the Middle East war. OAPEC reversed itself because it was now "convinced the Dutch government's attitude toward the Middle East had changed," Saudi Arabian Petroleum Minister Sheik Zaki al-Yamani said.

In another action taken at the OAPEC meeting, a resolution was approved establishing an Arab investment corporation with assets of $1.1-billion, with Arab states to purchase shares.

The Dutch Foreign Ministry July 10 expressed satisfaction with OAPEC's decision, saying it would "stimulate the development of good relations between the Netherlands and the Arab world, as the Netherlands has always wanted."

The Dutch port of Rotterdam, which depended heavily on oil traffic, had lost about $9.4-million in port duties during the embargo. Oil reserves during that period, however, had increased 30%, reaching 5.1 million tons in May as opposed to 3.5 million tons in 1973. The increase in reserves was attributed by the Economics Ministry to a 15% drop in oil consumption as a result of mild weather, highway speed limits and gasoline price rises.

Libya ends U.S. oil embargo. Libya had lifted its 14-month oil embargo against the U.S. without making an official announcement, the Times of London reported Dec. 31.

Foreign oil companies in Libya said that although they had not been officially informed of the government's action in ending the embargo, the decision would have little bearing on the resumption of shipments. They cited the high cost of Libyan oil, about $12.50 a barrel, $2.50 more than the foreign companies had to pay for Persian Gulf oil. As a result, Libyan production had dropped to about 800,000 barrels a day in 1974 from an average of 2.2 million in 1973.

OPEC says '73 output up. The Organization of Petroleum Exporting Countries (OPEC) reported Aug. 22 that its member nations had increased their oil production by 13.5% in 1973 despite Arab cutbacks.

According to an OPEC bulletin, the 1973 output more than doubled the previous year's figures and compared with a worldwide production increase of 8%. The increase "could have been even higher had it not been for the fact that some member countries had decided, for political reasons, to curtail their production during the last quarter of 1973," the bulletin said.

Saudis pressured U.S. on Israel. Saudi Arabia had used U.S. oil companies in May 1973 in an attempt to pressure the U.S. government to change its pro-Israel policy, according to testimony released Aug. 6, 1974 by the Senate Foreign Relations Subcommittee on Multinational Corporations. The subcommittee had been investigating the Arabian American Oil Co. (Aramco) and its four U.S. shareholders, Exxon, Mobil, Texaco and the Standard Oil Co. of California.

Subcommittee Chairman Frank Church (D, Ida.) said the seven-month probe showed "the extent to which the companies that are hostages of the Saudis are forced to operate at their beck and call. The companies followed these instructions and reported on their activities to the king [Faisal]. The capstone of their efforts" was a joint memorandum to President Nixon sent by the chief executives of the owner companies Oct. 12, 1973, warning Nixon "that any actions of the U.S. government at this time in terms of increased military aid to Israel will have a critical and adverse effect on relations with the moderate Arab producing states." The memo was sent six days after Egypt and Syria had attacked Israel.

Company memorandums portrayed Faisal as warning the firms' executives that they would "lose everything" unless they intensified their efforts on behalf of the Arabs.

Subcommittee chief counsel Jerome Levinson said that although the U.S. had become friendlier to the Arabs, there was no evidence that this change in policy had been brought about by the companies' efforts.

One document supporting the subcommittee's charges was a cable sent to Aramco owning companies shortly after the outbreak of the war by Frank Jungers, Aramco's chairman and chief executive. It said that James E. Akins, U.S. ambassador to Saudi Arabia, through an Aramco official had "urged that industry leaders in the USA use their contacts at highest levels of USG [U.S. government] to hammer home point that oil restrictions are not going to be lifted unless political struggle is settled in manner satisfactory to Arabs."

According to another aspect of the investigation record, Saudi Arabia, after it realized it could not administer the oil embargo against the U.S. smoothly, called on Aramco and its owners to run

PRODUCING GOVERNMENT OIL REVENUES

[In billions of dollars]

	[1] 1974	[2] 1972	[2] 1970
Total of listed countries	89.1	15.1	7.9
Middle East	56.6	8.9	4.2
Saudi Arabia	[3] 18.9	3.1	1.2
Iran	[3] 16.4	2.4	1.1
Kuwait	[3] 7.6	1.7	.9
Iraq	[4] 7.5	[5] .6	.5
Abu Dhabi	[3] 3.3	.6	.2
Other	[3] 2.9	.5	.3
Africa (nations listed)	18.1	3.5	2.0
Libya	[6] 7.9	1.6	1.3
Nigeria	[3] 6.6	1.2	.4
Algeria	[6] 3.6	.7	.3
Venezuela	[7] 10.0	1.9	1.4
Indonesia	[8] 4.4	.9	.3

[1] Based upon non-Communist world consumption in 1974 equal to the 1972 level of 44,700,000 barrels daily and per-barrel revenues as shown below.

[2] Actual revenue as reported.

[3] Assumptions: (i) one-third of production sold at 93 percent of January 1974 posting; (ii) government revenue on the remaining two-thirds based upon 12.5 percent expensed royalty plus 55 percent income tax. Average per barrel government revenue works out as follows: Abu Dhabi, $8.78; Kuwait, $8.22; Iran, $8.43; Nigeria, $10.61; and Saudi Arabia/other Arab Middle East, $8.28.

[4] Assumes government revenue from nationalized Kirkuk production of $12 per barrel; government revenue at the Persian Gulf averages $8.93 per barrel.

[5] Excludes government revenues from nationalized Kirkuk and North Rumaila output.

[6] Assumes average government revenue in Libya and Algeria to be equal. Based upon Libyan revenue half-way between posting ($15.77) and tax-paid cost ($10.07), or $12.92.

[7] Assumes average revenue of $8.85, based upon 16⅔ percent expensed royalty plus 58 percent income tax.

[8] Based on a $10.80 realized price, yielding government per-barrel revenues of about $8 in the production-sharing formula.

the embargo for them. Sen. Church said the Saudis received total compliance with their wishes, "including the operations of a primary and secondary embargo aimed at the U.S. military." Jungers Oct. 21, 1973 had sent the owning companies a cablegram in which he had informed them of the Saudis' intention of "looking to Aramco to police" the embargo and reported the embargo on shipments to the U.S. armed forces.

Sources of U.S. oil imports. The Federal Energy Administration released figures Oct. 23 showing that Canada and Nigeria supplied the U.S. with twice as much crude oil as all of the Arab oil producing countries combined during the first eight months of 1974.

Of the 6 million barrels of petroleum products imported to the U.S. on a daily basis, crude oil comprised two-thirds of that total and refined products about one-third.

Canada was the largest single supplier of foreign oil, but its shipments had been decreasing from 35.8% of U.S. imports (883,800 barrels a day) in 1972 to 24.7% in 1974 (896,400 barrels a day).

Nigeria had nearly tripled its exports to the U.S. Shipments were up from 235,800 in 1972 (9.6% of U.S. imports) to 633,800 (17.5%) in 1974. Nigeria replaced Venezuela as the second largest exporter of oil supplies to the U.S.

Iran ranked third in imports received by the U.S., moving up from fifth place in 1972. It more than doubled its exports in that two-year period—from 140,200 barrels (5.7%) to 561,900 (15.5%) in 1974.

Venezuelan exports to the U.S. fell from a daily average of 534,700 barrels in 1973 to 405,100 barrels daily in 1974, a reflection of Venezuela's decision to freeze production as part of its pricing strategy.

Cooperation Plans Advance

COMECON plans energy cooperation. Delegates of the 10 member nations of the Council for Mutual Economic Assistance (COMECON) met in Sofia, Bulgaria June 18–21, 1974. In a communique issued June 21, the Communist-country group announced plans for implementing energy co-operation plans. They agreed that the Eastern European members should participate in the exploitation of future natural gas strikes in the Soviet Union and decided to formulate a project to unify their electrical energy systems and develop a mutual system of atomic power stations.

Arabs set up Energy Institute. The Organization of Arab Petroleum Exporting Countries (OAPEC) agreed at the end of a two-day meeting in Cairo July 11 to establish an Arab Energy Institute. In addition to a search for new uses for oil, the institute would conduct nuclear research and study the possible use of solar energy.

The conferees also signed an agreement to form the Petroleum Investments Co. to finance the development of oil industries

U.S. Average Daily Crude Oil Imports

	1974 (Jan.-Aug.)		1973		1972	
	Thousands of barrels	% of U.S. imports	Thousands of barrels	% of U.S. imports	Thousands of barrels	% of U.S. imports
Canada	896.4	24.7	1,094.1	30.8	883.8	35.8
Nigeria	633.8	17.5	410.0	11.6	235.8	9.6
Iran	561.9	15.5	207.5	5.9	140.2	5.7
Venezuela	405.1	11.2	534.7	15.1	450.6	18.2
Saudi Arabia	301.0	8.3	434.2	12.2	201.3	8.2
Indonesia	283.7	7.8	207.3	5.9	156.8	6.4
Algeria	195.7	5.4	127.6	3.6	85.3	3.5
Ecuador	78.7	2.2	57.3	1.6	—	—
Trinidad	78.6	2.2	—	—	34.2	1.4
United Arab Emirates	62.6	1.7	68.2	1.9	32.9	1.3
Angola	47.3	1.3	44.6	1.3	—	—
Others	80.1	2.2	360.9	10.1	247.8	10.4

in the OAPEC countries. OAPEC pledged an $80 million fund to help compensate Arab oil-importing countries for the increase in oil prices. The beneficiaries would be Jordan, Lebanon, Mauritania, Morocco, Sudan, Somalia and both Yemens.

ECG adopts oil-sharing plan. The 12-nation Energy Coordination Group (ECG) reached formal agreement after a two-day meeting in Brussels Sept. 19–20 on a detailed plan under which member countries affected by any oil-producer embargo would receive supplies from a common pool when their oil imports dropped by more than 7%.

The plan was to be managed by a new agency within the Organization for Economic Cooperation and Development (OECD) and would be open to all ECG members. The plan was subject to official approval by member governments. The ECG hoped to put the proposed new agency to an OECD vote in November.

One of the ECG members, Norway, had declared at the Sept. 19 meeting that it could not approve the plan until its Parliament debated the issue. Norway had begun to develop vast oil reserves in the North Sea and there was opposition in Oslo to any sharing arrangement that would conflict with its sovereignty over its oil supply.

The plan had been proposed by the U.S. in Brussels June 17–18 at a meeting that also heard and took under consideration a working group report on multinational oil firms proposing that the companies provide the 12 member nations with more information about their operations.

The need for government surveillance of international oil companies and multilateral reduction of energy consumption had been discussed at a previous ECG meeting in Brussels April 3–4.

Preliminary agreement on the plan was then reached at a July 29–31 session.

According to details of the plan disclosed Aug. 5 by U.S. Assistant Secretary of State for Economic Affairs Thomas O. Enders, the U.S. and the other ECG members would reduce their oil consumption during a crisis by 7%–10% by such measures as high gasoline excise taxes, rationing or government intervention in allocation of oil supplies. ECG oil producers (the U.S., Britain, Canada and Norway) would share equally their output with other members in time of need. The accord stipulated that a majority of the group could override a request by one or two members for putting the emergency measures into effect.

Enders said the ECG also had prepared a comprehensive plan for developing alternative energy supplies, for dealing with monetary dislocations caused by sharp increases in oil prices and for monitoring the activities of international oil companies.

EEC action. The EEC (European Economic Community) foreign ministers, meeting in Brussels July 23, 1974, were blocked by Great Britain from adopting a joint EEC energy policy plan for 1975–85.

The plan, formulated by the European Commission, had been disclosed May 30. The program aimed at reducing EEC dependence on imported energy resources during 1975–85. The proposals included a 10% reduction, through rationalization programs, of the community's energy consumption; a 10% increase in electricity use, to reach 35% of total consumption; increase in the use of nuclear energy in electricity production, from the current 5.7% to 50% (by the year 2000 nuclear energy and natural gas should supply 80% of the community's energy needs); maintenance of coal production at its present level and an increase in coal imports; and an increase in natural gas internal output and imports. The Commission proposed that oil companies be required to provide more information to public authorities on their imports, investments and operations.

But the ministers Sept. 17 approved a "declaration of intent" committing the community to a joint energy policy. The statement, approved at a meeting in Brussels, called for energy conservation, reduction of energy imports from outside the community, more nuclear energy production, diversification of non-community supplies and joint research. Britain lifted its two-month veto of the plan after slight textual changes were made to soften emphasis on the energy autonomy of the community.

The EEC energy ministers Dec. 17 then worked out agreements on their joint objectives for energy use by 1985. They agreed to reduce the 10-year growth in their nations' domestic energy consumption from a projected 50% to 35% and to cut dependence on imported energy supplies from the 1973 level of 63% to 40%–50%. The plan called for reducing the share of oil in energy production from 61% to 41%–49% through increased use of nuclear power and natural gas.

Among other actions, the ministers agreed to require oil companies to report twice a year on their energy imports and decided to create a joint fund of nearly $50 million to encourage oil exploration and development in Europe. Britain stressed it would cooperate on a joint energy policy but insisted it would retain full national control of its North Sea oil and natural gas resources.

European Economic Community development aid ministers agreed Oct. 3 to jointly contribute $150 million to aid the 25 developing nations worst hit by the quadrupled price of oil imports. The decision was reached at a meeting in Luxembourg. The EEC was committed to give $30 million to the United Nations emergency fund set up for this purpose, while the remaining $120 million would go directly to the affected countries. Claude Cheysson, the EEC development cooperation commissioner, said a further $350 million would be granted before July 1, 1975 if other industrialized and oil-producing nations made proportional donations.

Five days later the European Commission and the Organization of Arab Petroleum Exporting Countries (OAPEC) agreed Oct. 8 to maintain regular contacts and exchange technical information. The decision came at the end of two days of talks in Brussels between OAPEC Secretary General Ali Attiga of Kuwait and two Commission members—Henri Simonet and Claude Cheysson, commissioners for energy and development cooperation respectively.

Simonet Oct. 25 announced the Commission's approval for member nations to join the oilpool of the 12-nation Energy Coordination Group (ECG), to which all community partners, except France, belonged. He said, however, that the ECG's oil pooling plan in the event of a new energy crisis must not violate the community principle of the free movement of goods throughout the nine nations. According to the London Times Oct. 26, France would benefit from the oil pool without any of its commitments.

Commonwealth ministers meet. The finance ministers of 34 Commonwealth nations Sept. 26 called for low-interest loans and other measures to aid developing nations whose economies had been strained by the increases in oil prices. The action was endorsed at the end of a two-day meeting in Ottawa, Canada.

The ministers urged a faster and more controlled "recycling" of oil profits from petroleum exporting nations through inexpensive loans granted by the International Monetary Fund to the developing nations. Canadian Finance Minister John Turner, chairman of the conference, later told reporters that without such action all the oil profits would end up in the countries with the strongest economies, creating a "desperate situation" for developing nations.

5-nation meeting on oil, money problems. Foreign ministers and finance ministers of the U.S., Great Britain, West Germany, France and Japan met Sept. 28–29 in Washington to discuss joint efforts that could be undertaken to meet the oil-price crisis. No communique was issued after the two-day meeting, but the London Times reported Sept. 30 that another purpose of the conference was to draw up a contingency plan in the event that Middle East oil producers imposed a new embargo because of mounting tensions in the area.

The meeting was called by U.S. Secretary of State Henry Kissinger because of fears that unless the major industrialized nations acted decisively to coordinate their policies in the face of the threat posed by higher oil prices and a possible cutoff of supplies, Western nations faced bankruptcy or political collapse.

According to an unofficial report Sept. 29, no agreement was reached on how to force down rising oil prices, although it was accepted that a lower price should remain a long-term goal. French Foreign Minister Jean Sauvagnargues said at a news conference that day that France was opposed to an "economic war" between oil producing and oil importing nations.

One of the strategies reportedly discussed at the meeting was a Kissinger proposal that concerted action be taken to reduce oil consumption—a 15% cutback was mentioned, sources said. Kissinger also reportedly praised the French decision to impose a payment ceiling of 51 billion francs on oil imports for 1975.

Officials reportedly were agreeable to the plan, but further study was sought. According to the Kissinger plan, reduced consumption would be paired with a major fuel conservation program in each country, sources said.

IMF establishes oil facility. The consequences of a series of sharp increases in the price of oil were felt throughout the international money market network.

An "oil facility," proposed in January by H. Johannes Witteveen, managing director of the International Monetary Fund (IMF), was established by that organization Aug. 22, 1974 to help member nations meet their greatly increased balance of payments needs, which had been affected by escalating import costs for petroleum and petroleum products.

Seven oil producing nations agreed to lend the IMF the equivalent of $3.4 billion until Dec. 31, 1975. The funds borrowed by the IMF would then be made available to nations hard pressed to meet their international obligations. (Interest paid by the IMF would be at an annual rate of 7%, with repayment of the loans scheduled to be made in eight equal semiannual installments beginning after three years and to be completed not later than seven years after the transfer of currency to the IMF.)

The seven lending nations participating in the oil facility and their contributions (expressed in SDRs and dollars): Abu Dhabi, 100 million SDRs or $120 million; Canada, 257,913,900 SDRs or $300 million in Canadian currency; Iran, 580 million SDRs or $700 million; Kuwait, 400 million SDRs or $480 million; Oman, 20 million SDRs or $24 million; Saudi Arabia, 1 billion SDRs or $1.2 billion; Venezuela, 450 million SDRs or $540 million.

It was announced Sept. 6 that drawings on the oil facility totaled 142.6 million SDRs. Countries drawing on the loan program were Bangla Desh, with 12.43 million SDRs; Chile, 41.47 million; Haiti, 1.15 million; Kenya, 9.21 million; Korea, 21 million; Pakistan, 30.61 million; Sri Lanka, 11 million; Sudan, 9.41 million; and Tanzania, 6.32 million.

Interest on the loans, which would be repaid in 16 equal quarterly payments to be completed not less than seven years after the date of the original borrowing, would be 6.875% annually for the first three years, 7% at the beginning of the fourth year and 7.125% beginning in the fifth year.

The IMF and World Bank held their annual joint meeting in Washington Sept. 30–Oct. 4. The interlocking problems of oil and money dominated the talks.

A number of proposals were debated dealing with the need to "recycle" the vast amount of money earned by oil exporting countries and the huge sums paid for oil by importing countries, but no conclusive agreements were reached. It was announced Oct. 4 that the IMF would draw up plans for a major new lending operation, administered by the IMF and funded principally by oil-producing nations, to benefit countries that were in financial disarray because of their inability to pay for the needed oil imports.

Participants at the Washington conference were divided over the need for a new recycling mechanism and over the scope of any proposed expansion in the credit program. According to Treasury Undersecretary Jack F. Bennett, the U.S. questioned the need for a larger oil facility. At a news conference Oct. 4, Bennett said the U.S. preferred to deal with the problems of massive transfers of funds by relying on Eurocurrency markets and an increase in investment by oil exporting countries through nonbanking channels. "There is no proof yet" that an official, expanded IMF aid program was needed, Bennett said. The U.S. view remained unchanged, Bennett said, that worldwide economic distortions caused by higher oil prices could be alleviated by a reduction in price.

World Bank President Robert McNamara told the conference Sept. 30 that the poorest nations "are the principal victims" of worldwide inflation. Large increases in aid from developed nations were required to prevent a worsening of their plight, he said.

Of the 2 billion persons living in developing countries, McNamara said, 20% had prospered because of inflation (mostly those in oil and mineral exporting countries). "But for the poorest of countries that represent fully half of the total population of all the nations we serve, countries containing 1 billion human beings—the situation is desperate," McNamara said.

The poorest nations—those with per capita incomes of less than $200 annually—were expected to "suffer an annual decline" in income by 1980 even if capital flows to developing nations increased to $33 billion from $23 billion in 1973, McNamara said. Oil exporting nations currently had committed an estimated $5.5 billion in aid to developing nations, McNamara said, but most of their oil revenues were expected to be channeled to high and middle income nations.

IMF oil facility borrowings—Thirty-three nations had borrowed a total of 1.716 billion in Special Drawing Rights (SDRs) (about $1.792 billion) from the International Monetary Fund's (IMF) oil facility since it was established, the IMF announced Dec. 31.

Purchases from the special credit fund by IMF member nations:

Honduras, 16.785 million SDRs; Korea, 33 million SDRs; Sudan, 19.3 million SDRs; and Uruguay, 29.015 million SDRs. Announced Nov. 15.

Costa Rica, 13.477 million SDRs; El Salvador, 13.49 million SDRs; Guinea, 3.51 million SDRs; Haiti, 2.675 million SDRs; Iceland, 15.5 million SDRs; Italy, 412.5 million SDRs; Kenya, 22.82 million SDRs; Mali, 4 million SDRs; New Zealand, 85.68 million SDRs; Sri Lanka, 23 million SDRs; Tanzania, 22.03 million SDRs; Yemen, 9.306 million SDRs. Announced Dec. 4.

Central African Republic, 2.025 million SDRs; Chad, 1.095 million SDRs; Cyprus, 6.38 million SDRs; Korea, 36 million SDRs; Yugoslavia, 99 million SDRs. Announced Dec. 31.

Nigeria had agreed to loan the IMF currencies equivalent to 100 million SDRs (about $122 million) for use in its oil facility, and also had loaned the World Bank $240 million at 8% interest, according to reports Dec. 23.

The Nigerian loan to the World Bank brought its borrowings from oil producing countries during 1974 to $2.8 billion, including a $750 million loan from Saudi Arabia. It was the largest single World Bank borrowing in history, according to the announcement Dec. 17.

U.S. proposals. U.S. Secretary o State Henry A. Kissinger issued a ca Nov. 14, 1974 for "collaboration amon the consuming nations" to avoid "per petual crises" over rising oil prices.

Speaking at the University of Chicago Kissinger urged the nations of North America, Western Europe and Japan to pool their efforts to protect themselve from "the threat of a new oil embargo." He proposed a $25 billion international lending agency financed by the consume nations to help industrial states cope with the high cost of oil. This would serve to stabilize the financial system by providin support to banks already burdened by th impact of oil dollar investments, Kissinge explained.

The secretary also proposed: reductio of oil imports of the industrial nations to one-fifth of total energy needs rather tha the current one-third of total needs; development of new energy sources, including coal, solar power and uranium; a effort to bring down the price of oi through dialogue between the oi consumers and the oil producers; and creation of a special trust fund to assis developing nations, which had $20 billio in payments deficits.

Jackson proposes special council—U.S Sen. Henry M. Jackson (D, Wash.) Nov 11 urged major oil consuming nations to establish a special council of economic and finance ministers, authorized to work for a reduction of oil prices and to protect the international banking system from an unmanageable flow of investments by oi producing nations.

In a London speech before the Pilgrims Society, Jackson said the council's first act must be "to insulate the price of oi produced by consuming nations from the control of the producers' cartel." The Organization of Petroleum Exporting Countries (OPEC) should not be allowed to set worldwide oil prices, Jackson declared. If oil consuming nations which also produced oil, like the U.S., orderd a rollback in their prices, Jackson said, pressure would be brought on OPEC nations to reduce their prices.

In calling for "tough programs" aimed at lowering the price of oil, Jackson opposed the International Monetary Fund plan, favored by many European states, for dealing with the financial crisis caused by higher oil import costs—recycling petrodollars through oil consuming states in the form of loans and investments.

Recycling would make the U.S. "a lender of last resort and we would end up with all the funny money," Jackson said. One task of the new special council, as outlined by Jackson, would be to issue guidelines for petrodollar investments to insure that industrialized nations maintained "continued control" over "their own essential financial infrastructures" and were not overwhelmed by the volume of OPEC investments.

Simon details oil-facility plan—U.S. Treasury Secretary William E. Simon Nov. 18 provided details of the recycling plan, first proposed by Secretary of State Henry Kissinger, for dealing with the growing financial instability caused by high oil import costs and payments deficits.

In a New York speech, Simon proposed that the oil facility be associated with the 24-member Organization for Economic Cooperation and Development, representing the major industrial nations, rather than the 130-member International Monetary Fund, in which developing and oil producing nations were heavily represented.

Simon said the U.S. contribution to the loan facility would be financed through the Treasury Department's existing $5 billion Exchange Stabilization Fund, which engaged in international lending operations to stabilize the dollar.

(In testimony before the Joint Economic Committee of Congress Nov. 25, Simon said a final decision had not been made about the amount of the U.S. contribution, but he said it was his "preliminary view" that the U.S. would furnish 25%-30%, or about $8 billion, of the total funds needed. Simon also told the committee that if necessary, the U.S. could meet its payments through the purchase of U.S. government securities by oil exporting nations. Operation of the oil facility "would not require inflationary expansion of money and credit," Simon said, or "lead to an increase in the federal government debt.")

Other key elements in the U.S. proposal as outlined by Simon: participation "should be linked with a commitment to cooperate in reducing dependence on oil imports"; members of the mutual loan fund would be asked "to follow responsible adjustment policies and avoid the use of trade restrictive measures"; the oil facility should supplement and not replace ordinary credit channels within the private market or through official groups, such as the IMF; financial support to member countries would be decided on a weighted vote of participants and not be based on a specific criterion, such as oil import costs, but rather on the nation's overall economic condition; "all nations should share the credit risk on the basis of their share of participation."

U.S. participation would require Congressional authority. Simon repeatedly sought to assure committee members at the Nov. 25 hearing that the Administration would consult with Congress as negotiations to establish the fund were undertaken.

Simon had been on record earlier as opposing efforts to institutionalize the recycling process through establishment of another international group, instead favoring increased use of the private credit and investment markets to channel oil revenues to oil consuming nations.

International energy group formed. An International Energy Agency (IEA) to pool energy resources by 16 industrial nations in the event of an emergency was formed in Paris Nov. 15, 1974.

The body, established within the framework of the 24-nation Organization for Economic Cooperation and Development (OECD), was composed of the U.S., Austria, Belgium, Britain, Canada, Denmark, Ireland, Italy, Japan, Luxembourg, the Netherlands, Spain, Sweden, Switzerland, Turkey and West Germany. Twenty-one OECD nations voted to set up the agency, but five of them refused to join. France, which also refused to join, abstained along with Greece and Finland.

The IEA replaced the 12-nation Energy Coordinating Group (ECG), which had promoted the creation of the new agency and formed its nucleus. The ECG had held its last meeting Nov. 6 in Brussels.

Norway, an ECG member, refused to join the IEA as a full member because it feared giving up its sovereignty over its oil resources. Foreign Minister Knut Frydenlund had said in Parliament Nov. 2 that his country's interests did not coincide with other members, who were oil importers, while Norway was a potential exporter of oil.

France, which had boycotted the ECG from the start, also had decided to remain out of the IEA, arguing that the consumer nations should be negotiating with the producer states rather than forming a bloc that would antagonize them.

A key clause in the charter of the IEA was a pledge by members to come to the assistance of other participants whose oil supply was reduced by 7% even if that meant a reduction of consumption by the donor states.

OPEC aid to developing nations. Members of the Organization of Petroleum Exporting Countries (OPEC) gave $8.6 billion in official economic assistance to developing nations during the first nine months of 1974, the International Monetary Fund (IMF) reported Nov. 18.

The aid was in addition to the $3.1 billion made available to the IMF's oil facility for use by nations with balance of payments problems, and $1 billion loaned to the World Bank.

Bilateral commitments made by OPEC nations over the nine-month period totaled $6.2 billion. Major donors were Iran ($2.8 billion) and Saudi Arabia ($2.4 billion). Multilateral aid totaled $2.4 billion, with the largest grant ($900 million) earmarked for the Islamic Development Bank.

According to the IMF, OPEC nations had given between $380 million and $530 million annually during 1970-73, with 83% of the aid directed to Egypt, Jordan and Syria. Until 1973, only three OPEC nations—Kuwait, Libya and Saudi Arabia—had made significant contributions to developing nations.

Western leaders confer. U.S. President Gerald Ford and West German Chancellor Helmut Schmidt conferred in Washington Dec. 5-6, 1974 on economic problems and the related world energy crisis. In a joint declaration issued at the end of the talks, both leaders said they were "determined not to permit a serious deterioration in their economies to occur."

According to the declaration, Ford and Schmidt agreed to consider establishing "a supplementary financial safety net" in the framework of the 25-nation Organization for Economic Cooperation and Development to help oil consuming nations cope with the soaring costs of petroleum.

Both leaders, with Secretary of State Henry Kissinger present, had agreed Dec. 5 to try to persuade France to join the joint effort of Western industrial nations to cope with the world energy crisis.

Later Ford and French President Valery Giscard d'Estaing conferred on the French Caribbean island of Martinique Dec. 15-16 and reached compromise agreements on energy, gold and other issues. The accords, reached, according to the communique, in an atmosphere of "cordiality and mutual confidence," represented a sharp departure from the conflicts that had marked U.S.-French relations in recent years.

In the joint communique issued Dec. 16, the U.S. agreed to participate in a French-proposed conference of oil producing and exporting nations "at the earliest possible date." France agreed to "intensive consultations among consumer countries in order to prepare positions for the conference," a concession to the U.S. demand for a unified front among oil consuming nations. Both leaders stressed the importance of "solidarity" among the oil consumers. They agreed that prior to the preparatory consumer-producer meeting, consumers would strengthen their cooperation on energy conservation and development and on "setting up of a new mechanism of financial solidarity," the last a reference to U.S. Secretary of State Henry Kissinger's proposal for a $25 billion emergency fund to help oil consuming nations hurt by soaring oil costs.

At the end of the talks, Kissinger told newsmen that France would participate in the $25 billion fund, but not within the framework of the new 16-nation International Energy Agency. He added that the U.S. had abandoned its efforts to persuade France to join the group.

Consumers Woo Producers

Many of the developed oil-consuming nations began intensive efforts during early 1974 to cement relations with oil-producing states and thus, they hoped, assure themselves of the supplies of petroleum on which their economies depended. It was reported that envoys of the developed nations were offering trade incentives, technology, advanced weapons and political cooperation in attempts to work out bilateral deals that would keep the oil flowing.

French mission to Arab states. French Foreign Minister Michel Jobert visited Saudi Arabia, Kuwait and Syria Jan. 24–29 in a move aimed at establishing French economic, political and military influence in the region.

In his meetings with Saudi Arabian King Faisal Jan. 24–26, Jobert discussed a proposed 20-year agreement in which France would receive 800 million tons of oil in return for sophisticated arms and industrial equipment.

Jobert was said to have sought a similar oil-arms agreement in his discussions in Kuwait Jan. 27–28. Kuwait government sources reported Jan. 28 that France was ready to supply fighter planes, tanks and anti-aircraft missiles without political conditions. Jobert, the sources said, offered French participation in petrochemical and oil refinery projects in Kuwait in return for a yearly guarantee of oil delivery.

The French foreign minister conferred in Damascus Jan. 29 with Syrian President Hafez al-Assad and Foreign Minister Abdel Halim Khaddam.

Jobert had defended his government's policy of bilateral arms-oil agreements with Arab countries Jan. 21. He charged that this policy was being criticized in some quarters "because France might take someone else's place or ask that a little room be made for her."

Jobert visited Baghdad Feb. 6–8 and held talks with Iraqi leaders on oil-economic aid arrangements.

France Jan. 9 had confirmed an oil agreement with Saudi Arabia. The government announcement said France would receive 27 million tons of crude oil from Saudi Arabia over the next three years.

The Paris statement did not say what the Saudis would receive in return, but Saudi Petroleum Minister Sheik Ahmed Zaki al-Yamani had said Jan. 7 that his country would get French warplanes and other military equipment in exchange. Other sources said the French also would provide industrial machinery and technical advice.

France signed energy agreements with Iran and Libya in February.

The accord with Iran, signed in Paris Feb. 9 by French and Iranian Economic Ministers Valery Giscard d'Estaing and Hushang Ansary, provided for collaboration in energy development and industrialization expected to be worth $3 billion–$5 billion.

Negotiations were being conducted for a contract under which France would build up to five nuclear power plants in Iran and a natural gas pipeline to France. Both nations would jointly construct a gas liquefaction plant and a fleet of gas tankers. Other negotiations would be held for French companies to construct a petrochemical complex, steel factory and electric plant.

Oil deliveries to France were not mentioned in the agreement, although French objectives were seen as assuring long-term energy supplies and offsetting increased oil costs with sales of industrial equipment and know-how.

In signing the accord, Ansary noted that Iran was "the first oil-producing country to adopt nuclear power as a source for its energy, taking the view that oil is too valuable to be used for lighting and heating."

In the deal with Libya, France agreed to provide technical and economic aid in exchange for Libyan oil in an agreement in principle signed in Paris Feb. 19 at the end of a week-long visit by Libyan Premier Abdel Salam Jalloud. The agreement, also signed by French Premier Pierre Messmer, mentioned no overall amount of the total value of exchanges nor named specific projects. Possible projects included French construction of nuclear power plants in Libya, development of transport, harbors and telecommunications, and joint French-Libyan investments in third countries.

Jalloud said the agreement "opened great prospects, because we want to prove to Europe that a real economic force can be created in an association of the Mediterranean countries, Europe and the Arab world in general."

The Defense Ministry announced April 16 that France had agreed to sell an undisclosed number of Mirage F-1 fighter planes to Kuwait.

It was reported that the accord called for providing Kuwait with 32 of the aircraft with spare parts as well as construction of a radar network and early-warning system at a cost of $300 million. The aircraft were to be equipped with air-to-air and air-to-ground missiles, with delivery starting at the end of 1974. No conditions had been imposed.

French-Iranian deals—The shah of Iran paid an official visit to France June 24–26, 1974 and placed orders for about $4 billion to $5 billion worth of items over a 10-year period under the framework of the February accord.

Under an agreement signed June 27 by foreign ministers of both countries, Jean-Pierre Fourcade and Hushang Ansari, Iran agreed to make a $1 billion advance deposit with the Bank of France to cover the orders, a payment that would substantially ease France's oil-related balance of payments deficit.

French President Valery Giscard d'Estaing, who conferred several times with the shah during his visit, said June 27 that the agreements were a sign that "in international affairs, Iran and France have parallel attitudes since both intend to maintain their independence and at the same time, to cooperate in the advent of a new international order."

Under the contracts, France would supply Iran with five 1,000-megawatt nuclear reactors worth about $1.1 billion. Delivery would be completed by 1985. France would also supply the atomic fuel, train Iranian scientists and technicians and establish a nuclear research center. The French Foreign Ministry said the agreement had implied safeguards, apparently a reference to the clause pledging both parties to respect their "international obligations." Iran had signed and ratified the 1968 nuclear nonproliferation treaty which provided for a safeguard system maintained by the International Atomic Energy Agency.

The other agreements cited in a joint communique issued June 27 provided for France to build a subway in Teheran, supply a steel plant to be built by Creusot-Loire, build a liquified natural gas plant with other nations, participate in construction of a natural gas pipeline and build 12 large tankers through a French-led consortium. Other potential projects were envisaged, including the creation of a petrochemical industry in Iran and electrification of Iranian railroads. The communique also said Iran would increase its oil shipments to France, but did not specify the amount. (Iran currently supplied 12% of France's oil.)

At a press conference in Paris June 27, the shah confirmed that the French accord also involved military sales but declined to give details other than the purchase of fast motor boats. The shah also called for nationalization of the entire oil industry, which would permit transactions on a state-to-state basis. Defending the sharp increases in oil prices, he said "We're trying to defend ourselves against your rampant inflation. You are going to blow up, and you're going to blow us up with you."

The shah also reaffirmed Iran's readiness "to turn our area into a non-nuclear zone, that is, an area where no nuclear weapons should be used or stored." (The Iranian government had denied June 24 that the shah had said Iran intended to have nuclear weapons soon. The shah had been quoted in the French business magazine Les Informations June 23 as saying Iran would have nuclear weapons "undoubtedly and sooner than is believed.)

At a luncheon with French businessmen June 26, the shah had proposed construction of a natural gas pipeline from Iran to Europe, saying that Iran could supply 50% of West Europe's gas needs.

The shah returned to Teheran June 2[illegible] in the anglo-French Concorde supersonic jet, for which Iran had three purchase options.

French Premier Jacques Chirac visited Teheran Dec. 21–23 and announced that h[illegible] had signed economic agreements with Iran worth $6 billion.

One of the major agreements called for installation of the French color television system, which had been vying with the West German system for the contract. Others involved construction of a Teheran subway system, a steel plant, a Renault car factory and liquefied gas tankers. The $6 billion in contracts did not include the value of two nuclear power plants for which Creusot-Loire, the French heavy industry group, had, according to a Nov. 20 report, received firm orders.

Chirac also said he had assured the shah that French President Giscard d'Estaing and U.S. President Gerald Ford, who had met in Martinique, had not decided to revalue gold but had agreed only to let central banks value their gold holdings at market prices. He called this an "accounting-system" reform that would not reduce the purchasing power of the oil countries' petroprofits.

French Finance Minister Jean-Pierre Fourcade had announced Dec. 20, a few days after the Martinique summit, that France would revalue its official gold holdings at market prices in 1975, a move that reportedly would increase the dollar value of France's gold reserves from about $4.2 billion to $18–$19 billion.

The shah, in a statement reported by the Associated Press Dec. 19 and subsequently disputed by Iran, had warned that upward revaluation of gold would lead to steep oil price increases and could spark "the collapse of the whole monetary system."

Other deals—The French government announced Dec. 3 that France and Iraq had reached agreement on business contracts worth over $1 billion during a visit to Baghdad by Chirac Dec. 1–2. But the government denied reports issued Dec. 2 that Iraq had granted France a $1 billion loan to help the French pay for Iraqi oil.

The contracts provided for delivery by the French corporation, Pechiney, of an aluminum plant and a military hospital and for Iraq's adoption of the French color television system. An agreement in principle was also reached for construction by Creusot-Loire of a petrochemical complex at Basra.

A deal between France and the Persian Gulf emirate of Qatar provided for Qatar to lend France the equivalent of more than $110 million and also deposit more than $55 million with the Bank of France against the future delivery of French goods, it was reported Dec. 19. The accord provided for creation of a joint subsidiary by France's state-run CDF-Chimie Co. and Qatar Petroleum Co. which would construct an oil refinery complex in Qatar and one in France within two years.

British deals with Iran. Iran would supply Great Britain with five million additional tons of crude oil in exchange for about $240 million worth of industrial goods under an agreement concluded Jan. 25, 1974. The deal was made in St. Moritz, Switzerland, where British Chancellor of the Exchequer Anthony Barber and Trade and Industry Secretary Peter Walker conferred with the shah of Iran, who was on vacation.

The British goods would include textile fibers, steel, paper and petrochemicals, but no arms, according to British spokesmen. However, Iranian officials had recently discussed possible weapons purchases in London.

On their return to London, Walker and Barber said the extra oil, representing about 5% of Britain's oil imports in 1973, would mean that Britain would now receive 25% of its oil from Iran. The price of the five million tons would be about $7 a barrel, less than Iran received from concessionary oil firms.

Iran July 22 agreed to extend to Britain a loan of $1.2 billion in three annual installments. Chancellor of the Exchequer Denis Healey said the funds would be used to assist Britain's failing government-owned industries.

(Two British construction companies, Costain Civil Engineering and Taylor Woodrow International, had announced a contract to build a major dry dock complex in Dubai on the Persian Gulf, it was reported Jan. 2. The complex, valued at the equivalent of $218 million, would have facilities for three supertankers and eight crude oil carriers. The project was scheduled for completion in four years.

(London merchant bankers Morgan Grenfell and the Arab Bank established a new joint company, the Arab and Morgan

Grenfell Finance Co., for development of business in the Middle East and other areas, it was reported Jan. 14. The bank would organize "project finance" and help attract Arab funds into regional investment opportunities.)

Britain gets more Saudi oil—Saudi Arabia had given the Arabian American Oil Co. permission to increase the quota of oil to Britain, it was reported in London Jan. 15, 1974. The size of the new quota was not immediately disclosed, but a New York Times report from Beirut Feb. 3 said Britain was receiving an immediate supply increase of 200,000 barrels a day.

Japanese-Iraqi deal. Under a deal made in Tokyo Jan. 17, 1974, Japan was to lend Iraq $1 billion to construct oil refining and other industrial facilities in exchange for crude oil and petroleum products. Japan also was to provide Baghdad with technical and training assistance and tanker ships.

The accord was signed in Tokyo Aug. 16 by Iraqi Economy Minister Hikmat el-Azzawi and Japanese Foreign Minister Toshio Kimura.

The announcement said Iraq would receive a $250 million Japanese government credit to run for 18 years at an annual interest rate of 4% and $750 million in Japanese bank loans. The development projects in Iraq would include an oil refinery, a petrochemical complex, cement and fertilizer plants and an aluminum smelter.

Iraq was to supply Japan with 1.12 billion barrels of crude oil and oil products over the next 10 years.

West German oil agreements. West Germany announced Jan. 27, 1974 that it had agreed in principle to finance and build a $1.2 billion oil refinery in Bousair, Iran, in return for 182.5 million barrels of Iranian oil for five years. The refinery, to have an annual capacity of 25 million tons, would be totally owned by Iran after 15 years.

The agreement was reached after talks in St. Moritz between the shah and West German Economics Minister Hans Friderichs. They also discussed Iranian proposals for a German-financed billion-dollar chemical plant in Iran.

A West German firm and Algeria's government oil company, Sonatrach, was reported Jan. 22 to have signed Sonatrach's largest joint venture agreement for oil exploration in the Sahara.

The West German government-backed combine Deminex signed a 2-4 year $103 million prospecting plan for six parcels totaling over 11,000 square miles. Algeria would develop any fields discovered, with production to be shared.

(Two other Algerian deals were reported—with Petrobras of Brazil, and Copex of Poland, worth $39 million and $32 million respectively.)

Iran to supply India. Under an agreement announced Feb. 22, 1974, Iran was to supply India with oil on credit and to invest $300 million in a joint project to develop India's iron ore export.

The announcement was made at the end of Foreign Minister Swaran Singh's visit to Teheran, where he sought Iranian aid in meeting his country's difficult balance of payments situation resulting from oil price increases.

Italian deals. An agreement under which Libya would increase its oil exports to Italy from 23 million tons to 30 million tons annually in exchange for technological aid was signed in Rome Feb. 25, 1974 after a four-day visit by Libyan Premier Abdel Salam Jalloud. Italy would help construct petrochemical, steel and shipbuilding works in Libya, assist Libyan farmers and provide oil tankers and freighters. The price of the oil would be decided later.

Italian Foreign Minister Aldo Moro had visited Middle Eastern oil nations in late January and early February. He was reported to have received Saudi Arabia's tentative agreement to a sharp increase in crude exports to Italy in exchange for Italian participation in construction of industrial and agricultural projects in Saudi Arabia.

Italy's Istituto Ricostruzione Industriale later negotiated an agreement to supply and build various public works and manufacturing plants in Iran, it was reported June 11. The projects, valued at over $2 billion, included construction of a major steel plant and development of the infrastructure of the Bandar Abbas area in Iran, between the Persian Gulf and Gulf of Oman.

Italy and Iran Dec. 19 signed an accord for Iran to receive technological aid in exchange for providing Italy with credits of $5 billion, part of them in oil, to help reduce Rome's large balance of payments deficit. The agreement was signed at the end of a four-day visit to Iran by President Giovanni Leone and was announced Dec. 20.

Both countries would engage in joint development projects in Iran that would include construction of a steel mill in the port of Bandar Abbas, a $250 million chemical fertilizer plant, an aluminum factory and a facility for desalination of sea water.

Italy and Iraq July 17, 1974 signed a 10-year economic and technical accord under which Italy would supply industrial plant, public works, agricultural development aid and technology to Iraq in exchange for increased supplies of Iraqi oil.

U.S.-Saudi arms agreement. Riyadh radio April 14, 1974 announced that the U.S. had agreed to develop and modernize the Saudi Arabian national guard under an accord signed by U.S. Ambassador James E. Akins and Prince Abdul Ben Abdul Aziz, the guard commander. The U.S. was to provide the weapons and training valued at $335 million. The broadcast said the arms would include armored vehicles, antitank weapons and artillery.

The signing of the pact followed an announcement in Riyadh and Washington April 5 that the countries had agreed to strengthen their cooperation in the economic, industrial and military areas. Negotiations were to be held to set up the mechanism for bilateral exchanges.

According to an accompanying State Department statement, "Despite the strain that arose during the oil embargo, we remained in close touch with our Saudi friends." The U.S.' aim was "to broaden and deepen the entire range of Saudi-American relations in ways that will enhance stability in the Middle East."

The two countries agreed to wide-ranging economic and military cooperation under an accord signed in Washington June 8 by Secretary of State Henry A. Kissinger and Prince Fahd Ibn Abdel Aziz, second deputy premier and half-brother of King Faisal.

Fahd, who had arrived in Washington June 6 for the negotiations, said the pact was "an excellent opening in a new and glorious chapter in relations" between the two countries. Kissinger called the agreement "a milestone in our relations with Saudi Arabia and with Arab countries in general."

The agreement established two joint commissions on economic cooperation and on Saudi Arabia's military requirements. The economic commission, which would hold its first meeting in October in Saudi Arabia, would be comprised of members of the U.S. State Department, the Treasury, the Commerce Department and other agencies as well as comparable Saudi government agencies.

The military commission "will review programs already under way for modernizing Saudi Arabia's armed forces in light of the kingdom's defense requirements, especially as they relate to training," the agreement said.

U.S. Treasury Secretary William E. Simon visited Saudi Arabia and Kuwait July 19–21 during a Middle East trip whose purpose was to discuss financial investments, oil and trade developments.

(Simon denied at a Cairo news conference July 16 a report July 15 by The American Banker, a banking industry publication, that he had said he was bypassing Iran on his current mission because Shah Mohammed Riza Pahlevi "is a nut." Simon said the term was taken out of context, that he was "using the vernacular in the same way anyone would describe himself as a nut about tennis or golf." He intended to depict the shah as having "very firm ideas about oil," Simon explained. The American Banker had further quoted Simon as being critical of the shah, saying he "wants to be a superpower. He is putting all his oil profits into domestic investments, mostly military hardware.")

Simon's meetings with King Faisal and other Saudi officials July 20 were aimed at establishing the procedures to implement the U.S.-Saudi agreement. Petroleum Minister Sheik Ahmed Zaki al-Yamani said July 20 that his country would invest a major portion of its oil income in the U.S., particularly in government securities and possibly in the American stock market.

Simon met with government officials in Kuwait July 21. At a news conference following a dinner for the secretary, Kuwait Petroleum Undersecretary Sheik Ali Khalifa Al Sabah announced that his country was planning to participate in nuclear energy programs in partnership with some of the developing countries.

It was disclosed that Simon had sought to persuade the Saudis and Kuwaitis to invest in special U.S. Treasury bonds. Treasury aides accompanying Simon remained behind in those countries to work out the terms.

Simon said in Paris July 24 that he was convinced that oil prices would drop in August and that the Middle East oil producers were "responsible investors" who recognized a self-interest in preserving stable financial conditions.

Libyan-Rumanian accord. The official Libyan news agency Arna announced that Rumania would get 12 million tons of crude oil over the next four years in exchange for technical assistance in expanding Libya's road and rail networks and in school and hospital construction projects (reported Sept. 7, 1974).

EEC bid to Arabs. The nine foreign ministers of the European Economic Community (EEC) agreed June 10, 1974 on the first day of a two-day meeting in Bonn, to pursue its previously proposed offer of economic, technical and cultural cooperation with 20 Arab nations. Progress on the decision reached in early March had been delayed because of U.S. objections.

Under the plan, initiatives for exploratory talks would be launched immediately. Only if these proved fruitful would the Arabs and EEC hold a ministerial-level conference. An unidentified German official said the range of cooperation with the Arabs would cover culture, education, technical assistance and aid, adding "this is not an 'oil dialogue.'"

U.S. Secretary of State Henry A. Kissinger, after meeting June 11 with West German Foreign Minister Hans-Dietrich Genscher at Bad Reichenhall, near the Austrian border, said the U.S. still objected to the prospect of foreign ministers of the EEC and Arab nations meeting "without a very clear idea of what's on the agenda."

(Thirty-five members of parliament of seven EEC nations March 25 had formed an association for "long-term economic, technical and cultural cooperation" between Europe and the Arab nations, called the Parliamentary Association for European-Arabic Cooperation. Political parties, conservative to Communist, were represented from all the EEC nations except the Netherlands and Luxembourg. The group would be financed by "sympathetic" business concerns and international foundations, not by governments, according to association spokesmen.)

EEC finance ministers agreed in Luxembourg Oct. 21 to seek a joint loan of up to $3 billion, mainly from oil-producing countries, to help members cover balance of payments deficits caused by soaring costs of oil imports.

The $3 billion figure, including principle and interest, would be for a minimum of five years. The nine members would jointly guarantee the loans, with the three big powers—France, West Germany and Britain—each assuming 22% responsibility for loans by any member. In case of multiple defaults, each member would guarantee a maximum of double its normal share.

Prices & Production

Controversy flared during 1974-75 over claims that the petroleum producers manipulated production in order to charge unconscionably high prices.

World oil surplus seen. A U.S. government study predicted that reduced demand for oil and a modest increase in production would result in a world-wide oil surplus, estimated at 1.5 billion barrels a day by the second half of 1974 and rising to 2.7 million barrels a day during the first half of 1975.

According to the study, which was submitted to the 12-nation Energy Coordination Group meeting in Brussels and reported in the New York Times July 1, 1974, the surplus would put downward pressure on world oil prices and create cracks in the unity of the Organization of Petroleum Exporting Countries (OPEC).

Oil industry estimates generally confirmed the U.S. study, but anticipated a slightly larger surplus in future months than did the government report. Oil companies, whose stocks had been severely depleted during the Arab oil embargo, had completely resupplied in recent months, spokesmen told the Times. One unnamed company official said, "You can almost say that we have nowhere else to put the stuff."

Oil prices already had dropped in European markets because of the increase in supplies—the wholesale price of a metric ton of gasoline ranged from $134–$139 in Rotterdam June 28, compared with a $174–$183 price spread two months earlier.

A similar downward movement in price had not occurred in the U.S., however, because the nation relied more heavily on domestic supplies of oil, which was increasing in price as oil from OPEC nations was declining, according to industry spokesmen.

"The present combination of free world oil production and price levels is unsustainable. Reduction in output, prices or both seems likely this summer," the U.S. study declared. Saudi Arabia had based its objections to an OPEC proposal to raise prices on this ground, contending that severe economic dislocations in oil importing countries caused by dramatic price increases necessitated a reduction in price, and they threatened to increase production to force down prices in order to achieve this goal.

Other OPEC nations had rejected the Saudi argument, saying that oil prices should keep pace with worldwide inflation. They also threatened to cut production in an effort to negate the effects of the Saudi boost in production levels.

The Beirut-based Middle East Econic Survey reported Sept. 9 that the Middle East's crude oil output in the first half of 1974 averaged 21,820,000 barrels a day, a 1.1 million barrel a day increase over the previous year. The highest increase was registered in Saudi Arabia, whose production rose by 870,000 barrels a day to 8,078,000 and in Iran, whose output was up 314,000 barrels a day to 6,165,000, the journal reported. Production in Kuwait and Qatar dropped.

U.N. raw materials parley. Oil prices and supply were taken up at a special U.N. General Assembly session held in New York April 9–May 2, 1974 to discuss the whole matter of raw materials and world development.

The session had been sought by Algeria, which was host to preparatory talks among nonaligned nations in March. It was widely accepted that Algeria sought to deflect mounting criticism of Arab oil producers for production cutbacks and price increases which had hurt poor as well as prosperous nations, the New York Times reported April 10.

U.N. Secretary General Kurt Waldheim opened the session with a call for collective action by U.N. member states to establish the basis for "a more equitable and workable global economic system." He warned that a "global emergency" faced both rich and poor countries in the form of mass poverty, population growth pressure, food and energy shortages, military expenditures and inflation.

Waldheim said the main theme of the session was to obtain "optimum use of the world's natural resources" to promote social justice. He stressed the "interdependence" of nations and warned against the pursuit of "short-term national interests."

Algerian President Houari Boumedienne addressed the session April 10, calling on underdeveloped countries to take control of their resources from foreign interests. However, he also urged continued foreign assistance to poor nations and proposed that the $80 billion owed by underdeveloped countries to their creditors be canceled or renegotiated.

Boumedienne minimized the damage done by rising oil prices, asserting: "The impact of oil in overall cost makeup has always been ridiculously small." He said current inflation was caused by "overconsumption and gadgetization," and waste that was "rampant throughout the developed countries."

The head of the Chinese delegation, Deputy Premier Teng Hsiao-Ping, followed Boumedienne with a scathing attack on the U.S. and the Soviet Union. "The two superpowers are the biggest international exploiters and oppressors of today," Teng said. The Soviet Union, he charged, was an "international merchant of death" that was selling outmoded arms and equipment in exchange for raw materials, and peddling munitions for Arab oil, which it then sold at high profits.

Other speakers April 10 included West German Foreign Minister Walter Scheel, who stressed the damage done by recent changes in the policies of oil-producing nations. "Trade relations, currency conditions, the development of countries has been seriously upset," he declared. "All this together has had the effect of an earthquake, leaving nothing unscathed."

Soviet Foreign Minister Andrei Gromyko told the session April 11 that the improvement in U.S.-Soviet relations would bring economic benefits to underdeveloped countries. He charged that oil companies were responsible for the energy crisis in the West, and said they should be curbed.

U.S. Secretary of State Henry Kissinger appeared before the session April 15. He pledged a major effort by the U.S. to help developing countries, but warned against use of "the politics of pressure and threats."

Kissinger warned commodity producers against jointly raising prices as had members of the Organization of Petroleum Exporting Countries. "Large price increases coupled with production restrictions involve potential disaster: global inflation followed by global recession from which no nation could escape," he asserted.

"The great issues of development can no longer be realistically perceived in terms of confrontation between the haves and have nots," Kissinger said.

Kissinger listed the issue of expanding energy supplies at equitable prices as one of six problem areas troubling world economy and development. Pledging U.S. action in each area, he said the U.S. would seek to expand energy supply by helping oil-producing countries to diversify and to reduce dependency on non-renewable resources.

Before adjourning, the special session May 1 adopted two documents—an action program and a declaration of principles for a "new economic order." They had been drafted largely by underdeveloped nations.

The documents, approved without a vote, were considered weak, but they signaled a change in relations between the world's rich and poor countries, according to many delegates.

The action program included establishment of an emergency operation to help the world's poorest countries maintain essential imports over the next 12 months (with a call on industrialized nations and other potential contributors to announce their aid by June 15); and establishment of a special fund under U.N. auspices, with voluntary contributions from industrialized countries and others, to provide emergency relief and development assistance beginning Jan. 1, 1975.

The declaration of principles, diluted by a negotiated compromise between European and developing nations, included provisions upholding the right of countries to nationalize their industries (without any reference to prompt and adequate compensation); a demand that developing nations be compensated for past exploitation by colonialists and others; a call for development of cartel-like associations of poor nations; and link of any rise in the price of manufactured goods to the price of the raw materials produced by the poorer countries.

After the two documents were adopted, a series of speakers from the industrialized nations rose to say there were parts of each they could not accept, despite agreeing with their general principles. The strongest criticism came from U.S. Ambassador John Scali, who denounced what he called "steamroller" tactics used in the Assembly.

Scali had introduced a U.S. proposal April 30 for a $4 billion program to help the nations in greatest need, to which the affluent countries and oil-producing nations would contribute, with the U.S. donating a "fair share." The proposal was quickly turned aside by the Assembly, with many delegates noting it had been introduced too late to be given proper consideration.

Some delegates accused the U.S. of intentionally planning the timing to force a rejection and have an excuse for refusing to participate in the emergency program that finally was adopted.

The timing was attributed to U.S. Secretary of State Henry Kissinger, who delayed approval of the proposal until the last minute, ostensibly because of objections by outgoing U.S. Treasury Secretary George Shultz. Schultz had said the economic predicament of the poorer countries had been caused by the sharp rise in oil prices, and the solution was to get oil producers to lower their prices, rather than try palliatives such as emergency assistance.

Oil prices hurting third world. Several black African nations were reported questioning the wisdom of siding with the Arabs on the Middle East conflict in exchange for the promise of an Arab boycott of Rhodesia, Portugal and South Africa, according to the New York Times March 3, 1974.

Petroleum prices in the undeveloped countries had risen 30% and the Arab states had rejected requests for price reductions. Furthermore, a promised grant of $200 million for development had not been received, the African Development Bank said.

The subject provoked criticism of Arab policy at meetings held by the OAU (Organization of African Unity) Council of Ministers in Kampala, Uganda April 1–4, by the OAU foreign ministers in Mogadishu, Somalia June 6–11 and by the OAU heads of state in Mogadishu June 12–16.

Both the Kampala and Mogadishu conferences revealed a growing uneasiness over Arab affiliations and, in a number of actions, expressed disillusionment with Arab attempts to exert more influence on the continent.

Economic issues generated the most serious confrontations, beginning at the foreign ministers' meeting June 8 when an OAU committee reported that Arab oil-producing states had refused to lower oil prices for African states. The Arabs maintained that prices were fixed by the Organization of Petroleum Exporting Countries (OPEC) and were therefore not negotiable. The ministers voted June 9 to accept "in principle" an Arab offer to set up a $200 million fund for soft loans to African states at about 1% interest. A number of countries including Ghana, Ethiopia, Nigeria, Tanzania, Kenya, Uganda and Zaire, were reported to have demanded that larger sums of money and better loan facilities be provided.

Senegal President Leopold Senghor had said in a June 5 interview published in the French newspaper Le Monde that the Arabs had to be "shaken" over the oil issue. A Kenyan official suggested June 19 that African countries reconsider their severence of relations with Israel in face of "the Arabs' uncompromising attitude toward Africa on oil."

Mahmoud Riad, secretary general of the Arab League, defended the Arab offer June 14 in Mogadishu where he was attending the OAU conference as an observer. He said the combined total of the grants, loans and investments that Arab oil-producing countries were providing or planning to provide to poor African nations was "far higher" than the estimated annual increase they faced in the cost of petroleum. He noted that establishment of at least three special funds plus a number of bilateral grants "would come to many, many millions of dollars."

However, increased oil prices were said to be costing Black African countries more than $1 billion, the New York Times reported June 9. An OAU report stated that 33 of the organizations' 42 member states had been hard hit by the oil crisis. Not affected were Algeria, Egypt, Libya, Morocco, Sudan, Tunisia, the Congo, Nigeria, and Gabon.

OPEC bars price hike, boosts royalties. The 12-nation Organization of Petroleum Exporting Countries agreed June 17, 1974, after a three-day meeting in Quito, Ecuador, to continue oil's posted price

RETAIL PRICE OF REGULAR MOTOR GASOLINE

Including Duty & Taxes

[U.S. dollars per U.S. gallon]

Country	As of July 31, 1974			As of July 31, 1973			As of July 31, 1972			As of July 31, 1971			As of July 31, 1970		
	Retail price	Duty	Tax	Retail price	Duty	Tax	Retail price	Duty	Tax	Retail price	Duty	Tax	Retail price	Duty	Tax
Austria	$1.190	$0.048	$0.450 [1] +16	$0.851	$0.051	$0.433	$0.585	$0.037	$0.339	$0.520	$0.035	$0.319	$0.495		$0.315
Denmark	1.351	.562	.177	1.033	.568	.137	.752		.519	.693		.482	.658		.477
Finland	1.208		.542	.821	.014	.429	.694	.014	.429	.657	.014	.420	.599		.321
France	1.318		.756	1.062		.817	.817		.594	.756		.552	.736		.547
Greece	1.701	.014	.050	.881	.005	.183	.756	.014	.446	.756	.014	.446	.643	$0.014	.401
Ireland	.967	.416	.058	.866	.520	.045	.677	.432	.033	.675	.430	.033	.617	.415	.0389
Italy	1.690	[1] 7	1.152	.988	[1] 6	.761	.988	[1] 6	.771	.919	[1] 6	.717	.780	[1] 6	.612
Luxembourg	1.115	.506	.052	.877	.540	.015	.605	.370	.029	.607	.370	.030	.593	.375	.023
Netherlands	1.368		.858	1.142		.779	.773		.578	.692		.476	.608		.420
Norway	1.470		.631 [1] +20	.841		.596	.841		.596	.782		.578	.661		.454
Portugal	1.410	.500	.463	.846	.500	.097	.792	.466	.090	.750	.444	.084	.737	.446	.083
Spain	1.160		.380	.783		.393	.595		.326	.570		.310	.530		.310
Sweden	1.179		.628	1.012		.665	.809	NA	.534	.731	NA	.497	.643		.424
Switzerland	1.120	.657	.050	.896	.542	.036	.700	.440	.020	.580	.360	.020	.520	.330	.010
United Kingdom	1.250	NA	.630	.780		.480	.676	.461	NA	.652	.450	NA	.625	.450	
West Germany	1.265	[1] 6	.648 [1] +11	1.180	[1] 6	.715 [1] +11	.751	[1] 6	.467 [1] +11	.660		.380 [1] +11	.578		.362 [1] +11

NA=not available.

Source: U.S. Bureau of Mines. Prepared by IPAA. February 1975.

at current levels for three months, from July 1. In another action, all OPEC members, except Saudi Arabia, announced a 2% increase in royalties levied on Western oil companies.

The dual decision on prices and royalties would have the effect of curbing the oil companies' profits unless they could pass the added cost on to consumers.

Saudi Petroleum Minister Sheik Ahmed Zaki al-Yamani, who had blocked OPEC attempts to raise prices, explained that his government was opposed to increasing royalties because of its forthcoming negotiations with the Arabian American Oil Co. (Aramco) "to establish a new relationship. Therefore, we cannot associate ourselves with any fiscal changes which might contradict the future arrangement we will have with Aramco."

An OPEC communique said the organization's economic commission had complained of "the excessive profits earned by the international major companies" and of their failure during January–June "to contain the alarming rate of inflation while the level of posted prices was kept constant by OPEC countries." OPEC's decision not to raises prices for at least three months was designed to give the companies "another opportunity to adopt the necessary measures in these respects," the communique said.

The OPEC meeting June 16 had rejected an appeal by the European Economic Community (EEC) against new taxes on oil exports. An EEC note circulated among the delegates had argued that the taxes to absorb excess profits would only increase oil prices in Europe and harm the world economic situation.

Iran, Kuwait, Qatar and Libya raised royalty rates paid by oil companies operating in their countries two percentage points to the 14.5% level. Abu Dhabi the previous week had been the first OPEC member to increase royalties to the new level.

Since royalty rates were based on posted prices, the latest round of increases would mean a boost of about 24¢ a barrel paid to Iran and Qatar, about 23¢ to Kuwait, and about 32¢ to Libya.

Another OPEC member, Indonesia, had raised the export price of its oil July 1 to $12.60 from $11.70 a barrel. Although OPEC had agreed June 17 to freeze prices for three months, an Indonesian spokesman said his government was not bound by that decision. "Our oil prices are determined on a bilateral basis with buyers," he said.

In its last previous increase, Indonesia April 1 had raised its oil price from $10.80 a barrel. A spokesman for the state-owned Petamina Oil Co. had said that the April 1 increase, the fifth raise since April 1973, was intended largely to increase the government's revenue and was not related to the current energy crisis.

Libya reduced the price of its oil from $16 a barrel to $14.50–$14.80 a barrel, the Middle East Economic Survey of Beirut reported April 6.

Representatives of 10 OPEC states met in London and agreed Aug. 16 to exchange data with each other on prices and sales of government oil in order to stabilize prices at current high levels. The discussions examined recent price trends and ways of insuring smoother marketing of OPEC's crude oil.

An unidentified Arab delegate was reported to have said during a recess of the Aug. 15 session that world oil production was running at a surplus of up to 4 million barrels a day. OPEC's daily output amounted to 30 million barrels. Oil storage facilities were then virtually full in both consuming and producing nations, as well as in tankers on the high seas, the delegate confirmed.

Arabs cut oil output. Oil production in Saudi Arabia was sharply cut in August, the Saudi government and the Arabian American Oil Co. (Aramco) confirmed Aug. 26.

Aramco said the drop in output from the August target of 8.5 million barrels a day to less than 8 million was the result of bad weather disrupting freighter loadings at the company's terminal at Ras Tanura, Saudi Arabia. Other sources, however, attributed the production cut to the fact that Aramco's customers were purchasing less oil because of a world surplus and to the fact that many of the firm's storage tanks in Europe and elsewhere were filled to capacity.

Kuwait had cut output in August and planned further cuts during the next two months, Oil Minister Abdel Rahman Atiki said in an interview published Aug. 21 by the Beirut weekly magazine As Sayyad.

Atiki did not disclose the extent of the production drop, but current output was about 2.55 million barrels a day, 15% less than in the pre-embargo days of September 1973. The minister asserted that "if prices are determined by supply and demand, then we shall reduce the supply of our crude oil to increase the demand on it."

Kuwait, Atiki said, was not convinced by the arguments of consumer nations that lower oil prices would end chaos in the world economy. Referring to a visit that U.S. Treasury Secretary William E. Simon had made to Kuwait July 21, Atiki said "our talks with Simon were difficult and exhaustive but he did not convince us of his viewpoint at all." The minister added: "Why should we be responsible for helping America solve her economic problems? When the Arab states were poor and our oil used to be sold cheaply, what assistance did the United States give us?"

Ford's comment. The problem was mentioned briefly by President Ford at a televised press conference Aug. 28, 1974.

Ford was asked if the U.S. could take any action against oil exporting nations, which recently had decided to reduce production in order to maintain high prices, through cartels such as Aramco, in which U.S. oil firms participated.

Ford gave no direct answer but urged acceleration of Project Independence—increased oil and natural gas drilling, expansion of geothermal and solar research, and greater efforts toward expediting the licensing of new nuclear reactors.

The situation also required short-term action entailing international cooperation, the President continued. "High oil prices and poor investment policy" for these nations' greatly increased petrodollars could result in "adverse repercussions" for the industrial world, Ford said, and underscored the need for oil consuming industrial nations to "meet frequently and act as much as possible in concert."

It was reported by Arab informants in Beirut Dec. 28 that the U.S. had sent messages to major oil producers in July expressing displeasure with the sharp increase in petroleum prices. One note was said to have stated that the U.S. "believes the tendency to continue to increase the price is equally unfortunate and unjustified" and was contrary to the interest of the oil producers themselves. The messages were said to have been sent to at least five oil-producer states, including Iran, Kuwait, the United Arab Emirates and Venezuela.

OPEC to raise oil taxes. The Organization of Petroleum Exporting Countries (OPEC) agreed at the end of a two-day meeting in Vienna Sept. 13 to raise by 3.5% taxes and royalties paid by foreign oil companies to the oil-producing states. Saudi Arabia was the only OPEC member which did not approve the decision.

The tax on a barrel of oil would rise by 33¢, effective Oct. 1. An OPEC communique said the government price for oil would remain frozen for the last quarter of 1974 but that "as of January 1975, the rate of inflation in the industrialized countries will automatically be taken into account with a view to correcting any future deterioration in the purchasing power of oil revenues."

In imposing the tax increase, the OPEC countries argued that price increases in industrial countries for machinery and food that the oil countries must import required a boost in government revenue from oil.

Saudi delegate Sheik Ahmed Zaki al-Yamani said his country did not oppose the tax increase, but held that it should be coupled with a reduction in the price of oil. His country would not impose the tax increase until it completed negotiations with the Arabian American Oil Co., Yamani said.

U.S. Federal Energy Administrator John C. Sawhill, commenting on the OPEC action, said Sept. 13 that any tax or price increase by the international oil cartel was "economic blackmail." He said some of the oil firms might absorb the rise, "but they won't absorb all of it."

Saudis raise oil price—Aramco confirmed Sept. 16 that Saudi Arabia had increased the price it charged oil companies for its petroleum by 22¢ a barrel, or 2%. The price change applied to "buyback" oil, petroleum accruing to the Saudi government from its current 60% interest in Aramco which was sold back to the oil companies.

Saudi Oil Minister Yamani explained Sept. 17 that the increase did not conflict with his government's desire to see the price of oil go down generally. He said U.S. oil executives had been informed by Saudi Arabia in June of its decision to boost the price of buyback oil from 93% of posted price to 94.864%, reflecting the government's policy of keeping participation rates at the same level charged by other Persian Gulf producers.

Ford & Kissinger issue warnings. U.S. officials asserted Sept. 23, 1974 that the high prices set by the oil-producing nations imperiled the world's economy and could lead to "confrontation" and "a breakdown of world order and safety." The warnings were contained in speeches delivered by President Ford to the opening of the World Energy Conference in Detroit and by Secretary of State Henry A. Kissinger to the United Nations General Assembly.

The hardening of the U.S. stance on the high oil prices prompted an angry reaction from Arab oil producers and predictions from other quarters of a possible military showdown.

In his address, Ford asserted that the U.S. recognized the oil producers' need to develop their own economy, "but exorbitant prices can only distort the world economy, run the risk of a worldwide depression, and threaten the breakdown of world order and safety." "Sovereign nations," Ford added, "cannot allow their policies to be dictated, or their fate decided, by artificial [price] rigging and distortion of the world commodity markets."

Stressing the principle of interdependence among states, Ford said "There is no way in today's world for any nation to benefit at the expense of others—except for the very short term and at a very great risk." With representatives of Arab oil nations in the audience, Ford said "the whole structure of our society rests upon the expectation of abundant fuel at reasonable prices," and that expectation "has now been challenged." The President noted that "throughout history, nations have gone to war over natural advantages, such as water, or food or convenient passages on land or sea." But he pointed out that in this nuclear age "war brings unacceptable risks for all mankind."

Ford's warning against the use of oil as an economic weapon was coupled with an appeal to oil-consuming nations to take these steps to avoid its consequences: expand domestic oil production; "resolve not to misuse their resources"; "fully utilize their own energy resources"; and "join with others in the cooperative efforts to reduce their energy vulnerability."

In response to newsmen's questions prompted by the Ford and Kissinger warnings, U.S. Defense Secretary James R. Schlesinger said Sept. 25 that the U.S. had no intention of seeking military action against oil-producing nations in the Middle East, but instead was seeking a solution to rising prices through "amicable discussions." While the U.S. "regards the problem of oil prices as detrimental to the world economy, we expect to have a solution through negotiations," Schlesinger said. The inflation that was being caused in part by high oil prices, he said, could reduce the military power of the U.S. and the Western alliance.

Schlesinger ruled out as "inappropriate at this time," the extensive arms sales by the U.S. to Persian Gulf countries "as a lever" to persuade them to lower their oil prices.

In an address to the U.N. General Assembly Sept. 18, Ford had challenged oil-producing nations to stop using oil as a political and economic weapon.

Ford told the General Assembly that the U.S. "recognizes the special responsibilities we bear as the world's largest producer of food," but he said "it has not been our policy to use food as a political weapon despite the oil embargo and recent oil price and production decisions." He emphasized that "energy is required to produce food, and food to produce energy."

Ford challenged oil-exporting nations to "define their policies to meet growing needs" without "imposing unacceptable burdens on the international monetary and trade system." He said that by "confronting consumers with production restrictions, artificial pricing and the prospect of ultimate bankruptcy, producers will eventually become the victims of their own actions."

Ford urged all countries to join in a "global strategy for food and energy" and warned that "failure to cooperate on oil and food and inflation could spell disaster for every nation represented in this room."

Kissinger's analysis—Kissinger told the General Assembly Sept. 23 that "strains on the fabric and institutions of the world economy," caused largely by artificially high petroleum prices, threatened to "engulf us all in a general depression."

In a gloomy assessment of the world economic situation, Kissinger said "the early warning signs of a major economic crisis are evident." Inflation rates "unprecedented in the past quarter century are sweeping developing and developed nations alike," he declared. "The world's financial institutions are staggering under the most massive and rapid movements of reserves in history. And profound questions have arisen about meeting man's most fundamental needs for energy and food."

In the past, Kissinger noted, "the world has dealt with the economy as if its constant advance were inexorable." Now, he said, "we continue to deal with economic issues on a national, regional or bloc basis at the precise moment that our interdependence is multiplying."

Kissinger noted that the U.S. had launched programs with such oil producing nations as Saudi Arabia and Iran to help them diversify their economies. He implied that if such cooperation failed to ease oil prices, the U.S. might change its policy. "What has gone up by political decision can be reduced by political decision," he said.

Kissinger also expressed concern about the spread of nuclear explosives following atomic tests by India. "Political inhibitions" about atomic weapons were "in danger of crumbling," he said, and "nuclear catastrophe looms more plausible whether through design or miscalculation, accident, threat or blackmail."

Kissinger pledged the U.S. would soon make new proposals to tighten safeguards to prevent countries from transforming nuclear materials meant for peaceful purposes (plutonium and its by-products) into explosives.

Kissinger noted that "the shadow of war remains" over the Middle East, and he urged the Assembly to be "realistic" in its debate on the issue. "The art of negotiation is to set goals that can be achieved at a given time and to reach them with determination," he said. Progress could be thwarted "by asking too much as surely as by asking too little."

Arab and other reaction—Arab anger over the comments by President Ford and Kissinger was expressed Sept. 24 by political leaders, public figures and the press, who accused the U.S. of waging a war of nerves against the Arab countries.

Kuwaiti Petroleum Minister Abdel Rahman al-Atiki said if the U.S. went ahead with plans to form a bloc of oil-consuming nations, a confrontation with the oil producers was inevitable.

Clovis Maksoud, who had recently completed a six-month tour of the U.S. as a special Arab League representative, called for cancellation of Kissinger's scheduled visit to the Middle East and accused Ford of pro-Israel bias.

An official of the Organization of Petroleum Exporting Countries (OPEC) took issue with Ford's view that oil prices were exorbitant. He said: "Inflation did not begin with the increase in oil prices. It is rather because of inflation that oil prices have had to be adjusted."

A representative of the Arab League said in Cairo Sept. 24 that his organization had read the Kissinger and Ford statements with "great surprise coupled with concern." "If the United States wants cooperation with the Arabs, one would expect different words" from U.S. leaders, he asserted.

Saudi Arabian Oil Minister Sheik Ahmed Zaki al-Yamani said he "heard no threats" in Ford's speech. "What I heard from him was a tone of cooperation rather than confrontation," he said. Yamani warned the U.S. in a later statement of the danger of "economic imperialism" in dealing with friendly oil-producing states.

French Foreign Minister Jean Sauvagnargues said Sept. 24 that U.N. delegates and their governments seemed "apprehensive about the implicit threats" in the Kissinger and Ford statements. He warned against a confrontation between oil producers and consumers, and noted that Western European nations and Japan were reluctant to push toward such a showdown because they, unlike the U.S., were almost totally dependent on oil imports.

Shah Mohammed Reza Pahlevi of Iran reacted to Ford's address by rejecting the President's bid to cut oil prices. He called instead on industrial nations to reduce prices of their exports first. The shah, who was on a week-long visit to Australia as part of an Asian tour, said in a speech in Canberra Sept. 26: "No one can dictate to us. No one can wave a finger at us, because we will wave a finger back." The Iranian leader added: "If the world prices go down, we will go down with oil prices. But if they go up, why should we pay the bill?"

The U.S. State Department reacted to the shah's statement by saying "there is no spirit of confrontation at all." The U.S., the department said, was attempting "to solve a problem which affects all of us, developed and less-developed countries, consumers and producers."

Arabs defend stance—Representatives of Kuwait and Egypt defended Arab oil policies before the General Assembly and denied they had contributed to world inflation.

Sheik Sabah al-Ahmad al-Jaber, Kuwait's foreign minister, said Sept. 30 that inflation had plagued the industrial nations since the end of World War II, and he attributed it to these nations' "inability to properly manage their domestic affairs." The countries "which are heaping blame on the oil producing countries now," he asserted, "are the ones that started the practice of classifying goods, treating some of them as strategic materials subject to special trade rules."

Sheik Sabah charged that the major oil companies had deliberately frozen oil prices for more than 25 years, while the prices of all basic commodities, manufactured products and services exported by advanced countries were rising. "Raising the price of oil was in essence the correction of an inequitable situation," he asserted.

Egyptian Foreign Minister Ismail Fahmy echoed Sheik Sabah's statement Oct. 1, saying it was "regrettable" to hear it "even claimed that the fragile framework of international economic cooperation would be exposed to danger if the oil producing countries continue their present pricing policies."

Fahmy asserted that the Arabs had used their oil only to secure their "legitimate rights" and had imposed the 1973 embargo only after "warning the countries which assist Israel" in maintaining control over Arab territory.

China backs Arab oil policies—Chinese Deputy Foreign Minister Chiao Kuan-hua told the General Assembly Oct. 2 that the Arab nations' use of oil as a political weapon was an "historic pioneering action" in the Third World's struggle against imperialism.

"The profound significance of the oil battle lies in the fact that the developing countries have united themselves and independently exercised control over their natural resources and fought against plunder, exploitation and the shifting of crises onto them," he said.

Chiao noted that some non-producing countries were suffering temporary hardships due to high oil prices, but he said this should not negate the "historic significance" of the exporting countries' action.

Saudi ties price cut to peace. Saudi Arabia Petroleum Minister Sheik Ahmed Zaki al-Yamani told newsmen in Washington Oct. 2 that oil prices would decline if there were a political solution to the Arab-Israeli conflict. Yamani warned, however, that if the Israelis did not withdraw from occupied Arab territories, "this would produce a war" that would "have a very dangerous effect on prices, as well as on the supply of oil." "Any solution that will stop the fighting," he said, "is in the hands of the American government."

Asked why he was linking a political solution of the Middle East situation with a cut in oil prices, Yamani said, "If you give them [the Arabs] an incentive, they will be on your side."

He described the Sept. 23 remarks by President Ford and Secretary of State

Henry A. Kissinger on oil prices as an "obstacle" to a dialogue between producer and consumer nations.

Kuwait raises tax, royalty rate. Gulf Oil Corp. and British Petroleum Co. were to pay Kuwait $1.14 a barrel more for one million barrels of oil a day during October–December as a result of a tax and royalty hike announced by Kuwait Oct. 1.

A Gulf spokesman said the royalty rate for the fourth quarter would rise from 14.5% to 16.7%, and that the tax rate would be boosted from 55% to 65.75%, bringing Kuwait an additional 3.5% total revenue.

Kuwait's action wiped out a slight price reduction for buyback oil sold to Gulf and BP announced Sept. 30. Kuwait at that time had agreed to sell the two firms 900,000 barrels a day during the fourth quarter at 93% of the posted price, or $10.74 a barrel, a drop from 94.9%, or $10.95 a barrel.

Indonesia bars price increase. Pertamina, the Indonesian state oil company, announced Oct. 1 it had decided not to raise petroleum prices for the current fourth quarter. The decision was believed to reflect a world surplus estimated at 1 million–3 million barrels a day.

Kissinger seeks oil price stability. U.S. Secretary of State Kissinger continued his travels in the Middle East during October and November 1974 in efforts to promote Arab-Israeli peace and to achieve lower and stable oil prices.

In talks with Kissinger in Riyadh Oct. 13, Saudi Arabian King Faisal acknowledged that high oil prices could wreck the world's economy and pledged to help bring them down. In an airport statement before leaving, Kissinger noted that the king's "attitude was constructive and enlightened" after he had informed Faisal of the U.S. concern about high prices and "the impact this can have on the whole structure of the world economy and the stability of the whole international system."

Saudi Foreign Minister Omar Saqqaf, who was seeing Kissinger off, echoed the secretary's remarks, saying that Saudi Arabia was "following a policy on oil which bespeaks a sense of responsibility toward the welfare of the world communities."

Kissinger flew to Algiers Oct. 14 to up the oil situation with Algerian President Housari Boumedienne. Kissinger said they had "discussed our differing approaches to oil prices" and "reviewed ways and approaches to reconcile these different points of view in the months ahead." Boumedienne, an advocate of high oil prices, did not pledge to get them lowered.

Kissinger visited Teheran Nov. 1–2.

In his meetings with Shah Mohammed Riza Pahlevi, Kissinger expressed U.S. concern over high oil prices but received no assurances from the shah that they would come down. A joint communique issued Nov. 2 established a U.S.-Iranian economic commission.

In Riadh Nov. 6, King Faisal again assured Kissinger that he would seek stable oil prices. Speaking for Faisal, Foreign Minister Omar Saqqaf said the Saudi policy was "to keep the oil prices as they are and to try to reach a reduction, albeit a symbolic reduction, or if we can, a greater reduction."

OPEC raises oil prices. The Organization of Petroleum Exporting Countries (OPEC) decided at a meeting in Vienna Dec. 13, 1974 to raise oil prices and adopt a new uniform pricing system.

Petroleum prices were to be increased about 38¢ a barrel, or almost 4%, Jan. 1, 1975 and were to continue at that level until Oct. 1, 1975. The decision to increase the government revenue per barrel for oil marketed through their foreign companies to a maximum of $10.12 from $9.74 followed a formula that had been adopted by three OPEC members Nov. 10—Saudi Arabia, United Arab Emirates and Qatar.

An official communique indicated that OPEC's action was aimed at further reducing the profits of the major international oil companies.

Under the new pricing system, OPEC for the first time based actual prices on government revenues instead of the posted price, the traditional artificial price upon which tax and royalties were estimated.

U.S. Interior Secretary Rogers C. B. Morton said Dec. 13 that the OPEC price increase would add an additional $4 billion to world oil costs "and further depress economic activity." Morton said "the OPEC government's take has now risen over fivefold since last year—$1.70 then compared to $10.12 now."

In their earlier action of Nov. 11, Saudi Arabia, the United Arab Emirates and Qatar reduced the posted price for a barrel of oil from $11.65 to $11.25. The royalty rate on the posted price was increased from 16.67% to 20%, while the tax on the companies' net income was boosted from 67.5% to 85%. The new rates were retroactive to Nov. 1 and were to continue to July 1, 1975, according to the communique.

The three states said the purpose of their move was to reduce the companies' "excess profits" and to help the world's oil customers.

Saudi Arabia Petroleum Minister Sheik Ahmed Zaki al-Yamani said: "We have taken away some of the excess profits that the companies have been making and we are giving it to the consumers in lower prices."

United Arab Emirates Petroleum Minister Manna Saeed al-Oteiba said: "It is up to the consumer to keep an eye on the companies and prevent them from passing on the higher taxes and royalties through higher prices."

Mideast oil output down. A drop in oil production during January 1975 in Kuwait Iran, Saudi Arabia and other Middle East countries was reported Feb. 19. The reduction was attributed by industry sources in London to a decline in world demand, which in turn resulted from a slowdown in world economic activity, energy conservation and mild winter weather in Europe and elsewhere.

The Kuwait Oil Co. said it had cut back January production to 1,806,846 barrels a day, compared with an output of 2,579,-544 barrels a day for the same period in 1974.

Iran's output in January averaged 5,067,000 barrels a day, down from 5,457,-000 barrels in December 1974 and 5,676,-000 barrels in January 1974, an Iranian consortium official reported.

The Arabian American Oil Co. (Aramco) reported that its production in Saudi Arabia was about one million barrels a day less than the production limit set by the Saudi government.

The Trans-Arabian Pipeline Co. (Tapline), an Aramco subsidiary, announced Feb. 19 that it had suspended pumping Saudi crude oil to its terminal at Zahrani, Lebanon Feb. 9 because it had no contracts to ship from there for the next six months. Tapline carried an average of 450,000 barrels a day, but the average had dropped to 50,000 barrels a day in early February, a company official said.

Lower demand for oil was reflected in a further cutback in production by major Arab producers in February.

A survey by the Organization of Petroleum Exporting Countries showed that the daily output in February had totaled between 26 million and 27 million barrels, down more than 10% from the 13 OPEC members' daily average production of 30,-773,000 barrels in 1974, it was reported Feb. 28. In addition to a decline in global demand by industrial consumers, the production cut was attributed to oil company refusal to pay premiums that had been common for high-grade, low-sulphur crude rather than to any production cuts affecting supply.

The world's largest oil producer, Saudi Arabia, had cut its February crude oil production to 6.5 million barrels a day, down 23.5% from the 1974 output, the Middle East Economic Survey reported March 8. The cutback was said to indicate that although the Saudis had refused to agree to a formal production quota within OPEC, they were going along with a drop in output despite decreasing demand, a move that was expected to maintain prices at their current level.

Qatar disclosed March 3 that its production had been reduced to 480,000 barrels a day in February, more than 7% below its normal level. The resultant decline in revenue was said to threaten the country's development plans.

Abu Dhabi's oil minister, Mana Said al-Otaiba, disclosed March 1 that his country, with OPEC authorization, had reached an arrangement with Western oil companies to reduce the price of a barrel of oil by about 55¢ to spur sales; the previous price was $11.20 a barrel. Abu Dhabi's output in October-December 1974 had totaled 1.6 million barrels a day but dropped to 400,000-700,000 barrels a day in January-February, Otaiba said.

Consumer Policy, Producer Response

U.S. hints military move over oil. Secretary of State Henry A. Kissinger warned that the U.S. might use military force in the Middle East "to prevent the strangulation of the industrialized world" by the Arab oil producers. Kissinger made the statement in a Business Week interview in the magazine's Jan. 13, 1975 issue (made public Jan. 2). It aroused angry world reaction, particularly among Arab oil-producing nations.

Kissinger said the use of force would be "considered only in the gravest emergency." "We should have learned from Vietnam that it is easier to get into a war than to get out of it," he said. "I am not saying that there's no circumstances where we would not use force. But it is one thing to use it in the case of a dispute over [oil] price; it's another where there is some actual strangulation of the industrialized world."

As for possible counteraction by the U.S.S.R., Kissinger said, "Any President who would resort to military action in the Middle East without worrying about the Soviets would have to be reckless. The question is to what extent he would let himself be deterred by it. But you cannot say you would not consider what the Soviets would do."

Kissinger said "the only chance to bring oil prices down immediately would be massive political warfare against countries like Saudi Arabia and Iran to make them risk their political stability and maybe their security if they did not cooperate." He ruled out this action, however, as being too risky since it entailed the possible destruction of those countries' systems and their take-over by extremists, which would defeat the "economic objectives" of the West.

Kissinger discounted reports of a possible outbreak of war in the Middle East as "exaggerated." If there were such a conflict, it is not certain the Arabs would impose an oil embargo as they did in 1973, Kissinger said. "It would now be a much more serious decision than it was the last time."

Kissinger reiterated his call for cooperation among the oil-consuming nations of the West to make them "less vulnerable to the threat of embargo and to the danger of financial collapse."

Commenting on statements made in December by Shah Mohammed Riza Pahlevi that Iran would side with the Arabs in any future war, Kissinger said this was indicative of "the trends in the Moslem world" toward "the direction of greater solidarity."

(The shah had asserted in a Cairo newspaper interview published Dec. 27 that Iran would not enter into any future war in the Middle East, although its sympathies would be with the Arabs. Al Ahram quoted the shah as saying: "Of course Iran is not thinking of participating in the fighting. You are aware of geographical and other obstacles. But our sentiment will certainly be on your side." He called for "close regional cooperation" between Iran and the Arab states to remove "the military presence of the great powers" in the Middle East. The shah said he had been misinterpreted by a Beirut magazine which reported Dec. 12 that he hinted of a possible Iranian military alignment with the Arabs in the event of any new war with Israel. The magazine Hawadess had quoted the shah as saying: "All Islamic countries would be involved in a new war with Israel. Of course, we would have no choice this time. The war would be ours.")

Questioned by newsmen about the interview, Kissinger Jan. 2 restated his belief that the use of U.S. military force in the Middle East was unlikely and that the circumstances that would warrant it were extremely remote. He expressed confidence that "the oil problem would be dealt with by other methods," but, repeating his remarks to Business Week, Kissinger declared that he was not saying "there's no circumstances where we would not use force."

The secretary repeated his assurances to newsmen Jan. 3 and said his statements in the interview reflected the views of President Ford. "I do not make a major statement on foreign policy on which I do not reflect his views," he said.

Ford's press secretary Ron Nessen Jan. 4 confirmed that Kissinger's statement on the possible employment of military action "did reflect the President's views."

Nessen had said Jan. 3 that Ford regarded Kissinger's statement on the possible use of military action "a highly qualified answer on a hypothetical situation involving only the gravest kind of emergency with the industrialized world."

The use of military force in the Middle East as a possible "option" was said to have been discussed by Ford's energy advisers at a meeting Dec. 14-15, 1974 to prepare policy recommendations for the President. A Ford spokesman said Jan. 3 that the President "knew of no plan for military action" discussed at the meeting. However, Ford conceded that such contingency plans might exist in the Defense Department or in other branches of government, the spokesman said.

Arab and other reaction—Iranian Premier Amir Abbas Hoveida warned that the use of force against the oil-producer states by one superpower would result in military intervention by the other superpower and cause "a catastrophe," according to an interview published Jan. 4 in the Egyptian newspaper Al Ahram.

Algerian President Houari Boumedienne said Jan. 6 that "occupation of one Arab state would be regarded as an occupation of the entire Arab world." U.S. military action, he predicted, would destroy the oil fields.

Egypt endorsed Boumedienne's position in a statement released Jan. 7 by Information Minister Ahmed Kamal Abul Magd. He asserted that Kissinger's declaration "did not serve the cause of American-Arab relations or the cause of peace in the area."

The Soviet press and television Jan. 6 carried a summary of critical reaction to Kissinger's remarks by newspapers in Asia, Africa and Europe. These comments, the press agency Tass said, showed "that the times of gunboat diplomacy and intimidation are gone."

The Soviet Communist Party newspaper Pravda charged Jan. 7 that "defenders of monopoly interests" in the West were employing "military blackmail" against the Arab oil producers in an effort to bring oil prices into line.

West German government officials and the press expressed anxiety about Kissinger's statement on force, it was reported Jan. 5. A government spokesman said: "I don't see the danger of [industrial] 'strangulation' at the moment. We are not interested in any kind of confrontation with the oil countries, but rather in cooperation. . . ."

OECD analyzes oil needs, payments. In an analysis of anticipated energy needs and demands, released Jan. 13, 1965, the OECD said that most Western industrial nations could achieve near self-sufficiency in energy supplies by 1985 if steps were taken to conserve energy supplies and if world oil prices remained approximately unchanged. There was a potential for saving 15%-20% of energy consumption by 1985, the report stated.

According to the forecast, OECD member nations in Western Europe, North America and developed Asia would be producing 80% of their energy requirements by 1985. In 1972, the OECD noted, these same nations had produced only 75% of their energy needs. Before oil prices had been increased, the OECD had expected the production figure to decline to 55% by 1985.

The report, the first comprehensive assessment of the group's energy needs prepared since oil prices had quadrupled, also stated that the transfer of new wealth to the oil-producing states represented a 1.5% tax on the domestic output of goods and services (the gross national product measure) in OECD countries.

In a later report March 5, the OECD noted considerable progress had been made among member nations in reducing their balance of payments deficits—red ink positions that reflected the impact of higher oil prices. The combined payments deficits, measured on the current account basis, of the 24 member nations were expected to total $27.5 billion at the end of 1975, and disappear entirely by 1980 as oil-producing nations increased their purchases from industrial nations and extended more aid to developing countries.

According to Otmar Emminger, an official of the West German central bank, the OECD nations' combined payments deficits for 1974 totaled $33 billion—that was $5.5 billion less that the OECD forecast made at the close of the year. Emminger's remark was reported March 5.

OPEC backs talks with consumers. The Organization of Petroleum Exporting Countries (OPEC) met in Algiers Jan. 24-26 and decided to hold a conference of their chiefs of state in about five weeks. The meeting would be held in preparation for a conference with oil-consuming countries on energy, raw materials and the world economy that was formally proposed in the OPEC parley.

A communique issued at the end of the discussions also denounced what it termed U.S. threats of force against the Middle East oil fields, declaring that such "threats create confusion and lead to confrontation."

Algerian Industry and Energy Minister Belaid Abdelsalam, opposing consumer cooperation, scored the U.S.-sponsored International Energy Agency as "a war against our people" and called the U.S. plan for a $25 billion fund for oil importers an attempt to extend American "hegemony." The statement was coupled with an Algerian recommendation of an OPEC agreement "on a global embargo of their oil against any aggressor" attacking any member of the organization.

The OPEC ministers agreed to retain prices at their current level for the remainder of 1975 and gradually increase them in 1976 and 1977 on the basis of an inflation index that might be agreed upon with industrial consumers.

Oil users back conservation plan. The 18-nation International Energy Agency of oil users met in Paris Feb. 5-7, 1975 and tentatively agreed on plans to continue the search for new fuel sources, to reduce their dependence on Arab oil and to eventually force down the price of petroleum. (IEA membership had grown to 18 nations with the inclusion of two new members—Norway and New Zealand.)

The IEA insisted that any meeting with the Organization of Petroleum Exporting Countries be limited to the discussion of the oil problem, despite recent OPEC demands that such a dialogue be held only if all raw materials were taken up.

The Paris conference reacted coolly to an American proposal for a price floor to assure development of non-Arab energy supplies. The proposal, submitted to the Feb. 5 meeting, had been unveiled by Secretary of State Henry A. Kissinger in a speech to the National Press Club in Washington Feb. 3. Calling for a system of floor prices for imported oil sold in the West, Kissinger suggested that the minimum price be considerably lower than the current price ($10-$11 a barrel), but high enough to make it economically feasible for investment in "long-range development of alternative energy sources." The floor-price plan might also encourage oil producers to agree on a long-term, lower-cost arrangement with the oil consumers, the secretary said. The oil producers, he pointed out, would be faced with two choices: "They can accept a significant price reduction now in return for stability over a long period, or they can run the risk of a dramatic break in prices when the program of alternative sources begins to pay off."

Kissinger saw the possibility that once the West began to develop energy sources of its own, the oil producers might reduce their prices to the old levels and make the alternatives non-competitive. "In order to protect the major investments in the industrialized countries that are needed to bring the international oil prices down, we must insure that the price for oil does not fall below a certain level," he said.

Kissinger proposed the formation of a consortium for the development of synthetic fuels that would pool the resources of Western nations and provide them with incentives for taking part in the floor-price plan. He also suggested the establishment of a consortium for energy research and development to coordinate national efforts in such fields as nuclear fusion and solar energy.

Kissinger proposed a five-point program for consumer-producer relations that included exploration of means of investing the oil producers' huge surpluses, ways to help them develop industries in their own countries and, for consumers, assurance of a steady oil supply.

Meeting March 20, the IEA agreed to set a minimum common price for oil imports to encourage investments in alternative energy sources.

Under the plan, which the U.S. had proposed, the value of imported oil would be allowed to fluctuate with international market demand. Consuming nations, however, would maintain prices on their domestic markets at a figure to be determined by July 1 through tariffs, quotas or other means. Once a minimum was established, national legislatures would have to approve appropriate measures to maintain the minimum price.

The U.S. had demanded the agreement as a condition for its participation in an April 7 preliminary meeting between consumers and producers. Energy-rich countries like the U.S. and Canada wanted the plan to protect investments in alternative energy sources such as oil shale and coal gasification. They feared attempts by the oil cartel to expand its markets by drastically increasing the supply of oil, thus causing prices to drop and making the development of alternative energy sources not viable. Energy-poor countries, such as Japan, had objected to the plan since it could keep oil prices artifically high.

According to IEA chairman Viscount Etienne Davignon, the decision was nearly unanimous, with an abstention by Sweden, which did not yet want to commit itself to a minimum price. Unanimity was thought essential for successful negotiations with producers.

Thomas Enders, U.S. Assistant Secretary of State, said he was "very satisfied." Other U.S. conditions, joint oil conservation programs and a $25 billion fund for mutual balance of payments, were also met. It was agreed that the U.S. would conserve one million barrels of oil a day, and the 17 other consumers would jointly do the same.

Consumers also agreed to cooperate in developing new energy sources. Britain would stimulate coal technology and Japan would experiment with solar energy. Research projects in utilization of waste heat, in use of municipal garbage, and in extraction of hydrogen from water were also approved.

OPEC offers consumers price talks. A three-day summit meeting of the Organization of Petroleum Exporting Countries ended in Algiers March 6, 1975 with an offer to negotiate with industrial consumer nations on the "stabilization" of oil prices on condition that such a conference not be confined to discussion of prices.

The OPEC bid to the consumers was contained in a 14-point charter declaration that also was to serve as the organization's position in its dialogue with the consumers. Among the other principal points of the document:

- Future oil prices must be based on "availability and cost of alternative sources of energy" and the use of oil for such nonenergy purposes as chemical products. The value of petroleum must be protected against inflation and monetary depreciation by linking oil prices to the prices of manufactured goods and services.
- The producers would "insure supplies that will meet the essential requirements of the economies of the developed countries, provided that the consuming countries do not use artificial barriers to distort the normal operations of the laws of supply and demand."
- OPEC was prepared to assure supplies to the industrialized nations to meet their "essential requirements" but would take "effective measures" against any grouping of consumer nations that sought "confrontation."
- The international monetary system must be reformed in order to provide a "substantial increase in the share of developing countries in decision-making, management and participation in the spirit of partnership for international development and on the basis of equality."

The conference failed to endorse Algeria's proposal for a special OPEC fund of $10 billion–$15 billion for aid to developing countries. It proposed instead coordinating grants and loans by funds under control of individual countries or by regional funds.

King Faisal of Saudi Arabia stayed away from the conference, reportedly because of his displeasure with other OPEC countries' refusal to lower prices. He was represented at the talks by Prince Fahd ibn Abdel Aziz, deputy premier and interior minister. The heads of state of Nigeria, Indonesia, Iraq and Libya also declined invitations and were represented by their ministers.

OPEC's 14-point declaration had been drafted at a preliminary meeting of the organization's oil ministers in Vienna Feb. 25–27.

Preparatory oil talks collapse. Oil producer and consumer nations opened preliminary talks in Paris April 7 to map plans for a full-scale world economic conference later in the year but the discussions collapsed April 15 after both sides failed to agree on the agenda and other terms for the proposed parley.

A statement issued after the talks said the delegates had "agreed to remain in contact . . . with a view to resuming, as soon as circumstances appeared favorable, the preparation of the conference."

The stalemate stemmed from demands by the U.S., the European Economic Community and Japan that the international conference be confined to energy and related economic problems, while oil-producing Algeria, Saudi Arabia, Venezuela and Iran, and three non-oil producing countries—Zaire, Brazil and India—insisted that the projected meeting be broadened to include the discussion of raw materials and development aid.

U.S. reverses raw materials stand. Secretary of State Henry A. Kissinger told a foreign ministers meeting of the 18-nation International Energy Agency in Paris May 27, 1975 that the U.S. favored discussion of international arrangements for greater price stability for raw materials at any future meeting between the oil producers and the industrial nations. He also expressed hope that further efforts would be made to deal with the problems of the developing nations and thus end the impasse between the oil-producing and oil-consuming nations and help avoid another energy confrontation.

The Kissinger statement was a reversal of the foreign economic policy of the U.S., which, with other IEA nations, had opposed discussion of raw materials at international oil conferences.

Kissinger said, "It has become clear as a result of the April preparatory meeting that the dialogue between the producers and consumers will not progress unless it is broadened to include the general issue of the relationship between developing and developed countries." As a means of resuming the conference with the oil producers, the secretary suggested the creation of three commissions to deal with critical areas such as energy, problems of the most seriously affected nations and raw materials.

A communique issued by the IEA ministers offered to resume "discussions with the oil producers at any time and in any manner found mutually convenient." The statement made no mention of the Kissinger proposal on raw materials, saying only that the agency's governing board would "examine the manner in which the dialogue should be continued."

(Kissinger had made a similar proposal May 13, in which he said the U.S. backed an international parley on raw material prices "on a case-by-case basis." Speaking at the Kansas City International Relations Council, the secretary warned that the U.S. would continue to oppose demands by the developing countries for a "totally new economic order founded on ideology and national self-interest." Kissinger said the U.S. also did not believe that "tying commodity prices to a world index of inflation is the best solution" because "it would strengthen those least in need of help because most raw material production still takes place in the industrial countries.")

At a meeting of the Organization for Economic Cooperation and Development (OECD) in Paris May 28, Kissinger again sounded the theme of cooperation between the developing and industrial nations. Warning that "economic issues are turning into central political issues," the secretary said the economic well-being of the West "depends on a structure of international cooperation in which the developing nations are, and perceive themselves to be, participants." He added: "The new problems of our era—insuring adequate supplies of food, energy and raw materials—require a world economy that accommodates the interests of developing as well as developed countries."

The OECD ministers later issued a "declaration on relations with developing countries" pledging greater cooperative efforts, including the resumption of talks on energy matters that also would deal with the subject of food and other raw materials.

OPEC holds prices. At the conclusion of a three-day meeting in Libreville, Gabon June 11, the Organization of Petroleum Exporting Countries (OPEC) announced it would retain until the end of September the nine-month freeze on oil prices it had agreed on in January. The OPEC communique also said the organization would lessen its dependence on the dollar by adopting a price structure based on Special Drawing Rights (SDRs), the weighted basket of currencies established by the International Monetary Fund.

The meeting had been opened June 9 by Gabonese President Omar Bongo, who complained that because of U. S. dollar inflation, oil-producing nations were "witnessing, powerless, the constant and appalling deflation of our already meager incomes." Mohamed Yeganeh, the chief Iranian delegate, told newsmen there was "basic agreement that in order to smooth fluctuations in exchange rates, the SDR would be more suitable than other proposals." He acknowledged that the proposed switch to SDRs would mean "a small increase in the price of oil." Referring to one of the topics discussed at the meeting, Yeganeh said recent overtures by U. S. Secretary of State Henry A. Kissinger on reconvening the interrupted producer-consumer dialogue were "a

The June 11 communique said the forthcoming price increase had been decided because of "increasing inflation, the depreciation of the value of the dollar, and the consequent erosion of the real value of the oil revenues of member countries." News reports said pressure for an increase beginning July 1, rather than at the end of September, had been exerted by Algeria, Nigeria and Libya but that the move had been opposed by Saudi Arabia, Venezuela and Iran.

Ford sees 7%–8% price rise limit. Any increase in world oil prices should be limited to 7%–8%, a spokesman for President Ford said June 10. Some OPEC members reportedly had been considering raising oil prices by as much as 38%, or $4 a barrel. Such an increase would be "totally without economic justification," Ford said through his press spokesman, Ron Nessen, because the price of industrial goods were expected to rise only 7%–8%. (The 7%–8% figure was based on projected inflation rates in oil-consuming countries during the first nine months of 1975, Nessen said.)

Action to raise oil prices 38% "would not be in the interests" of oil-exporting nations, Nessen said, adding, however, that the remark should not be viewed as a threat. Ford believed a price cut could be justified, Nessen said, since oil-exporting nations currently were operating at only two-thirds of capacity.

U.S. Energy Situation

Shortage Claims Disputed

The oil industry was accused by press and political sources early in 1974 of deliberately causing the oil shortage, in collusion with the oil-producing countries, for the purpose of raising prices and increasing profits. U.S. government officials were accused of acting in the interests of the oil companies rather than of the public.

Fuel shortage estimates conflict. There were apparent contradictions between the American Petroleum Institute's (API) report of a worsening fuel shortage and information released by major oil companies showing few perceptible drops and many large increases in their reserves of petroleum products compared with supplies a year earlier.

The API said Jan. 16, 1974 that in the week ended Jan. 11, oil imports had fallen sharply, causing crude oil and gasoline stocks to hit low levels. Imports had been declining since Nov. 2, 1973, when imports were a record 7.1 million barrels a day, according to API officials. Shipments averaged 4.9 million barrels in the week ending Jan. 11, a 12.2% decline from the previous week.

According to the API Jan. 14, the nation had averted a critical energy shortage in 1973 because large amounts of oil had been imported before the Arab embargo took effect. Overall oil imports were up 30% from 1972 but the increase in demand had slowed to 5.4% in the 12-month period.

Domestic crude oil production dropped 2.7% in one year, but the 1973 daily average of 10,963,000 barrels was the highest rate in the world.

Extent of shortage challenged—The Conference Board, an independent economic research group in New York, Jan. 2 accused William Simon, director of the Federal Energy Office, of overestimating the U.S. oil shortage by up to 1.3 million

In contrast to the Administration's prediction that supplies would run 4.5 million barrels a day below demand in each of the first two quarters of 1974, the group estimated there would be a shortage of 2.4 million barrels a day and 2.6 million barrels a day in the first and second quarters if the economy remained "strong." In a "weak" economy, the shortfall could drop to 2.2 million barrels a day in each quarter, the group said.

According to the Conference Board, these figures differed from those of the Administration because Simon ignored an "unusual degree of inventory building," or stockpiling of supplies, during the summer and fall of 1973 by distributors and industrial consumers. The research group said its figures also took into account the rapid rise in retail costs, which had caused demand to drop sharply.

At a briefing later, Simon Jan. 3 denied reports that oil tankers were collecting offshore waiting for prices to rise before making deliveries. A Coast Guard check found no "unusual concentration, hovering or bunching up of tankers," Simon said.

Simon also reported that oil imports for the 30 day period ending Dec. 21 remained 400,000 barrels a day above expected levels. In an interview that was published Jan. 10, Simon said the Central Intelligence Agency was providing his office with information about tanker operations throughout the world.

Public polled. As major oil companies announced another round of price increases in early January, a Gallup Poll released Jan. 9 indicated that the public held oil companies and the Administration largely to blame for the current energy crisis. Consumers also were held responsible for the nation's energy problems, but very little blame was placed on Arab nations and none on Israel. Only 6% of those surveyed believed there was no shortage.

Nixon said 'tied' by oil firms. Rep. Les Aspin (D, Wis.) charged Jan. 1 that 413 directors, senior officials and stockholders of 178 oil and gas companies had contributed $4.98 million to President Nixon's re-election campaign. Because of these donations, nearly 10% of Nixon's total campaign receipts, Aspin said, "President Nixon's hands are tied, preventing him from dealing effectively with the current energy crisis. After their massive contributions, there is little he can do to control them."

The list, which did not include large gifts from the Rockefeller family (Exxon stockholders), was compiled from public information on file with the General Accounting Office and from a secret donors list obtained after a court fight by Common Cause, the public interest lobby.

Refiners audited. Federal Energy Office Director William Simon announced Jan. 10 that price, profit and supply records of the nation's 140 petroleum refiners would be audited by a task force composed of FEO officials and Internal Revenue Service agents.

The investigations would have a double purpose, Simon said, in making certain that price increases reflected only cost increases and not higher profits. The check also would "verify the accuracy of refiner reports on crude oil and product supplies and is a major step toward establishing an independent reporting and information system on refinery inventories at FEO," Simon said.

Texaco Inc. was the first major oil company to make full disclosure of its petroleum inventories. The firm reported Jan. 10 that its supplies of crude oil (on Jan. 1) were 17.6 million barrels, off 1.4 million from a year ago; gasoline stocks were 23.3 million, up .1 million; middle distillates, which included home heating oil and aviation and diesel fuels, were 22 million, up 6.3 million; residual or heavy fuel oil supplies were 7.6 million, up .2 million.

Texaco officials said the firm had released the data to counter widespread suspicion that oil companies had been manipulating shortage figures to increase profits and eliminate independent competitors. No information had been withheld from the government, they said.

Four more oil companies released supply data Jan. 11. According to their reports:

As of Jan. 1, Shell Oil Co. had 20.2 million barrels of crude oil reserves, or 3.5% more than in the previous year; gasoline and other motor fuels were 17.4 million barrels, up 1.2 million; middle distillate were 15.1 million barrels, up 2.2 million.

Mobil Oil Corp. reported crude oil stocks of 16.2 million barrels, down from 17.2 million last year; 15 million barrels of gasoline, down from 15.2 million; distillate stocks were 23.7 million barrels compared with 16.2 million. (Inventories were based on Dec. 31, 1973 supplies.)

Standard Oil of Indiana reported, as of Dec. 31, 1973, crude oil stocks of 20 million barrels, down from 20.5 million; gasoline stocks of 21.3 million barrels, up from 19 million; distillate reserves of 21.5 million barrels against 17.9 million.

Gulf Oil announced its crude oil supplies totaled 18.5 million barrels Jan. 1, up from 16.7 million; gasoline supplies were 16.9 million barrels against 17.2 million; distillates were at 16.4 million, up from 14 million;

Exxon USA announced Jan. 14 that as of Jan. 4, petroleum stocks totaled 20.8 million barrels, or 1.5 million barrels higher than in the previous year. Refinery operations also were above January 1973 levels and gasoline inventories, set at 17.2 million barrels, equaled last year's levels. Heavy fuels and distillate also were higher.

Congress probes oil supply. The chairmen of four Congressional subcommittees announced Jan. 11 that their groups would initiate hearings to determine whether the energy problems facing the country were "fact or fiction."

Sen. Henry Jackson (D, Wash.), chairman of the Senate Permanent Subcommittee on Investigations, charged that there is a "total lack of public confidence in the oil industry, in the federal agencies charged with regulating the industry, and in the validity of the spiraling costs of gasoline and heating oil. People are not going to make sacrifices unless they get some straightforward answers about the extent of the shortage and who is benefiting from the shortage."

U.S. PETROLEUM SUPPLY & DEMAND
(In billions of barrels per year)

Year	Domestic demand	Domestic production	Imports	Imports as a percent of demand
1950	2.38	2.19	.19	8
1955	3.10	2.73	.37	12
1960	3.59	2.93	.66	18
1962	3.80	3.04	.76	20
1964	4.03	3.21	.82	20
1966	4.41	3.47	.94	21
1968	4.90	3.86	1.04	21
1970	5.37	4.12	1.25	23
1971	5.55	4.12	1.43	26
1972	5.99	4.34	1.65	28
1973 [1]	6.30	4.24	2.16	35
1974 [1]	6.64	3.97	2.67	40

[1] Based on preliminary figures supplied by FEA.

Federal Energy Office Director William Simon testified Jan. 14 before a subcommittee of the Joint Economic Committee. He acknowledged the inadequacy of data supplied to his office by the oil industry to verify the extent of the fuel crisis, but he added that aides were drafting legislation to compel oil, coal and uranium companies to provide needed information.

Simon was responding to Subcommittee Chairman Sen. William Proxmire (D, Wis.), who declared that a "shockingly large proportion of our people, perhaps most of our people, doubt the existence of the energy crisis . . . [and] believe it is a government-oil industry sponsored 'put-on' to raise prices and increase profits at the expense of the consumers." Proxmire had asked Attorney General William Saxbe the previous day to investigate the major oil companies for possible antitrust violations, particularly the industry's pattern of vertical integration which also allowed refiners to control oil exploration, transportation, distribution and retail sales.

In testimony Jan. 16 before a House Select Committee on Small Business' subcommittee, chaired by Rep. John D. Dingell (D, Mich.), four energy policy experts said the oil industry had been able to reap advantages from the fuel shortage by keeping the government "in the dark" about their supply and price information.

They also criticized tax credits given multinational oil companies for royalties paid to foreign governments. The effect of the tax rulings, according to the experts, was to stimulate oil exploration and production abroad at the expense of domestic ventures, and to provide multimillion dollar tax advantages to the firms when royalties were increased sharply in 1973.

An official of Exxon USA also appeared before the subcommittee, saying that his company would welcome a "government-sponsored system for collecting and publishing timely petroleum data." He conceded, however, that there was "some basis" for the "broad generalization" that oil companies had been unwilling to voluntarily release information about their petroleum reserves.

Interior Department officials admitted at a Jan. 17 session that statistics supplied by the oil and natural gas industry had formed the basis of government projections and policy on fuel production. The government had agreed to keep the records secret in order to insure voluntary compliance with its request for information, officials added.

As a result of this information policy, officials said they believed there was evidence that the industry was "seriously underestimating" gas reserves. They also conceded they were unable to gather evidence to determine whether oil and gas wells on government land had been capped to drive up prices.

Subcommittee members questioned Federal Energy Office (FEO) officials Jan. 18 about the Cost of Living Council's recent approval of a $1 barrel increase in "old" oil prices.

Rep. Henry Reuss (D, Wis.) termed the action "unjustified." "The drilling equipment is in place, the derricks are pumping, and oil is flowing out of the ground," Reuss said. He won admissions from two FEO officials that they had no firm evidence that oil companies were reinvesting the amount of increase in new production sites.

Rep. Dingell charged that the price increase guaranteed oil companies a $4.75 billion annual "bonanza."

Hearings Jan. 21–23 of the Senate Permanent Investigations Subcommittee provided a forum for airing of charges that the fuel shortage had been contrived, that oil industry profits were excessive, particularly during the period of the Arab embargo, and that at least one major oil company had cut off supplies to U.S. military forces during the Middle East crisis at the order of Saudi Arabia. Oil executives from seven major companies—Exxon, Texaco, Mobil, Gulf, Shell, Standard Oil (California) and Amoco, a subsidiary of Standard Oil (Indiana)—testified in rebuttal over the three-day period.

The subcommittee had laid the groundwork for the hearings, which opened as Congress reconvened to begin work on the stalemated emergency energy bill, by asking the oil companies to respond to questionnaires about their business operations.

Sen. Henry M. Jackson (D, Wash.), chairman of the subcommittee and sponsor of the emergency energy legislation, said at the opening of testimony that Mobil and Gulf had been the only companies to comply in full. Jackson said he was "flabbergasted" that Exxon had stamped "proprietary" to the answer of every question. Exxon officials based their refusal to answer questions on "how many stations they own and operate, how many stations are independent customers and how many have closed over the past year" on the fear that the disclosures would jeopardize trade secrets, Jackson said.

Despite the lack of cooperation from the industry in providing complete information on the extent of the fuel shortage, the subcommittee concluded that at the end of 1973, inventories in all petroleum products held by the seven companies were 5.5% higher than at the end of 1972.

Oil officials, who did not challenge the assertion, insisted that the fuel crisis was real. They attributed their favorable position to "prudent management" and mild winter weather, but cautioned the public not to be misled. The full effects of the Arab embargo would be felt within a few weeks as crude oil stocks dropped rapidly, industry officials warned.

Under harsh questioning from several openly skeptical senators, the oil executives denied they had withheld fuel from the market to force up prices during the shortage period. With pressure mounting in Congress for passage of some form of excess profits taxation, the oil spokesmen indicated general support for a tax that would encourage capital investments in exploration, but they added that the tax should apply across the board to U.S. industries rather than to the oil industry alone.

In response to questioning, oil officials suggested that the excess profits tax could be triggered when profits rose to 12%–15% of investments.

Oil profits cited—During the hearing Jan. 22, Jackson disclosed that a subcommittee staff report showed that the seven major oil companies increased their profits by 46% during the first nine months of 1973 when sales rose by only 6%, a figure which prompted Sen. James Allen (D, Ala.) to inquire, "Would it be improper to ask if the oil companies are enjoying a feast in the midst of famine?"

Oil officials took exception to Jackson's comparison of 1973 earnings with 1972 figures, claiming that earnings had been depressed in 1972. Jackson countered with data showing an average 20% annual increase in profits over the two-year period.

According to the oil executives, their profits had increased during a period when the Administration's price controls were in effect because recent spectacular increases in the price of crude oil could be passed on to the consumer as raw material costs.

Sen. Abraham Ribicoff (D, Conn.) charged that the nation's tax structure permitted the oil industry to reap windfall profits. According to Ribicoff, Texaco had paid U.S. income taxes of only 1.7% and Gulf only 2% because of the depletion allowance and foreign tax credit contained in the U.S. tax code.

Texaco and Gulf spokesmen said that their total worldwide taxes amounted to 51% and 52% of their entire income. Sen. William Proxmire (D, Wis.) charged that the tax provision allowing such wide discrepancies between foreign tax payments and U.S. payments was a "gigantic loophole" costing citizens $2 billion–$2.5 billion a year.

He explained: A U.S. oil company earning $100 million in annual revenues in Saudi Arabia would pay that government $50 million in royalties, or 50%. The remaining $50 million in earnings was subject to a U.S. tax of 48% on corporate income. At that rate, the oil company would owe the government an estimated $24 million, but because the $50 million royalty payment was considered a tax paid abroad, the company received credit for the $50 million, thereby canceling any U.S. taxes owed on the remaining sum.

A spokesman for Mobil told the subcommittee that the company was "very close to going into a loss position of refining and marketing," yet data compiled by the subcommittee showed that Mobil's cost of producing one barrel of Saudi Arabia crude oil was 10¢. When Saudi Arabian royalties and taxes were added to the initial cost, Mobil's price rose to $7.12 a barrel.

Jackson told reporters after the hearings the the "big lesson" to be learned from the testimony was that "our tax policy, instead of encouraging us to be self-sufficient in energy," promoted "self-sufficiency for the Arabs, period."

Oil cutoff to military charged—Jackson said Jan. 23 that he had documentary evidence that Exxon had cut off fuel supplies to U.S. military forces abroad at the bidding of the Saudi Arabian government.

Jackson said he based his charges on the Dec. 1, 1973 issue of Business Week and corroboration from "independent [government] documentation." According to the magazine, Saudi Arabia's King Faisal ordered U.S. companies participating in the Arabian American Oil Co. (Aramco) to cease supplying U.S. troops with fuel derived from Saudi crude oil. Faisal threatened to "retaliate against any breach by extending the oil embargo to the company involved and the country in which the violation took place," Business Week stated.

According to Jackson, the order was issued in October 1973 during the U.S. military alert, but Business Week said the incident occurred during a meeting of the consortium in Jeddah Nov. 4, 1973.

Spokesmen for Aramco members—Exxon, Mobil, Standard Oil (California) and Texaco—denied personal knowledge of the alleged shutoff but said they would check further on the charge. Jackson labeled the action a "flagrant act of corporate disloyalty to the U.S. government."

The Pentagon had admitted publicly in October 1973 that Navy ships had been denied their usual source of Arab oil, a fact which had prompted the Administration to invoke the Defense Production Act to insure an adequate supply of fuel for national security purposes.

Sen. Jacob Javits (R, N.Y.) deplored the oil companies' "quasi-governmental" role in the production of oil, setting of prices and distribution of supplies. "Obviously, it is to their [the Arabs'] interest to raise prices," Javits said. "They determine even your profit. They determine how much of the deal you may own. And yet what they are doing imperils this country and the world."

Javits added: "The only persons talking to these governments are companies who have their own profits, their own stockholders and their own private interests to protect. That is an impossible situation."

In a statement issued Jan. 24, Exxon Corp. Chairman J. K. Jamieson did not deny that fuel shipments to the U.S. military were shut off but he refuted Jackson's charge that the incident constituted a "disloyal act."

"As is generally known," Jamieson said, "the Saudi Arabian government in late October [1973] imposed an embargo on the export of crude oil and products to the U.S. and certain other countries. Included in this embargo were deliveries to the U.S. military of products derived from Saudi Arabian crude. These developments and actions taken by Exxon were promptly reported to the Department of Defense and the Defense Fuel Supply Center."

Z. D. Bonner, chairman of Gulf's U.S. affiliate, issued a similar statement the same day:

"The oil industries in the Middle East were simply told by the Arab governments that 'You will not be allowed to supply the U.S. military overseas from our oil and if you do you'll be cut off completely,' " Bonner said. "The U.S. government of course was completely informed. . . . And the decision was made to increase military supplies out of the United States."

Aramco officials conceded Jan. 25 that fuel had been withheld from the U.S. military since Oct. 21, 1973, when King Faisal issued the embargo instructions. "The embargo action was taken by a sovereign state and Aramco's compliance came as the result of a direct order which had nothing to do with patriotism," officials said, adding that the threat of harsh, but unspecified reprisals, was explicit in the king's order.

Bonner also reacted angrily to the treatment the oil industry had received at the public subcommittee hearings, which he said were conducted like a "criminal trial." The sessions "opened with a bunch of accusations but we never got a chance to face our accusers," he said.

Bonner and other oil officials who refused to be identified accused Jackson of allowing his presidential ambitions and need for publicity to interfere with a dispassionate search for answers to the oil crisis. Congress "got politics mixed in with the shortage issue," Bonner said, "and it shouldn't be there."

Companies sued re shortage. New York State filed suit Feb. 6 against Shell Oil Co. and three other members of the Royal Dutch/Shell Group, accusing them of diverting and withholding home heating oil in a scheme to inflate prices and eliminate competition during a period of contrived winter fuel shortages.

According to the civil complaint, the defendants arranged to divert 1 million barrels of oil normally imported by Shell and store it in New Jersey under customs bonds. Fuel stored under bond technically was not regarded as imported until the bond was removed. The supply was never reported to the American Petroleum Institute, an industry group whose assessment of the extent of the shortage until recently gave federal officials their only measure of the supply crisis. The stored fuel was not reported to the Federal Energy Office until Jan. 15, New York charged, adding that one of the defendants "indicated . . . its intention to issue a fraudulent invoice" and conceal its true inventory position while the state investigation was taking place.

The fuel remained "hidden in New Jersey" until after the Arab oil embargo was imposed and then was sold in November 1973 at "exorbitant prices," New York charged.

While this oil was in bond, Shell customers were unable to receive their full supplies of oil, according to New York. The state also charged that when the disputed oil was sold to distributors, they paid up to 49.5¢ a gallon and passed on the increase to homeowners. Shell charged other wholesalers 18.35¢ a gallon for oil imported through other terminals, the complaint stated.

Named as defendants with Shell were the Asiatic Petroleum Corp., Cia. Shell de Venezuela, Ltd., and Shell Curacao, N.V.

The suit ended in a pretrial settlement July 10. The four companies agreed to offer the metropolitan area enough of its imports during the fall of 1974 and keep the state attorney general's office informed on a monthly basis of its inventories and sale of home heating oil for two years.

New York dropped its claim to fines and restitution for the added costs paid by distributors for the oil in 1973 when the oil was finally released from storage.

In other legal action related to the fuel crisis, it was reported Jan. 22 that the federal government had brought civil suits against 13 gasoline stations in cities from Chicago to New York. The dealers were accused of price gouging. One criminal indictment was obtained Feb. 13 against a station owner.

The National District Attorneys Association was investigating possible wrongdoing by oil companies during the energy crisis, it was reported Jan. 24. It was the first time that the 5,000 member organization with representatives in 50 states had undertaken a unified inquiry. A spokesman said the investigation was launched because "there is a feeling in this country" that the federal government "might not have been doing as much as it could be" in examining possible violations by the oil industry. The government's "close association" with oil interests could impair "its ability to make proper judgments," he added.

16% petroleum shortfall seen. Deputy FEO Administrator John Sawhill told Congress Feb. 19, 1974 that if the oil embargo persisted through June, the nation's supply of petroleum products would fall 16% (2.8 million barrels a day) below demand in the second quarter, although conservation efforts could somewhat alleviate conditions.

If the supply situation did not improve during the April–June period, Sawhill predicted shortages of 12% for gasoline, 27% for jet fuel, 32% for residual fuels and 6% for middle distillates.

A lifting of the embargo would not immediately end the shortages, Sawhill added, warning that demand would exceed supplies by 8% in the second quarter. Shortages expected under optimal conditions were 4% for gasoline, 13% for jet fuel, 16% for residual fuel and 3% for middle distillates.

Effects of Crisis Widespread

Reports of job layoffs. A flurry of announcements of further job layoffs resulting from the energy crisis abated in mid-January 1974, but statistics released Jan. 11 by the Labor Department indicated that up to 100,000 workers had been laid off in mid-December 1973 because of production cuts related to the shortage of fuel, especially gasoline.

According to a payroll survey for the week of Dec. 9–15,1973, employment was off sharply for service stations, automobile dealers, hotels, motels, recreation and entertainment facilities and transportation facilities. Furloughs for nearly 75,000 workers had been announced since Dec. 15, 1973 in the automaking and airline industries.

As car manufacturers were in the process of switching from a heavy emphasis on the production of large cars to an expansion of their small car operations, the industry reported a 26.9% slump in sales during the first 10 days of January (compared with the same period of 1973). According to the Jan. 15 announcement, General Motors Corp. sales were down 42.5% in the 10-day period, while American Motors Corp., which produced a larger proportion of small cars, showed a 33% jump in sales. Sales figure for Chrysler were off 7%, Ford 14%. Ford announced Jan. 9 that 7,000 workers in two midwestern assembly plants for large cars would be laid off, 2,500 of them indefinitely. Another plant in New Jersey, also making big cars, would be closed temporarily, officials said Jan. 10.

Disney World in Orlando, Fla. laid off 700 employes Jan. 10 in reaction to declining attendance figures for the last quarter of 1973. Other tourist attractions in Florida also undertook personnel cutbacks.

(In one of the few positive spinoffs of the energy crisis, the National Highway Traffic Safety Administration reported Jan. 14 that the highway death toll dropped 15%–20% during November 1973 in the 16 states which had voluntarily reduced speed limits to conserve gasoline. The 34 states which did not require lower speed limits at that time showed only a 2% drop in auto fatalities. But the National Safety Council announced Feb. 2 that the number of persons killed in auto accidents in 1973 totaled 55,600 compared with 56,600 in 1972. In December 1973 when the impact of the fuel shortage was severe, there was a 19% reduction in deaths over the previous December. By Dec. 10, 1973, 24 states had adopted lower highway speed limits. They reported a 22% drop in fatalities while areas with unchanged speed limits reported a 16% decline, according to the council.)

In a letter Jan. 8, the Airline Transport Association asked Federal Energy Office Director William Simon to impose "effective fuel price control" on aviation jet fuel before rapidly rising petroleum prices put the cost of air travel "beyond the reach of millions of citizens to whom it is essential."

The industry's concern over rising fuel costs also was evident in a warning Jan. 4 from Eastern Airlines President Floyd D. Hall who said airlines faced bankruptcy or the need for massive government subsidies unless fare increases were authorized.

Cost saving devices already adopted by carriers, in addition to extensive employe layoffs and curtailments in flight schedules, included grounding of the jumbo jets and other expensive equipment and abolishing cut rate and special fares.

United Air Lines announced Jan. 11 that another 490 employees would be laid off, effective Jan. 26, bringing the airline's total number of furloughed workers to 1,440. Pan American World Airways Jan. 16 announced plans to furlough an estimated 400 employes, bringing the cutback in the carrier's work force to "nearly 2,000," according to officials.

Industrial production off. Production in the nation's factories, mines and utilities fell .5% in December 1973, according to the Federal Reserve Board Jan. 16. It was the first significant monthly decline in industrial output since August 1971. Analysts attributed the decline to a sharp drop in automobile production and to reduced consumption of electricity and gas—factors directly related to the energy crisis.

As 1973 ended and 1974 began, the auto industry continued to suffer the most immediate and severe effects of the fuel crisis: production in January declined 31% from 1973 levels, U.S. auto makers reported Feb. 1; the figure worsened for February when production was 34.4% below the level set 12 months earlier, according to reports March 1; total sales of domestic and foreign cars fell 26% in February from year-earlier levels, after dropping 22% in January and 18% in December 1973, the Wall Street Journal reported March 6.

Big cars with low gasoline mileage were the principal casualties of production cutbacks and sales drops as consumers exhibited a sudden and marked preference for the more economical small cars. Industry leaders admitted that changes in the nation's car buying habits had forced manufacturers to undertake what Ford President Lee Iacocca called the "greatest industrial conversion in history, at least in peacetime."

According to the New York Times March 3, auto makers were planning to increase production of small cars from 3.4 million in the 1974 model year to 5.2 million for the 1975 model year.

General Motors (GM) reported the worst decline in production; in the period from February 1973 to February 1974, 12-month period from February 1973–74, production dropped nearly 47%. A 43% cutback (compared with a comparable period in 1973) was planned through March, officials said Jan. 24 as plans were announced to close 14 of GM's 22 assembly plants for several weeks at a time. The action would idle 75,000 workers.

In a report Feb. 25, GM said conversion of a large California assembly plant from big car to subcompact production would begin during the spring. The four-month closing would idle 1.300 workers.

The number of Ford workers laid off because of conversion schedules or temporary shutdowns continued to climb. Officials said Feb. 9 that 2,600 employes would be laid off, but by Feb. 15 the number on temporary or indefinite furloughs had risen to more than 19,000.

Chrysler officials said Jan. 24 that two plants would be closed for part of February and March, causing 10,000 workers to lose their jobs temporarily.

According to estimates published March 4, more than 100,000 auto industry workers faced the indefinite loss of jobs and 53,000 were temporarily unemployed. Not included in the figures were those employes encouraged to take early retirement options or salaried workers now employed in hourly wage jobs.

Truck strike. Most of the U.S.' independent truckers went on strike Jan. 30, 1974 in grievances related to the energy crisis. Most of them ended their stoppage

Feb. 11 on the basis of an agreement negotiated with the aid of government officials and the mediation of Pennsylvania Gov. Milton J. Shapp.

Acceptance of the pact had been uncertain because the truckers' basic demands for a fuel price rollback and a public audit of oil companies were not met.

The strike, which was loosely organized by a coalition of 18 regional groups, had been successful in halting deliveries of perishable goods and other items throughout much of the nation.

Major supermarket chains were reported to be airlifting supplies amid reports of panic buying, particularly in urban areas. A tight gasoline supply situation was made worse as protesting truckers set up pickets around oil depots and harassed nonstriking oil truck drivers.

According to the New York Times Feb. 6, the strike had caused at least 75,000 workers to be laid off as the flow of supplies dwindled. In addition to produce and meat-packing workers, others laid off included steel and auto making employes.

As the effectiveness of the strike mounted, reports of violence also increased. As in past protests by independent truckers, the level of violence was highest in Ohio and Pennsylvania—the nexus of the nation's trucking system and home territory for the most militant drivers—but incidents involving gunfire, stoning, beatings, burned rigs, and bomb threats also were reported in at least 20 states. Two deaths were reported—a driver was killed in Pennsylvania when a boulder was dropped on his cab, causing him to crash; another person was shot to death in Delaware.

The National Guard was called up in Ohio, Pennsylvania, Maryland, Michigan, Kentucky, West Virginia and Florida to patrol the highways and provide convoy protection to non-striking drivers. The strike was said to be effective in at least 28 states by Feb. 1, but by Feb. 6, protesting drivers claimed success in 42 states.

Pennsylvania Gov. Shapp, whose state was the center of the violent protest, had arranged the emergency meeting for the loosely organized truckers and governors of eight other mid-Atlantic states. William J. Usery Jr., a special assistant to President Nixon and director of the Federal Mediation and Conciliation Service, led the government negotiators.

As conditions worsened, President Nixon acted Feb. 6, ordering a 30-day freeze on fuel prices and promising to provide the truckers with all the fuel they needed. (Under the government's allocation program, they had been limited to 110% of their 1972 consumption.)

Truckers also rejected that offer, saying they "no longer can be made the scapegoat for absorbing the rapidly rising costs of moving goods."

Final accord was reached Feb. 7 on another key demand—the surcharge, although no agreement was concluded on the rollback or audit issues. The surcharge matter was settled when Congress acted that day to pass a bill allowing the Interstate Commerce Commission to waive a 30-day waiting period before truckers could pass through to shippers (and hence to consumers) their additional costs resulting from higher fuel prices. The FEO also renewed its pledge to provide the truckers with 100% of their current fuel needs and announced establishment of a truckers' hot-line for help with fuel price and supply problems.

Airlines face higher costs. The airline industry had been threatened by another aspect of the energy crisis—the rising cost of scarce jet fuel. According to the Civil Aeronautics Board (CAB) Feb. 20, the average cost of a gallon of jet fuel for domestic airlines climbed 20.5% from December 1973 to January 1974, or to 47% higher than for the year ending June 30, 1973.

Higher fuel cost was cited by the International Air Transport Association (IATA) in a March 1 announcement confirming that member nations had agreed to seek a 7% increase on passenger fares over North Atlantic routes (effective April 1). Agreement was also reached to raise passenger rates on flights between North and South America 7%, effective April 15.

It was reported March 1 that the CAB had approved a 7% hike on cargo rates across the Atlantic and a 7% increase on passenger rates for Mid-Atlantic and South Atlantic flights. The IATA proposals had been filed Feb. 6.

In testimony before Congress March 4, airline executives warned that the industry's fuel bill could double in 1974 to $2.4 billion. The $1.2 billion increase would be six times the industry's profits in 1973, they said.

The Civil Aeronautics Board March 22 authorized a temporary 4% increase in domestic air fares, effective April 16 through Oct. 31, citing a "precipitous" rise in the price of jet fuel. The CAB also indicated that airlines could raise fares an additional 2% at the same time if they wished.

Action came on a petition to raise fares filed by United Air Lines. Clearance on the increase was made effective April 1 but because of competitive pressures within the industry no action was expected until April 16 when increases took effect for all domestic carriers.

The IATA in Switzerland March 22 announced agreement on a decision to raise air cargo rates on North Atlantic routes by 5% and Mid and South Atlantic routes by 4%, effective June 1 and subject to government approval.

FEO drops jet fuel equalizer plan—The Federal Energy Office (FEO) March 26 dropped a proposal to equalize the prices of jet fuel used on domestic and international flights after the CAB, oil suppliers and domestic airlines expressed opposition to the plan.

In a letter to FEO Administrator William Simon March 21, CAB Chairman Robert D. Timm warned that domestic air fares would rise even higher than their present levels if carriers were forced to subsidize the cost of high-priced bonded fuel used by international airlines.

(Pan American World Airways Inc., an international airline which with Trans World Airlines was the principal supporter of the price equalizer plan, announced March 26 that it had incurred a $10.1 million net loss in February, compared with an $8.6 million deficit 12 months earlier. During the first two months of 1974, Pan Am lost $17.7 million, in contrast to a $16.2 million deficit in the same period of 1973. A 132% increase in jet fuel costs from February 1973-February 1974 was largely responsible for its worsening financial condition, Pan Am spokesmen said.)

Furloughs canceled—In the airline industry's first major cancellation of layoffs, TWA March 21 recalled 1,175 employes whose jobs had been eliminated when airlines were forced to retrench as jet fuel costs rose dramatically. An estimated 3,750 TWA workers had been laid off since October 1973. United also planned to recall all 650 furloughed flight attendants and Pan Am recalled 150 of 1,000 employes furloughed in late 1973, the Wall Street Journal reported March 22.

Tax Deals, Prices & Profits

Oil operations probed. The Senate Foreign Relations Committee's Subcommittee on Multinational Corporations began hearings Jan. 30, 1974 on relationships that had developed over more than 20 years among major oil companies, the U.S. government and oil producing states in the Middle East.

Earlier, subcommittee sources had disclosed Jan. 16 that subcommittee investigators had obtained 38 classified documents said to show that the Eisenhower Administration "helped set up" an international oil cartel in Iran in the 1950s. The papers "show how the antitrust legislation was interpreted to allow the use of foreign policy and national security to permit control of Middle East oil by seven major companies," sources told the New York Times Jan. 14.

Tax benefits—Testimony heard Jan. 30 concerned international tax aspects of the oil industry, including a secret Truman Administration decision and negotiations with Arab states over royalty payments, which gave the oil companies the opportunity to realize enormous after-tax profits.

Subcommittee Chairman Sen. Frank Church (D, Ida.) described the transactions, which had been detailed by tax analysts and former government officials in private subcommittee hearings.

In the summer of 1950, Church said, the National Security Council decided in "secret session" to make arrangements that would insure stable, pro-U.S. government in the Arab world. Accordingly, at the urging of the State Department, the

Treasury Department "agreed to a system in which the [oil] companies would increase their payments to the oil producing governments and the American government would permit them to reduce their U.S. tax payments correspondingly," Church said.

To encourage the tax switch, "Wall Street lawyers were sent to the Middle East to help these countries rewrite their laws to bring them within the purview of the tax credit provisions of the U.S. Internal Revenue Code," Church said.

The tax credit program proved mutually advantageous to the Arab governments and to the oil industry. According to Church, "the result of this arrangement was to abruptly reduce the taxes paid by the companies to the U.S. Treasury while dramatically increasing the tax revenues accruing to the oil producing governments." (If the royalty payments had been regarded as business expenses subject to deduction from gross earnings rather than as foreign taxes suitable for a dollar for dollar writeoff, the oil companies' after-tax profits would have been drastically reduced.)

Tax credits derived from the oil industry's highly profitable foreign operations had exceeded their U.S. tax liabilities since 1962, Church said. In 1973, when the five major U.S. oil companies—Exxon, Texaco, Mobil, Standard (California) and Gulf—made $4.5 billion abroad, they paid negligible amounts of U.S. taxes, he said, "Stockpiled" tax credits, amounting to more than $2 billion, would allow the oil firms to continue to avoid U.S. taxes for several years, even if the tax arrangement were abolished, Church added.

Church provided an illustration of the transfer of money from the U.S. treasury to one Arab state: In 1950, the Arabian American Oil Co. (Aramco) paid $50 million in federal taxes, but in 1951, the first year of the new royalty-credit arrangement, the same consortium paid only $6 million. Royalties paid to the Saudi Arabian government by Aramco totaled $66 million in 1950, but jumped $44 million one year later to $110 million.

The subcommittee also released private cable traffic from the London Policy Group, an oil industry association, and its New York advisory committee. In one cablegram, oil officials conceded that the "artificiality" of the posted price system used for setting royalty payments and computing tax credits was "obvious and well known, but it has not been challenged by the IRS."

The subcommittee staff pointed out that despite the artifically high price for Middle East oil, neither the producing states nor the oil companies had any incentive to reduce it because royalty payments enriched Arab treasuries and U.S. tax credits enabled the companies to write off their expenses.

Church also revealed that a study of the oil industry's tax setup had been made in 1958 by the Joint Congressional Committee on Internal Revenue Taxation, but the study had never been released to Congress or the public.

Sen. Walter F. Mondale (D, Minn.) had disclosed Jan. 27 that according to Treasury Department figures for 1971, U.S. oil companies used $2.4 billion in foreign tax credits to cut their federal taxes by more than 75%. The 1971 statistics showed that the oil companies' U.S. taxes were reduced from $3.2 billion to $788 million.

"No other U.S. industry uses this foreign tax credit to this massive scale," Mondale said.

It was reported May 27 that the U.S.' six largest multinational oil companies—Exxon, Mobil, Texaco, Gulf, Standard Oil of California, and Standard Oil of Indiana—had paid worldwide taxes of $25.4 billion in 1973, but only $642 million in U.S. income taxes. The group's 1973 net profits had totaled about $6.7 billion and gross revenues had nearly equaled $50 billion.

Origins of escalating price hikes—Sen. Church charged Jan. 31 that the oil companies' inability to present a united stand against demands from oil producing countries for a larger share of oil profits, together with "indifference" expressed by the State Department and Justice Department over consequences of escalating prices, permitted the recent trend toward "leap frog" price increases to reach chaotic and dangerous proportions.

Testimony from Henry M. Schuler, vice president of Hunt International Petroleum Corp., third largest independent oil producing venture in Libya, supplemented Church's charges.

Schuler described oil policy and negotiations since 1971 as an "unmitigated disaster" which had created the "unstoppable momentum" within the Arab world for higher prices.

Unless the oil companies and oil consuming nations stood firm in the face of these escalating prices, Schuler said, "there will be such serious economic and political dislocation for the entire world that demands for military intervention will become inevitable."

"If a political and economic monster has been loosed upon the world [by the oil crisis]," Schuler continued, "it is the creation of Western governments and companies. Together we created it and gave it the necessary push, so only we, acting in harmony, can slow it down."

Schuler began his testimony with the observation that since August 1970, the price of Libyan crude had increased 1,-100% from $1.50 a barrel to $16.

After the Libyan revolution in 1969, Schuler said, the government decided to seek a larger proportion of oil revenues than the 50-50 split, which had been in effect for 20 years, by raising prices and limiting production.

Occidental Petroleum Corp., the most vulnerable oil company for such a strategy because Libya was its only source of crude, was selected as the first target.

When Exxon refused to provide Occidental with an alternative source of petroleum to withstand the Libyan pressure, Occidental capitulated to Libya's demands and in September 1970 agreed to turn over 58% of its total profits.

With the Occidental precedent established, the oil "companies were picked off one by one," Church said. According to Schuler, the final breakdown in efforts to maintain industry unity before the "leap frog" strategy of negotiating separate price increases occurred in January 1971. At a meeting in London, the major companies and all but two independents—Hunt and a German firm—surrendered to the Middle East nations' price demands. (The London Policy Group evolved out of this unsuccessful effort to present a unified front in negotiations.)

On the basis of Schuler's testimony and numerous government documents on the issue, several senators charged that the oil companies' efforts were "undercut" by the U.S. government, which lacked even "the foggiest notion" of the price crisis' implications. The Justice Department had agreed to waive antitrust reprisals against the industry during its period of joint negotiations, but, Church declared, the Nixon Administration "waffled" the opportunity for presenting a united front when in January 1971 federal officials suggested that the price talks be divided into separate discussions with the Persian Gulf states and with Libya and Algeria. "We were seen to back off and off," Schuler said. "The oil producing states realized there was no way to stand against them. This is why we are in the position we are in today."

George T. Pierce, a senior official of the Exxon Corp., told the subcommittee Feb. 1 that his firm had underestimated the demand for petroleum since 1958, but he denied under questioning that the error was a deliberate tactic designed to boost prices during an ensuing shortage.

Although demand had risen annually since 1955, Pierce conceded that Exxon had made no effort to expand its spare capacity for crude oil since 1958. He also admitted that when Iran had asked Exxon to expand its production there in 1968 and 1969, the company had refused.

McCloy's antitrust efforts—John J. McCloy, a longtime presidential adviser, told the subcommittee Feb. 6 about his 10-year efforts to lay the groundwork for protection from antitrust prosecution for the oil industry in their joint bargaining with Middle East nations and the industry's stunning setback suffered in 1971 negotiations with oil producers.

McCloy, a Wall Street lawyer representing 23 oil companies, told of his talks with President John F. Kennedy in 1961 soon after producing nations formed the Organization of Petroleum Exporting Countries (OPEC). He mentioned the need for similar "concerted action" by the oil companies, McCloy said, adding that he personally had reminded every attorney general from 1961 to 1972 of the industry's problems in dealing jointly with foreign oil producing nations while constrained by U.S. antitrust laws.

McCloy said his efforts resulted in an informal agreement in which the government allowed the oil companies to bargain collectively. In 1971, McCloy continued, he sought and obtained a formal letter of enforcement intention disavowing any government challenge to the industry position as additional protection; his preparations proved fruitless, however, when presidential emissary John N. Irwin 2nd, and Douglas MacArthur 2nd, ambassador to Iran, agreed to a suggestion by the shah of Iran that the oil companies conclude separate price agreements with producing states. His fears that a "chain reaction inherent in separate [price] negotiations" would result were realized McCloy said.

(The Justice Department gave preliminary approval Jan. 29 to the oil companies' plan to negotiate jointly with OPEC. The department stressed that the action did not constitute immunity from future antitrust prosecution. The oil companies had sought an advisory opinion, called a business review letter, on the matter. The opinion they received allowing them to proceed with joint talks represented only an "expression of present enforcement intention," officials said. The ruling was similar to that won in 1971 by McCloy.)

In testimony released Feb. 9, John N. Irwin 2nd, currently ambassador to France and former undersecretary of state, described negotiations he attended in 1971 between representatives of oil companies and OPEC nations.

According to Irwin, he had been prepared to support the oil companies in their desire to negotiate as a single group, but he had agreed to split the talks between the Persian Gulf state producers and Libya and Algeria as another group after being persuaded by the shah of Iran. The U.S. ambassador to Iran, Douglas MacArthur 2nd, concurred in the decision, Irwin said.

The oil companies were not informed of the change in strategy, which effectively undercut their position, until after Irwin accepted the shah's proposal, he added.

Irwin was a lawyer with no background in oil when recommended for the negotiating post by the oil companies and selected by Nixon.

"The significance of Ambassador Irwin's testimony is not that he, as an individual, was ill-prepared for his policy-making role," Sen. Church concluded. "Rather, it is that the U.S. government had little institutional capability for dealing with complicated international oil negotiations. [It] reveals the appalling lack of governmental expertise in the area of international oil and the failure of the U.S. government to come to grips with the different bargaining strategies open to it in these negotiations," Church said.

Price hike tied to oil company dispute—Sen. Church charged Feb. 18 that a dispute between a major U.S. oil company and a large U.S. distributor forced up prices paid for petroleum by public utilities along the eastern seaboard. Customers, principally those served by Consolidated Edison of New York and Long Island (N.Y.) Lighting Co., paid $50 million more annually in higher prices as a result.

The charges were based on testimony taken in December 1973 from Edward M. Carey, president of New England Petroleum Co. (NEPCO), which sold fuel to the utilities.

NEPCO accused Standard Oil Co. (California) of breaking a long-term contract to supply it with crude oil, forcing the company to charter other tankers at a higher cost to import the fuel.

The dispute arose in 1973 when Libya seized 51% of Standard Oil (California) oil operations. Standard, which operated a refinery in the Bahamas with NEPCO to process the Libyan crude, notified NEPCO that it would suspend deliveries from Libya as part of its tough new policy aimed at denying Libya a market for the oil it had seized.

NEPCO, faced with the need to supply its customers and meet payments on the Bahamas refinery, negotiated a separate deal with the Libyan National Oil Co. and arranged for tankers to ship the "hot" oil Standard refused to deliver.

Carey explained the decision: "We did not feel that we could prudently sit and wait for the major oil companies to negotiate something with the Libyan government because [in the interim] ... you would have substantial blackouts and brownouts in the northeast part of the U.S."

After the deal was negotiated with Libya, NEPCO officials said they received two phone calls from Texaco and Standard executives warning the firm not to proceed with the import arrangement. (Texaco oil fields also had been seized.) Immediately thereafter, Frank A. Mau of the State Department's Office of Fuel and Energy, called NEPCO officials.

Mau's "conversation was essentially the same as the previous two except instead of saying that legal actions would be taken to preserve the rights of these two companies, we were told that the State Department felt this was the wrong thing for New England to do, that this would have repercussions in the Middle East and that the State Department opposed this action," Richard Manning, NEPCO's attorney, testified.

Manning said he returned the phone call to protest the government intervention: "If the State Department wanted us to stop picking up the oil and shut down the utilities on the East Coast," he said, "we would have to be notified by the elected representative of the people."

He also wrote Secretary of State Henry Kissinger about his protest: "It is indeed shocking that throughout this dispute none of the major oil companies involved (whose profits—on which practically no U.S. tax is paid—are at record high levels) have given any indication whatsoever of any concern as to the damage to the public which will result if their positions are sustained. Your department, in supporting the positions taken by these major oil companies, seems to be of the same disposition."

The oil industry pressure tactics proved unsuccessful over the long term as well. They were not able to shut off the world market in "hot" oil and Libya nationalized the remaining U.S.-controlled petroleum operations Feb. 11. The Wall Street Journal reported Feb. 15 that despite the embargo NEPCO continued to receive supplies of oil from Libya as a "gesture of gratitude" for defying the major oil companies.

By the latter part of 1973, however, electricity rates for utilities served by NEPCO rose 18.5% and their fuel prices soared 74%, according to Church. Standard had charged 40¢ a barrel to transport the oil to the jointly owned refinery but tankers chartered to ship the "hot" oil charged $1.90 a barrel, Carey said.

Secrecy reported—The subcommittee revealed that the Libyan Producers' Agreement, a private pact among major international oil companies to support each other in the event of nationalization efforts by the Libyan government, had been classified by the State and Justice Departments, the Washington Post reported Feb. 17.

Sen. Church said this was "the first time I know of when a government classification has been extended to a document between private companies."

There was a "second peculiarity" about the arrangements, Church said: the secrecy provision applied only to the U.S. companies although foreign firms involved in the agreement had copies. The document also was unclassified for 27 months after its execution in January 1971 but was hurriedly barred from publication after Ralph Nader filed suit to compel the Justice Department to disclose all agreements it had administratively exempted from antitrust challenges and to disclose the review letters which conferred such exemptions.

The document was declassified after Church's protests but he did not release it immediately when the Administration said that its disclosure "could damage foreign relations." Church asked a delay in its publication until after the Energy Action Group foreign ministers' meeting in Washington Feb. 11.

Testimony Feb. 20 revealed that the Eisenhower Administration had invoked "national security" to kill a 1953 criminal case against seven major oil companies accused of operating an international cartel to maintain artifically high prices.

The nature of the threat to national security by the antitrust suit was never explained by the National Security Council, the State Department or the Justice Department, witnesses said.

If U.S. interests were imperiled by the case, a former Justice Department official asked, why were they entrusted to a "private supranational government" which included two foreign based oil companies?

Another witness said Feb. 21 that President Truman had instructed him to drop a criminal action against oil companies "solely on the assurance of Gen. Omar Bradley that the national security called for that decision." Truman, however, urged vigorous prosecution of civil actions against the oil industry, Leonard J. Emmerglick testified. (He had headed the grand jury investigation in the criminal action.) That case was finally settled in the 1960s with consent decrees that were "cosmetic, and nothing more," Emmerglick said.

Oil tax subject of debate. According to a report released Feb. 3 by Sen. Henry M. Jackson (D, Wash.), tax breaks granted the oil industry as incentives to promote exploration and development had failed to generate new supplies of domestic oil. The study, prepared by the Library of Congress, put the cost of the tax breaks at $1.5 billion annually. (Another $600 million–$1 billion was lost to the government in tax credits allowed on foreign oil operations, according to the report.)

A tax study prepared by the Petroleum Industry Research Foundation, a trade group, said Feb. 8 that the oil industry bore a larger domestic tax burden per dollar of sales than any other U.S. industry.

In 1971, according to the group, the industry paid 5.59¢ on every dollar of gross revenue, compared with 4.58¢ for other mining and manufacturing firms, and 4.17¢ for other businesses.

As the depletion tax rate dropped, the report continued, the oil industry also increased its effective federal income tax rate on domestic operations; in 1969 it was 19.1%, 1970—22%, 1971—23.5%, and 1972—25.6%.

But a Senate Government Operations subcommittee said later (Dec. 2) in 1974 that the seven largest U.S. oil companies paid (in the aggregate) an effective U.S. income tax rate averaging less than 5% for the five-year period ending in 1972. (An effective tax rate was defined as the percentage of net income actually paid in federal income taxes.)

Huge profits during Arab embargo. The major oil companies issued reports detailing dramatic 12-month increases in earnings that were paced by sharply higher gains during the fourth quarter of 1973 when the Arab embargo was in effect.

Treasury Secretary George P. Shultz Feb. 4, 1974 gave Congress a report rebutting the oil industry's contention that 1973 profits merely appeared high because earnings in previous years had been unusually low. Averaged earnings of the 22 largest oil companies were higher in 1973 than any of the preceding 10 years, based on an analysis of return on stockholders' equity, according to the report.

In 1973, the return was 15.1% compared with 9.7% in 1972, 11.8% in 1968 and 10.9% for the average over 10 years.

Cities Service Co. Jan. 22 had been the first oil company to make public its fourth quarter earnings: 1973 net income was $135.6 million, up 36% from $99.1 million in 1972; fourth quarter profits increased 49% to $42.1 million from $28.1 million in the same period of 1972; return on shareholders' equity was 9.2%.

Exxon announced Jan. 23 that profits were up 59% both in the final period and for all of 1973. Net income for the October–December period rose $291 million to $784 million. Over the entire year, net income gained $91 million to total $2.44 billion. In one year, the return on shareholder's equity was 19%, up from 13% in 1972.

In the U.S., earnings were 16% higher based on a 14% rise in sales. The largest "absolute and percentage" gains in earnings were derived from operations in the Eastern Hemisphere—primarily Europe, the Middle East and Africa—where profits were up 83%.

The statistics were released to reporters by Exxon Chairman J. K. Jamieson, who said the press conference was held to offset "attacks on us about our secrecy."

Union Oil Co. of California, the nation's 12th largest oil company in sales, reported Jan. 23 that net income for 1973 was $180.2 million, up 47% or $58.3 million. During the fourth quarter, earnings totaled $51 million, up 55% from $32.8 million in the same period of 1972. The return on shareholders' equity was 11%, up from the 7.6% average over the past three years.

Texaco announced Jan. 24 that fourth quarter earnings were a record $453.5 million, up 70% from the 1972 level of $266.6 million. Full year profits were a record $1.29 billion, up 45% from $889 million.

Revenues were up 47% in the final period and 32% higher for the entire year.

Mobil also issued a profit statement Jan. 24. Earnings were $271.6 million, a gain of 68% from $161.5 million. Revenues climbed 36% at the same time. For the entire year, profits reached $842.8 million, up 46% from $574.2 million. Revenues increased 23%.

Shell Oil Co. announced Jan. 24 that there had been a 1.5% decline in fourth quarter earnings (down from $80.6 million in 1972 to $79.4 million), but full year figures showed a 27% gain. Profits totaled $332.7 million, up from $260.5 million. Revenues were up 25% in the fourth quarter and 18% for the year.

Standard Oil Co. (Ohio) announced Jan. 27 that its operating profits in the final quarter of 1973 dropped 40% on a 7% increase in revenue, but operating profits for the entire year rose 24% on an 8% gain in revenue. Net income for the October-December period was $11.6 million, down from operating profits of $19.3 million and net income of $17.1 million in the final period of 1972. Total operating profits for the year were $74.1 million, up from operating profits of $59.7 million. Net income over the same period climbed from $57.5 million to $89.4 million, a 55% gain

Standard Oil Co. (Indiana) Jan. 28 reported a 53% gain in earnings during the fourth quarter of 1973 and a 36% increase for the entire year. Net income in the final period was $121.4 million, up from $79.5 million. Revenue totaled $1.9 billion, a gain of 26%. For the year, net income rose to $511.2 million from $374.7 million on an 18% gain in revenue to $6.5 billion. Earnings from U.S. operations were 32% higher than in 1972.

Phillips Petroleum Co. posted a 128% increase in earnings during the final quarter and a 55% gain in profits for the entire year, officials said Jan. 29. In the December quarter, Phillips earned $86.7 million, up from $38.1 million. For the year, net income was $230.4 million, up from $148.4 million. Domestic operations accounted for 19% of 1973 earnings, according to the company.

Sun Oil Co. reported Jan. 29 that its earnings for the final period were 60% above year earlier levels, reaching $75 million. Full year earnings were $230 million compared with $155 million in 1972—a gain of 48%.

Because of a change in accounting procedures, Marathon Oil Co. released only full year profits statistics. Officials said Jan. 29 that net income for 1973 was $143.3 million. Earnings before extraordinary items in 1973 rose 62.2% in one year.

Standard Oil Co. (California), the nation's fifth largest oil company, reported Jan. 30 that its fourth quarter earnings were $283.1 million, up 94% from $145.8 million. For the year, earnings increased 54% to total $843.6 million. The 1973 return on net investment was 15%, up from 10.8% in 1972.

Atlantic Richfield Corp., the nation's ninth largest oil company, Jan. 30 announced a 41% increase in fourth quarter profits (compared with the same period in 1972). Earnings totaled $91.7 million. Figures for the full year put earnings at $270 million, up 38% from $195.6 million.

Amerada Hess Corp. Feb. 6 announced fourth quarter profits: $104.5 million, compared with $18.3 million in the final period of 1972. Net income for the year showed an even greater increase: up from $26 million in the previous year to $245.8 million for 1973. Revenues were up 42.2% over 12 months.

Gulf Oil Corp. announced Feb. 12 that its operating revenues had increased 153% in the final quarter of 1973 (compared with year-earlier figures).

Data for the full year showed operating profits gained 79% to total $800 million. Revenues were up 29%, with the bulk of the increase from foreign operations. Earnings from U.S. operations were up 14% while foreign earnings climbed to $560 million from $150 million in 1972.

Occidental Petroleum corp. announced Feb. 20 that its fourth quarter earnings were up 172% in 12 months, reaching $24.3 million compared with $8.9 million in the final three months of 1972.

For the entire year, profits soared more than 300%. Earnings were $79.8 million.

In 1972 operating net had totaled $19.7 million and net income had been $10.4 million. Revenues rose 27% in 1973. Occidental was the industry's 11th largest oil company.

The second largest international oil company, the Royal Dutch/Shell Group of Companies, announced Feb. 28 that 1973 "was an extraordinary year" as earnings soared 153% in 12 months from $704 million to $1.78 billion. During the same period, revenue rose 32% from $14.4 billion to $19 billion. The rate of return on net assets for 1973 was 17%.

Officials cautioned that the recent depreciation of the pound had artificially inflated the year's earnings by about $345 million and that inflation had also distorted the value of company assets.

According to an unofficial estimate of 1973 fourth quarter earnings, profits surged 175% to total $703.8 million. (Earnings in the final three months of 1972 were $255.3 million.)

British Petroleum Co., 49% owned by the British government, announced March 14 that profits rose 332% in 1973 to a record $760 million, despite a drop in the sale of crude oil and refined products. Higher oil prices and changes in currency values accounted for the increase, officials said.

Fourth quarter profits increased an estimated 250% to total $289 million. Revenue rose 60% in the final three months of 1973.

Aramco profits probed. The Senate Subcommittee on Multinational Corporations resumed hearings on the oil industry March 27, 1974. Aramco, which was owned jointly by the Saudi Arabian government and four U.S. oil companies—Texaco, Exxon, Mobil and Standard (California)—was the focus of the Congressional investigation into the relationship between sharply higher industry profits and sharply reduced oil supplies.

According to preliminary figures made public by the subcommittee for the first time, Aramco's gross income in 1973 totaled $8.7 billion and profits equaled $3.2 billion. (Another $1.1 billion was paid in royalties and $3.9 billion went to the Saudi government in taxes. U.S. taxes came to $3 million in 1973, $5 million in 1972 and $4 million in 1971.) These figures compared with revenues of $4.6 billion in 1972 and $3 billion in 1971. Profits in 1972 had totaled $1.7 billion and $1.1 billion in 1971.

Aramco officials said the figures, which were based on documents submitted by company auditors, were "grossly misleading" because they were based on the "posted" price rather than on the actual market value of oil.

Senate investigators noted the close relationship between surging profits and dramatic increases in the posted price of oil. They charged that at the current posted level, Aramco realized a profit of $4.50 a barrel. As a result, Sen. Frank Church (D, Ida.) said, Aramco had little incentive to hold down the posted price of oil when prices rose at a corresponding rate.

The subcommittee also accused the consortium of using various "control mechanisms" to hold down production during the 1960s when a supply glut threatened price stability. An Aramco chart showed that the group's production capacity far exceeded its actual production from 1963 through 1973. Other consortium documents revealed that the group imposed penalties on any owner taking oil in excess of its ownership share; that Aramco did not begin to accumulate earnings for use in expansion projects until 1970; and that Standard California proposed limiting production from 1969–73 to deal with "the problem of accommodating a large potential [oil] surplus."

An official of Continental Oil testified that the four U.S. owners of Aramco had joined with Gulf to limit production in Iran from 1962–68.

Other Aramco financial data, analyzed for the subcommittee by an official at the General Accounting Office, showed that the operating and general expenses for producing a barrel of oil was about 12¢ a barrel until 1973, when the price rose a few pennies; profits per barrel increased from 62¢ in 1963 to $4.50 at the current level; dividends paid by Aramco to its owners rose from $810.5 million in 1971 to $1.57 billion in 1972 and $2.59 billion in 1973; Aramco's $8.7 billion cash flow in 1973 was the largest in history for a single firm in international commerce.

Howard W. Page, a former Middle East coordinator for Exxon Corp., told the subcommittee March 28 that five major U.S. oil companies joined an international consortium in Iran in 1954 because Secretary of State John Foster

Dulles thought the action was necessary to prevent Iran from falling under Soviet domination.

Dulles also decided that the smaller U.S. independent oil firms could collectively hold no more than a 5% interest in the consortium, Page testified in response to Sen. Church's charge that the majors had prevented independents from obtaining an equitable share in the operation.

The subcommittee also prepared charts intended to show that until 1972, Exxon, Texaco, Gulf and Standard (California) proposed low production volumes for Iran in order to limit operations there.

Sen. Church made public an internal Standard (California) memo written Dec. 8, 1968 which indicated that the company would increase production in Saudi Arabia and Iran, where political pressure was being exerted to increase operations, but decrease their production levels elsewhere in a trade-off designed to protect the Persian Gulf concessions by retarding operations in other areas.

A company spokesman said the memo was of little import. "Nobody paid any attention to it [the memo] as far as operations were concerned," he testified.

Burmah Oil buys U.S. firm. The sale of the Texas-based Signal Oil & Gas Co. to Burmah Oil Inc., the U.S. subsidiary of the British-owned Burmah Oil Ltd., was completed Jan. 28. The purchase price of $480 million, to be paid to the parent Signal Companies, a California-based conglomerate, included $60 million in debt forgiveness and a cash payment of $420 million.

The take-over followed an abortive last-minute appeal to U.S. Attorney General William B. Saxbe by Sen. Floyd K. Haskell (D, Colo.) to seek a temporary restraining order against the transaction. The Justice Department, which was conducting an antitrust investigation into the arrangement, did not act on the request. Previously the department had asked for a 60-day delay in the takeover, but the firms rejected the request.

Haskell, chairman of a Senate Interior special subcommittee, held hearings on the merger Jan. 29. He expressed concern that the deal might result in cancellation of crude oil contracts to independent refiners and gasoline stations in the U.S. because of the possibility that most of Signal's crude would go to Standard Oil of Ohio, which was associated with Burmah Oil. Haskell had previously said the take-over would enable Burmah to obtain Signal's 20% interest in the huge Thistle oil field in the North Sea.

N.J.D. Williams, managing director of Burmah Oil Ltd., told the subcommittee that the acquisition would not lessen competition in the U.S. because Burmah was an "insignificant factor in the exploration for and development of crude oil in the U.S." and Signal Oil's domestic net crude production was only 52,000 barrels a day in 1972. Williams said Burmah had "no intention of breaching" any of Signal's existing contracts, which committed three-fourths of Signal's production for sale to domestic refiners through 1980. However, there would be no restraints on Burmah sales when the contracts expired.

The sale came after the Delaware State Supreme Court sustained a lower court's dissolution of a temporary injunction against the deal on the grounds that the plaintiffs, a group of Signal stockholders led by Loeb Rhoades & Co., the New York investment bank, and Cemp Investments Ltd., a Canadian company, had failed to post a $25 million bond. The plaintiffs had said the $480 million price was grossly inadequate.

Signal Oil & Gas was one of the U.S.'s largest suppliers of crude oil to independent refiners. It had reported sales of $267 million in 1972 and $185 million for the first nine months of 1973. Burmah had more than 20% stake in British Petroleum, 48% owned by the British government.

Government Controls

FEO fuel directives. Federal Energy Office Director William Simon Jan. 3, 1974 urged major oil refiners and independent oil distributors to restrict retail sales of gasoline to 10 gallons per customer, just as he had earlier asked motorists to limit their purchases to 10 gallons.

Simon received agreement to such a pledge from most oil companies but Gulf Oil and Standard Oil (Ohio) said Jan. 5 they would not comply.

Simon sent telegrams Jan. 7 to 26 refiners asking them to sell more domestic oil to New York and New England area independent wholesalers of heating oil who had been forced to rely on supplies of the higher priced imported fuel. They had complained of losing customers to the big oil suppliers whose use of the cheaper domestic oil had allowed lower retail prices.

The FEO's final fuel allocation rules were published Jan. 15 and became effective immediately.

Noting that industrial and public service users were favored over those requiring fuel for home heating and private automobiles, Deputy FEO Administrator John C. Sawhill said that the allocation program's aim was "to preserve jobs rather than to have homes at 75 degrees."

"There will have to be changes in the American lifestyle over the next several years," Sawhill said. Among the regulations issued, two would have immediate effects on the living habits of individual Americans:

■ Private homeowners could obtain only enough oil to heat their residences at temperatures that were six degrees lower than average settings in 1972. If homes were heated at 70 degrees in 1972, a 64-degree reading would be required for 1974; however, hardship cases could be appealed to state allocation boards, which would maintain 1.5%–4% reserves of each petroleum product, based on an estimate of use in each state. Oil suppliers would be allowed to deny deliveries to those refusing to cooperate. Violators of the regulations could be fined up to $2,500.

■ Gasoline supplies during the first three months of 1974 would be 20% below projected demand, according to Sawhill, necessitating limits on pleasure driving. But FEO officials said Jan. 15 that no cutback in gasoline production was being ordered. According to previous official announcements, refiners would be asked to limit gasoline production to 75%, 89% or 95% of 1972 production levels.

Officials said a mild winter and successful voluntary effort to conserve fuel had eased the home heating oil shortage so the Administration was not forced to ask refiners to concentrate on the production of middle distillate fuels at the expense of gasoline production.

With the exception of the gasoline production change, the final allocation regulations generally resembled previously published proposals. (The Pentagon was allowed 2% increase in fuel for the first quarter over 1972 use.)

Sawhill also released details of the Administration's contingency plan for gasoline rationing, insisting, however, that no decision had been made on whether to adopt the plan and voicing hope that rationing could be avoided.

Under the standby plan, top priority users in public service and mass transit would get all the fuel required. Businesses which had made bulk purchases of gasoline would be entitled to 100% of their 1972 consumption levels.

Licensed drivers, 18 and older, would apply for authorization to receive ration coupons, available in three-month supplies at local distribution points for $3 payments. No gasoline value had been assigned to the coupons because the value was expected to vary with the severity of the shortage. Coupons would be "transferable" and could legally be bought and sold.

Rural areas and dense population centers with poor mass transit facilities would be favored over other areas in distribution of coupons.

In another development, the Commerce Department imposed export quotas Jan. 23 on motor and aviation gasoline, heating and diesel oil, and heavy fuel oils. An export licensing system was replaced with quotas, officials said, because the volume of exports, although slight, had showed an upward trend.

In a followup order, the Commerce Department Feb. 8 placed new export quotas on kerosene, jet fuel, carbon black feedstock oils, butane, propane and natural gas liquids.

The FEO Jan. 30 issued proposals to control retail propane prices by limiting refiners' pass-through increases to actual crude oil price hikes. No provision for a price rollback was included in the program, despite charges by Georgia Gov. Jimmy Carter (D) that the cost of propane, which was widely used in the rural South, had risen an average 310% in one year.

Simon announced a plan Feb. 4 to equalize the price differences paid by heating oil customers in the eastern part of the country for cheaper domestic oil and the more expensive imported supplies. (Prices in December 1973 had ranged from 23¢ a gallon for domestic to 47¢ a gallon for imported.)

Under the plan which would take effect March 1, distributors were instructed to share their supplies of domestic and imported fuel in an effort to redistribute their individual supplies according to the total proportion of domestic and imported fuel available to all the region's 76 major suppliers.

The distribution of scarce gasoline supplies also was inequitable. Simon warned service stations Feb. 12 that they could not discriminate among buyers by selling only to regular customers or accepting large advance payments for fuel.

In another government-ordered adjustment of production, refiners were authorized Feb. 10 to begin increasing gasoline output as long as they maintained adequate supplies of distillate and residual fuels for the winter.

The FEO ordered refiners Feb. 14 to increase jet fuel production by 6% in order to supply the nation's airlines with fuel promised under the Administration's allocation program.

At a press conference Feb. 19, Simon scored Gulf Oil Corp. for its "completely resistive and uncooperative" position regarding the FEO regulation requiring refiners to share crude oil supplies.

Simon said the FEO was investigating complaints that Gulf was curtailing deliveries to independents and other customers it had serviced in 1972.

According to the New York Times Feb. 21, major oil companies were cutting back on their crude oil imports in an effort to circumvent the government's sharing order. Oil industry spokesmen argued that "disincentives" were built into the program, i.e., under allocation arrangements, high priced oil purchased abroad would be sold at lower levels to competitors in refinery operations. The price order was aimed at equalizing the spread between imported prices and the price of cheaper domestic oil.

Gasoline situation worsens. Calls mounted from governors and members of Congress for a nationwide gasoline rationing program but Federal Energy Office (FEO) Director William Simon continued to resist, suggesting instead that states adopt regional plans to deal with spot shortages.

As federal officials were admitting privately that the Administration's allocation system was riddled with problems, Simon announced Feb. 9 that gasoline allocations would be increased for 12 states and Washington D.C. Relief for states with severe shortages was promised Maine, New Jersey, Arkansas, Delaware, Illinois, Kentucky, Maryland, Mississippi, North Carolina, Texas, Virginia and Tennessee. Connecticut was added to the list after a personal appeal by Gov. Thomas Meskill.

Allocations were reduced for 10 states—Iowa, Kansas, Minnesota, New Mexico, North Dakota, Ohio, Oklahoma, South Dakota, Wisconsin and Wyoming.

The announcement highlighted a worsening gasoline shortage that was particularly severe in urban areas and in the Northeast, where a strike by independent truckers had interrupted the delivery of already scarce gasoline supplies.

Evidence of the uneven nature of the shortage and the allocation program's ineffectiveness in redistributing the dwindling supplies could be seen in New Jersey, where two–three hour waits at gasoline stations were common while motorists in much of the rest of the country faced no lines and few limits on their purchases.

By Feb. 15, five states and the District of Columbia (and adjacent areas of northern Virginia) had joined Oregon and Hawaii in implementing a form of rationing that based distribution on odd and even numbered license plates.

In Oregon, where the program was voluntary, and in Hawaii, where mandatory rules were in effect, the system had been effective in limiting weekday sales, eliminating panic buying and reducing long waiting lines and traffic jams.

Massachusetts, New York, Washington, Maryland, and Washington, D.C. used a voluntary version of the license plate plan. New Jersey adopted a mandatory program, and Virginia did so also Feb. 18. (More than 20 towns and cities in Connecticut also had adopted a voluntary form of the plan but there was no statewide program.) Pennsylvania had a program based on the serial numbers of state inspection stickers.

New York Feb. 26 abandoned its voluntary gasoline distribution system based on odd and even license plates and adopted a mandatory plan. Among other states joining in the practice on a statewide voluntary basis were New Hampshire, Vermont, Rhode Island, North Carolina and South Carolina. Delaware adopted a mandatory system. States using the odd-even plan on a local basis were California, Florida, Nebraska, Illinois, Kentucky and West Virginia.

A minimum purchase was included in several of the state programs. Simon had accused motorists of contributing to the chaotic situation by lining up to purchase only a few gallons of gasoline as panic buying conditions persisted.

Further adjustments were made in the government's allocation program for the statewide distribution of gasoline and dealers hardest hit by reduced gasoline deliveries were allowed to pass on a price increase to motorists.

In a ruling announced Feb. 16, the FEO allowed half the nation's gasoline stations to raise gasoline prices 1¢ a gallon, effective March 1. The markup was authorized only for those stations whose allocations had been reduced at least 15% from their supply level in the 1972 base period. The action, coming on top of the regular price adjustment allowed dealers at the beginning of each month, was expected to cost motorists $500 million a year.

According to Deputy FEO Administrator John C. Sawhill, the price increase was designed to compensate gasoline stations for their reduced profits resulting from the falloff in gasoline sales and the concomitant decline in sales of automotive accessories. "People just don't buy those accessories after they've been waiting in line [to buy gasoline] for 45 minutes or longer," he said.

Sawhill admitted that the agency acted in response to threats by many gasoline stations to shut down their operations entirely to protest what they considered the government's discriminatory restraints on profits. Few representatives of gasoline station organizations expressed satisfaction with the 1¢ increase, however, calling it "too little too late."

Dealers had intentionally exhausted their February allocations and closed by mid-month in the Tidewater area of Virginia, sections of Pennsylvania, Florida, Massachusetts, Oregon, Washington, Connecticut and North Carolina—areas where gasoline supplies were especially short. The dealers also were protesting a government regulation banning preferential treatment for their regular customers.

The number of gasoline stations closing permanently as a consequence of the fuel crisis had risen sharply, with independent dealers very hard hit by the dwindling supply.

According to the New York Times Feb. 16, the number of operating gasoline stations across the nation had dropped from 222,000 in 1969–70 to 216,000 currently. Non-brand independents were believed to number 36,000 but their share of the market was said to be 25%. Oil industry figures put the number of name brand station closings at 3,700 in 1973. Since spring 1973, when the gasoline shortage first became apparent to motorists, 9% of the majors and 12% of the independents had shut down their stations, according to an industry survey.

In a decision announced Feb. 19, the FEO ordered emergency gasoline allocations totaling 84 million gallons to be distributed to 20 states.

Ten states were scheduled to receive an increase of 5% over their February allocations "in a matter of days," FEO Administrator William Simon said: Alabama, Arizona, Georgia, Nevada, New Hampshire, Oregon, and Vermont. West Virginia, New Jersey and Virginia would receive an additional 3% to augment the 2% emergency supply authorized Feb. 9 but apparently never delivered.

States receiving an extra 2% were Connecticut, Florida, Indiana, Massachusetts, Missouri, New York, Pennsylvania and Rhode Island. Maryland and

Illinois, which had also been promised and never received an emergency supply of 2%, Feb. 9 were initially scheduled to receive another 3%, but after "clarification," the allocation was set at the original 2% level.

Simon said the emergency supplies would be drawn from oil companies' inventories rather than diverted from states where gasoline was more plentiful. He added that the emergency effort would be followed by an attempt to recalculate and equalize the allocation program in all 50 states.

The additional supplies for February, which augmented other statewide emergency reserves totaling 3% of available supplies, would be under each governor's supervision for distribution within the state.

Simon said he had been able to provide emergency gasoline supplies because of "the tremendous success of our [allocation] program" and the "tremendous response" of consumers in fuel conservation efforts.

FEO officials expressed irritation Feb. 20 at press reports noting that in many states the emergency supplies equaled less than a gallon of gasoline per motorist.

Maryland filed suit Feb. 20 against Simon, the FEO and 20 major oil companies to force the government to increase its allocation.

Simon again revised the gasoline allocation program Feb. 22 and 23.

Supplementary allocations to meet an emergency situation in February were expanded in a ruling announced Feb. 22. Gasoline deliveries for Washington, D. C. and 24 states were increased to total 10% of the month's basic supply. Two other states were allowed a total of 6% in additional gasoline. (South Carolina was the only new state added to the group already scheduled to receive extra gasoline supplies.)

The latest order would add 239.75 million gallons of gasoline to the 326.47 million gallons released for emergency use Feb. 19. The increases were mandatory and oil companies were ordered to supply the fuel from their own stockpiles.

Simon authorized a 2¢ a gallon increase in gasoline sold by independent retail dealers Feb. 23 when it appeared that many stations, especially in the Northeast, would continue their "pump out" protest.

The ruling was intended to give refiners an incentive to switch production emphasis from home heating oil to gasoline. A 6% increase in gasoline production was expected to result.

According to the FEO, the order, which would take effect March 1, "supersedes the 1¢ increase granted Feb. 16 for service station owners with less than an 85% allocation of their 1972 supply."

Simon also relented on another issue. Declining to impose a solution from Washington, he urged states to draw up their own plans for "priority customer treatment at gasoline pumps."

In another price decision, the FEO announced Feb. 25 that it had paired a 1¢ a gallon increase in the wholesale price of gasoline with a 2¢ a gallon decrease in the cost of middle distillate fuel oil, effective March 1. The decision reversed a gasoline price rollback and heating oil price hike authorized in December 1973.

But Simon told a Senate subcommittee Feb. 26 that gasoline prices would eventually "stabilize where they are," which, according to Simon, was at 50¢–55¢ a gallon. Simon also conceded, however, that prices could rise in short-term spurts to more than 70¢ a gallon.

FEO officials told the New York Times Feb. 28 that a serious gasoline shortage would affect wide areas of the country by late spring if crude oil supplies were not available to replenish depleted stockpiles.

According to FEO officials, unusually heavy winter demand for gasoline, particularly along the East Coast, had surprised refiners who were forced to release gasoline stockpiles normally held in reserve for spring and summer. (Other unpublished statistics, however, were said to show a sharp 9% decline in gasoline use during the first three weeks of February over the previous year's level, the Times reported March 7.)

An announcement Feb. 27 by Shell and American Oil Co. that March allocations would be cut back prompted fears that characteristic end-of-the-month gasoline droughts would reappear in March. Shell officials said national allocations to dealers would be cut from 85% of the supply level in February 1972 to 70% of the supplies received in March 1972. On the same basis, Amoco reduced allocations to its dealers from 80% to 75%.

The cutback was required, Shell said, "because we have 10% less gasoline available than in the base period of March 1972 and more customers to serve," and also because Shell had been forced to sell 77,000 barrels of crude a day to other refiners.

Other statistics released Feb. 28 by the American Petroleum Institute showed that gasoline production and stockpiles were nearly equal to the figures for the same period in 1972. Foreign crude oil imports were 9.6% below the daily average for the last week of January and the first three weeks of February. Crude oil stocks were 3.5 million barrels higher than 12 months earlier.

The American Automobile Association published the results of a nationwide random sample March 5 showing that 20% of all service stations were without gasoline. In the Southeast, 36% lacked supplies but in a six-state area of the central and mountain states, only 2% of the stations were without gasoline.

According to a Gallup poll conducted in early January, 53% of those surveyed opposed gasoline rationing, with 37% favoring the action. The poll was published Feb. 27.

In another FEO action, officials issued proposed revisions Feb. 27 in the government's controversial crude oil allocation program. The changes, which decreased the amount of forced sharing of supplies among refiners, were required to stimulate imports. Refiners, who objected to the regulation, had responded by reducing their imports to minimize their required redistribution with competitors.

According to the new rules, "additional quantities of crude, over and above the amounts estimated to be available this quarter," could be imported "without being allocated away to another refiner."

Small refiners with capacities of less than 75,000 barrels a day would get preferential allocation. The new system would raise their supplies "to 1972 levels subject to reductions across the board only when the nation's total supplies fall below those levels."

Under a new "neutral zone" designation, independent refiners producing 75,000–175,000 barrels a day "would not be required to allocate crude and they would be eligible to receive allocated crude only on the basis of increased crude oil acquisitions from outside the allocation system."

Refiners producing more than 175,000 barrels a day would be subject to "fixed allocation obligations."

Refiners had been subject to the sharing order because of a law passed by Congress in 1973 over the Administration's objections. (The law had been intended to prevent small refiners from being squeezed out of the market by bigger refiners.) FEO officials urged Congress to amend the existing law to minimize the unintentional consequence of import disincentives for big refiners.

Gasoline allocations increased. Federal Energy Office (FEO) Administrator William E. Simon told the National Governors Conference in Washington March 7 that "every state will receive a greater supply of gasoline in March than in February, even when the February emergency allocations to the states are included."

Under new allocation plans for March, total gasoline supplies available to the states would be about 8 billion gallons, or nearly 860 million gallons more than February's allocation. Every state would be guaranteed at least 85% of its gasoline supplies in the March 1972 base period, Simon said.

Simon said the adjustments reflected the growth of motor vehicle use within the different states. Eight states were scheduled to receive more than 100% of their March 1972 consumption levels because of this growth factor: Alaska, Kansas, Louisiana, Minnesota, North Carolina, Oklahoma, Texas and Wyoming.

Nationwide, the average state allocation would be 89% of the base period, but Simon promised that by April, the equalization process would be completed and no state would receive more than 95% of its 1972 base level. (February allocations had ranged from 61% to more than 100% of the consumption levels in 1972.)

By the time Simon had announced yet another readjustment in his much criticized gasoline allocation program, the supply situation had stabilized somewhat from the chaotic condition that characterized February. Emergency fuel allotions covered the critical spot shortages that were in evidence at the end of the month and new deliveries became available for March, but the price of gasoline continued to rise.

FEO allocation system faulted. The Federal Trade Commission (FTC) March 16 issued a staff report highly critical of the FEO's handling of fuel allocation programs.

According to the report, the federal government's efforts may have been instrumental in causing the near "chaos" that marked the nationwide distribution of fuel. Supplementary actions by the states, which were "more responsive to citizens' needs," probably "kept the [federal] program afloat," the study concluded. (Congressional actions, often the basis for the FEO's administrative decisions, also contributed to the supply problem, according to the report.)

Among the faults cited in the system were an Emergency Allocation Act that the FTC described as "vague and contradictory"; an inexperienced FEO staff that overextended itself in administering massive government programs; Congress' decision to use 1972 as the base year for calculating allocations (because small independent gasoline dealers also had little fuel in that period); government decisions, such as the refinery sharing order, that reduced incentives; a two-tier price system for "new" and "old" oil that resulted in more pronounced inequities between major dealers and the independent owners who relied heavily on the more expensive "new" oil; and "wide disparities" among state allocations.

Record natural gas price OKd. The Federal Power Commission (FPC) Feb. 2 allowed an interstate pipeline to buy natural gas at 55¢ per 1,000 cubic feet. It was the highest purchase price ever approved by the FPC and 22% above the previous record—45¢—set in May 1973, and itself a 73% increase over 1971 prices.

Coal miners strike. In an unusual side effect of the fuel crisis, more than 20,000 West Virginia coal miners launched a 10 day strike Feb. 26 to protest an order by West Virginia Gov. Arch Moore prohibiting motorists with more than a quarter of a tank of gasoline from purchasing additional fuel.

The strike persisted despite a Federal Energy Office order March 3 authorizing the release of 650,000 extra gallons of gasoline from the state's stockpile and Continental Oil Co.'s shipment of supplementary gasoline supplies to the area March 2.

Moore modified his order March 5 to permit motorists traveling more than 250 miles a week for business reasons to file for exemptions from the gas purchase limit. Most of the strikers, however, were dissatisfied with the exemption program, calling it "an incredible bureaucratic nightmare." Many miners lived at least 50 miles from their work.

The average price of gas sold interstate was 26¢. Between 1954 and 1972, the price of federally regulated natural gas had increased 200%. (Intrastate gas was not subject to federal regulation.)

Under the ruling, Southern Natural Gas Co. was authorized to buy Alabama-produced natural gas under a 20-year contract. The FPC staff had recommended a 38¢ purchase ceiling. The difference between the staff-approved rate and that set by the commission would total $1.8 million, an amount that could be passed on to distributors and customers.

In its 3–2 ruling, the commission appeared bitterly divided over the question of allowing further price increases as an incentive to producers to stimulate exploration and development.

In dissent, William L. Springer charged that the majority had "capitulated to the prescription of an industry-established price . . . rather than a just and reasonable rate set by regulatory review."

Another opponent of the price increase, Commission Chairman John N. Nassikas, joined Springer in denouncing the decision as a "travesty of regulatory justice." They noted, however, that because no interested parties had intervened to oppose the increase when it was before the FPC, a reversal on court appeal was not possible.

In another 3–2 ruling Feb. 5, the commission required natural gas producers to make public disclosures of the extent of their natural gas reserves. The decision reversed an earlier ruling holding that the information was a trade secret whose release would be "inimical to competition."

Elk Hills pumping halted. A federal judge in San Francisco Feb. 14 ordered Standard Oil Co. (California) to cease pumping operations at its leased lands adjacent to the naval oil reserve in Elk Hills, Calif.

The preliminary injunction was issued in response to a government charge that the company, which pumped 17,000 barrels a day from its property, also was drawing off 300 barrels a day in oil from the federal reserve set aside for national defense emergencies.

The injunction would be enforced until final determination of the charges was reached in a forthcoming trial.

Lt. Cmdr. Kirby Brant had resigned as deputy director of naval petroleum and oil shale reserves Jan. 11 because of his opposition to President Nixon's policy of allowing commercial production in the Navy's oil reserve fields.

Brant, who was subpoenaed by a subcommittee of the House Interior and Insular Affairs Committee, testified Jan. 17 about his policy differences with the Administration. He also accused Standard Oil of California of draining oil from the Elk Hills, Calif. field by drilling in leased areas around the borders of the tract.

Company officials said production, which had begun there in September 1973, would not be halted because oil on its property was a "completely independent oil accumulation unconnected in any way" with the nearby naval reserves.

In a letter to the oil company released Jan. 22, Assistant Navy Secretary Jack L. Bowers said "preliminary" findings had determined that Standard Oil was "drawing from pools in Elk Hills."

The Interior Department had admitted Dec. 19, 1973 that major oil companies had been draining oil from the naval reserves for 20 years in apparent violation of federal law. Officials said the drainage occurred because the department had allowed oil companies to lease lands in the previously proscribed buffer zones on the peripheries of the reserves.

Ex-oilmen in key FEO jobs. The Federal Energy Office (FEO) March 6 released a list of 58 former oil industry employes holding important jobs in that agency. The list was prepared at the request of Rep. Benjamin S. Rosenthal (D, N.Y.) who criticized the "incestuous game of musical chairs that is played so frequently by industry and government."

In an accompanying statement, the FEO said "several of these people haven't worked for oil companies for many years and some have held other government positions prior to coming to the FEO" (which employed a total of 2,030 persons).

Nixon Program

Energy message. President Nixon Jan. 23, 1974 sent Congress an energy message in which he proposed higher taxes on the foreign output of oil companies and a further easing of air pollution restrictions on automobiles. The President also urged Congress to enact some of his 1973 energy proposals.

His major tax proposal was elimination of the 22% depletion allowance for foreign crude oil production. Currently, oil and natural gas producers were permitted to deduct from their taxable income 22% of their production income, foreign and domestic. Nixon urged Congress to retain the allowance for domestic production "to encourage greater development of U.S. energy resources."

The President said the Administration was reviewing another tax break for the oil companies that permitted reduction of their U.S. income taxes by the amount of foreign taxes paid on crude oil produced in host countries. Nixon favored retention of this provision for "some reasonable portion" of the foreign taxes and treatment of the remainder as a normal operating expense. The latter would be deducted from taxable income instead of directly from the final tax. In essence, the revision under study would revise a portion of the tax break, currently equivalent to $1 deduction for $1 of tax paid to foreign countries, to the standard corporate tax rate of 48%, or 52¢ deduction for $1 foreign tax paid. Observers noted that the oil companies could still reinvest such profits overseas free from U.S. taxation until the profits were returned to the U.S.

A third tax recommendation was a renewed call for the Administration's December 1973 request for a "windfall profits tax" on crude oil prices, usually identified by others as an excise tax.

On the automobile emission item, Nixon called for a two-year delay on interim controls to allow manufacturers to focus on fuel economy instead. The Environmental Protection Agency had set an interim standard for 1976 models requiring that nitrogen oxide emissions be less than two grams per mile. The new proposal would reduce this to .4 gram for 1978 cars.

Other proposals in the energy message:

- "Efficiency labels" for automobiles and major appliances showing how much energy they used.
- An increased rate of federal oil and gas leasing on the outer continental shelf to more than 10 million acres a year, beginning in 1974 (about five million acres had been leased so far).
- Acceleration of the licensing and construction of nuclear power plants with a provision for advance approval of plant sites or "nuclear parks" (site clusters).
- A substantial increase in the federal commitment to energy research; coal-related research would be increased 160% in fiscal 1974 to the $427 million level.
- "Energy-related" unemployment insurance by extending the period of benefits and expanding coverage "in those labor-market areas that experience significant increases" in unemployment.

In the field of pending legislation, Nixon urged Congress to simplify the emergency energy bill currently before Congress with a host of amendments. He suggested a "basic bill" providing authority to institute mandatory fuel conservation measures, including gasoline rationing, which was again viewed by the President as a "last resort."

Improvement of urban transportation was put in the "special priority" category by the President, who endorsed greater "flexibility" in funding.

Nixon also renewed his legislative calls for sale of oil from the Elk Hills naval petroleum reserve in California, federal licensing authority for deep water ports, an end to federal regulation of prices on newly developed natural gas supplies and creation of a new energy administration, department and research agency.

The President disclosed that the Interior Department was anticipating competing applications for a natural gas pipeline to Alaska's North Slope, one from a Canadian-American consortium for a route along the Mackenzie River Valley to the upper Middle West, another from the El Paso Natural Gas Co. for a route "across Alaska."

In his message, Nixon gave this report on the current energy situation:

Last year the United States consumed roughly 18 million barrels of petroleum, in one form or another, every day. This represented about one-half of our total energy consumption. The level of petroleum consumption was also rising, so that we expected demands to reach about 20 million barrels a day in 1974.

While the country is rich in natural resources, our production of petroleum resources is far less than our demands. Last year we were producing approximately 11 million barrels of petroleum a day, and the level of production was declining.

The difference between our demands and our domestic consumption must be made up, of course, by imports from abroad, reductions in demand, or increased domestic production. Even before the embargo on oil in the Middle East, our foreign supplies were barely adequate. Since the embargo, the shortage has become a good deal more serious. The Federal Energy Office has estimated that during the first three months of 1974, our imports will fall short of our normal demands by 2.7 million barrels a day. If the embargo continues, shortages could exceed three million barrels a day during the rest of the year. That shortfall is the major factor in our current emergency.

With the Nation confronting a severe energy shortage, I appealed to the public eleven weeks ago to undertake a major conservation effort on a personal, voluntary basis. My appeal was repeated by public servants across the land. The

Congress acted quickly to pass laws putting the Nation on year-round daylight savings time and reducing the national highway speed limits to no more than 55 miles per hour. The Federal Government began moving swiftly to ensure that fuel supplies were allocated fairly and that conservation measures were undertaken within the Government. Most importantly, the people themselves responded positively, lowering the thermostats in their homes and offices, reducing their consumption of gasoline, cutting back on unnecessary lighting, and taking a number of other steps to save fuel.

Largely because of the favorable public response, I can report to the Congress today that we are making significant progress in conserving energy:

—Total consumption of gasoline in the United States during the month of December was nearly nine percent below expectations.

—Consumption of home heating oil has been reduced. A recent survey of 19,000 homes in New England showed they had reduced heating oil consumption by more than 16 percent under last year, after making adjustments for warmer weather.

—Utilities report that consumption of natural gas across the country has been reduced by approximately 6 percent over last year, while the consumption of electricity is down about 10 percent.

Beyond the progress we have made because of voluntary conservation, we have also been fortunate in two other respects. The weather in the last quarter of 1973 was warmer than usual, so that we did not consume as much fuel for heating as we expected. In addition, the oil embargo in the Middle East has not yet been totally effective, allowing us to import more oil than we first anticipated.

The Federal Government clearly has a major responsibility in helping to overcome the energy crisis. To fulfill that responsibility, several steps have been taken in the last three months:

—A major conservation program has been established and has cut consumption of energy by Federal agencies by more than 20 percent below anticipated demands in the third quarter of 1973.

—A sweeping investigation of fuel prices charged at gasoline stations and truck stops has been launched, putting an end to price gouging wherever it is found.

—A Federal Energy Office has been created to serve as a focal point for energy actions taken by the Government.

—Finally, a fuel allocation program has been set up to assure that no area of the Nation is subjected to undue hardships and to assure that in allocating fuel, the protection of jobs comes ahead of the satisfaction of comforts. As part of this allocation effort, refiners are being encouraged to produce less gasoline and more of the products that are needed in homes and industry, such as heating oil, diesel oil, residual fuel oil, and petrochemical feed-stocks. The Cost of Living Council has issued regulations to encourage the shift away from gasoline production. If necessary, additional steps will be taken to encourage shifts in refinery production.

The allocation program now underway will mean some cutbacks in travel, heating and other end uses of fuel, while uses which keep our economy operating at a high level will be permitted to remain at or above last year's levels.

Market forces are also at work allocating fuel. Due primarily to huge increases in prices for foreign oil, the price of gasoline has risen by 12 to 15 cents per gallon over last year. This obviously discourages the consumption of gasoline. Heating oil has also shown a comparable rise with similar effect.

There is a limit, however, to the amount of market allocation through higher prices which we will allow. We will not have consumers paying a dollar a gallon for gasoline. We must therefore seek to maximize the production of domestic oil at a price lower than the price of foreign oil. We will also carefully review requests for energy price increases, to ensure that they are genuinely needed.

All of the measures of conservation and allocation have greatly improved the Nation's chances of avoiding hardships this winter and gas rationing this spring. *Gas rationing, with its attendant bureaucracy and cost to the taxpayer, should be only a last resort.* Nevertheless, we are attempting to be prudent and therefore have developed a system of coupon rationing. The system is now on the record for public comment, and will be ready for use this spring should it prove necessary. . . .

The measures of allocation and conservation are, in the very short-run, the only actions which will have an effect in lessening the crisis. However, in the slightly longer term, we can and we are making efforts to increase domestic supplies of petroleum very rapidly.

Increases in supplies of domestic crude oil are necessary not only to assure supplies, but to keep the prices for consumers at a reasonable level. The prices charged by a foreign cartel for crude oil have risen so dramatically that U.S. oil prices are now greatly below the world market price

To ensure that domestic oil exploration continues and grows, the price of oil from new exploration and development has been removed from Economic Stabilization Act controls. Also, to compensate for increased production costs and to stimulate advanced techniques for recovering oil, we have permitted a $1 per barrel increase in the cost of petroleum under existing oil contracts.

As a result, domestic oil wells that had been abandoned because they were no longer profitable are being put back into production, and new American oil is now beginning to come into the market. We anticipate additional increases in the oil in the future.

As a greater domestic production fills more of our oil needs, we will be demanding less foreign oil, and the price for foreign oil will not be driven upwards by our demands. Our own domestic production will tend to put a cap on the prices foreign suppliers may charge.

To deal further with the world shortage of oil and its increasingly unrealistic price levels, I have invited major consuming nations to a conference in Washington on February 11. The conference will, I hope, eventually lead to greater international cooperation in the areas of energy conservation, research, pricing policy, oil exploration, and monetary policy.

Nixon reviewed his two-month-old Project Independence program:

Energy demand in the United States will certainly continue to rise. Were domestic oil production to continue to decline and demand continue to grow at over 4 percent annually, as it did before the embargo, imports would increase from 35 percent of U.S. consumption in 1973 to roughly half of U.S. consumption by 1980.

We must also face the fact that when and if the oil embargo ends, the United States will be faced with a different but no less difficult problem. Foreign oil prices have risen dramatically in recent months. If we were to continue to increase our purchase of foreign oil, there would be a chronic balance of payments outflow which, over time, would create a severe problem in international monetary relations.

Without alternative and competitive sources of energy here at home, we would thus continue to be vulnerable to interruptions of foreign imports and prices could remain at these cripplingly high levels. Clearly, these conditions are unacceptable.

To overcome this challenge, I announced last November 7 that the United States must embark upon a major effort to achieve self-sufficiency in energy, an effort I called Project Independence. If successful, Project Independence would by 1980 take us to a point where we are no longer dependent to any significant extent upon potentially insecure foreign supplies of enegry.

Project Independence entails three essential concurrent tasks.

The first task is to rapidly increase energy supplies—maximizing the production of our oil, gas, coal and shale reserves by using existing technologies and accelerating the introduction of nuclear power. These impotrant efforts should begin to pay off in the next 2 to 3 years. They will provide the major fraction of the increased supplies needed to achieve energy self-sufficiency.

The second task is to conserve energy. We must reduce demand by eliminating non-essential energy use and improving the efficiency of energy utilization. This must be a continuing commitment in the years ahead.

The third task is to develop new technologies through a massive new energy research and development program that will enable us to remain self-sufficient for years to come.

We cannot accept part of the overall program and ignore the others. Within the Federal sector, success will depend on a wide range of actions by many agencies. As an important part of that effort, the head of the Federal Energy Office, William Simon, will mount a major effort this year to accelerate the development of new energy supplies for the future.

Our strategy for Project Independence is reflected in urgent measures now pending in the Congress as well as many new legislative proposals and administrative actions I now plan to take.

Nixon also stressed the need for research and development:

Nowhere will the need for the combined efforts of industry and Government be greater than in energy research and development. If we are to see the successful culmination of Project Independence, the Federal Government must work in partnership with American industry.

For the last five years, I have provided for a continual expansion of our efforts in energy research and development. Federal funding increased almost 75 percent from $382 million in fiscal year 1970 to $672 million in fiscal year 1973 and was then raised to $1 billion for fiscal year 1974. Last June I announced my commitment to an even more rapid acceleration of this effort through a $10 billion Federal program over the next five years, and I asked the Chairman of the Atomic Energy Commission to develop recommendations for the expanded program.

Today I am announcing that in fiscal year 1975—the first year of my proposed five years, energy R&D program—total Federal commitment for direct energy research and development will be increased to $1.8 billion, almost double the level of a year ago. In addition, I will be requesting an increase of $216 million for essential supporting programs in basic and environmental effects research.

Regardless of short-term fluctuations in the energy supplies, our Nation must move swiftly and steadily on a course to self-sufficiency. The private sector clearly must provide most of the money and the work for this effort. We must also guard against Government expenditures which merely replace private sector investments. But the Federal Government does have a role to play in supplementing and accelerating private development and in filling major technological gaps where market incentives are lacking. The Federal expenditures which I am announcing today are designed to serve those purposes.

In pursuing our energy R&D program, we must maintain balance. We cannot afford to direct all our efforts to finding long-term solutions while ignoring our immediate problems, nor can we concentrate too strongly on finding short-range solutions. Our program must be structured to provide us with payoffs in the near, middle, and far term.

For the near term—the period before 1985—we must develop advanced technologies in mining and environmental control that will permit greater direct use of our coal reserves. We must speed the widespread introduction of nuclear power. And we must work to develop more efficient, energy-consuming devices, for use in both home and industry.

Beyond 1985, we can expect considerable payoffs from our programs in nuclear breeder reactors and in advanced technologies for the production of clean synthetic fuels from coal. By this time, we should also have explored the potential of other resources such as solar and geothermal energy.

For the far term, our programs in nuclear fusion, advanced breeder reactors, hydrogen generation and solar electric power appear to be the ultimate keys to our energy future.

In his conclusion, Nixon suggested that the challenge posed by the energy crisis could benefit the nation in the long run:

Although shortages were long in appearing, the energy crisis itself came suddenly, borne by a tragic war in the Middle East. It was a blow to American pride and prosperity, but it may well turn out to be a fortunate turning point in our history.

We learned at a stage short of the truly critical, that we had allowed ourselves to become overly dependent upon foreign supplies of a vital good. We saw that the acts of foreign rulers, even far short of military action, could plunge us into an authentic crisis. The Arab oil embargo will temporarily close some gasoline stations, but it has opened our eyes to the shortsighted policy we had been pursuing.

The energy emergency has shown us that we must never again be caught so dependent upon uncertain supplies. It is a lesson the American people must and will take to heart. By 1980, if we move forward with the proposals I have outlined today, I believe we can place ourselves in a position where we can be essentially independent of foreign energy producers.

America has half the world's reserves of coal. It has billions of barrels of oil in the ground, as well as convertible oil shale. It has vast natural gas reserves. We have the world's largest installed nuclear capacity and half the world's hydroelectric plants. This represents a truly enormous store of energy.

The United States also has the largest pool of highly trained scientific talent in

the world. Our managerial skills in the private sector are enormous. And our organized facilities for solving technical problems in universities, businesses, and government are unparalleled.

I have no doubt that the bringing together of these natural and human resources can propel us toward an era of energy independence.

It will take time. But along the way we will assure that no groups of Americans are better off because other groups are suffering. We will assure that the genius of the free enterprise system is maintained and not destroyed by its response to this crisis.

Years from now, let us look back upon the energy crisis of the 1970s as a time when the American spirit reasserted itself for the lasting benefit of America and the world.

Oil firms oppose tax change—Spokesmen for the oil industry immediately warned Jan. 23 that revision of their tax structure would puncture the vital push for energy supplies. Frank N. Ikard, president of the American Petroleum Institute, said "it would be a mistake at this point in time to disturb the tax treatment given to extractive industries."

He called foreign tax credits "absolutely essential to the petroleum industry and to American business generally" in order to remain competitive. Withdrawal of the depletion allowance on foreign drilling, he said, would have "a tremendous psychological impact" that could discourage increased production.

Nixon urges continued cooperation. In a nationwide radio address Jan. 19, President Nixon had said that with continued public cooperation in the face of the energy crisis, severe heating oil shortages and gasoline rationing could be avoided. Nixon cautioned, however, that energy shortages were "genuine," and that any slackening of conservation efforts by the public could result in the crisis being "brought home to America in a most devastating fashion."

Nixon said he could "assure" that the price of gasoline would not reach $1 a gallon, calling rumors of such possibilities ridiculous "scare stories." He recognized that some price increases were inevitable, however, and warned that rationing was a possibility that the government must prepare for.

Regarding the position of the oil companies, Nixon pledged to "do everything in my power to prevent ... unconscionable profits" from being made from the crisis. He noted that he had ordered a review of the international tax structure to "insure that American companies developing resources abroad are not permitted to avail themselves of special tax advantages abroad."

Nixon acknowledged the need for complete data in the formulation of energy policy and said he would ask for laws requiring "full and constant accounting" by oil companies of inventories, reserves, production and costs.

Nixon cited a degree of progress brought about mostly by voluntary cooperation by the public: gasoline consumption in December 1973 which was almost nine percent below expectations; a reduction in natural gas consumption of about six percent from the previous year; and a 10% cut in electric power use.

For the long term, Nixon said, the keys to energy stability would be the increasing independence of the U.S. by development of new supplies and international cooperation to halt the "self-defeating" use of politically-motivated embargoes and rapidly-increasing international prices.

Speaking for the Democratic Congressional leadership, Rep. John J. McFall (Calif.), House Democratic whip, called Jan. 26 for a cooperative approach to the energy crisis by the Administration and Congress. "The time has long passed where it does any good to assign blame for the current crisis," McFall said in a nationwide radio address in response to President Nixon's energy address Jan. 19.

McFall claimed the initiative in energy legislation for Congress and said many of the President's proposals the previous week were already pending in Congress in some form, many of them in the Emergency Energy Act bound for final action in Congress. "Throughout most of last year," McFall said, "the Administration appeared to prefer talk to action. We are gratified to note that the Administration is ready to join with Congress in a creative and cooperative approach to this problem."

McFall noted that legislation for energy labels on automobiles and major appliances had been introduced months ago and passed by the Senate despite Administration opposition. He also cited Administration opposition to a law approved by Congress for mandatory allocation of fuel supplies.

"The Democratic Congress, with the support of many of its Republican colleagues, is providing many of the answers," McFall said. The President was "asking the cooperation of the Congress. This he shall have. And in return we in Congress expect that same cooperation from his Administration."

Simon optimistic. Federal Energy Office Director William E. Simon Jan. 23 expressed optimism about the nation's energy and environment goals. Simon said that he did not believe that the Administration's energy program would impair the environmental programs except for a delay of a year or so.

Simon also spoke that day of the possibility of federal "seed money" to industries producing new forms of energy, such as oil from shale or coal-based synthetic crude oil.

In an appearance before the Senate Finance Committee studying tax revision, Simon said "some windfall profits have been made during the past few months and have contributed to the sharply increased overall profits of all oil producers." He expressed hope that the current "emotionally charged atmosphere" of hostility to the oil industry would not lead to "punitive legislation."

State-of-Union message stresses energy. In his State-of-the-Union message to Congress Jan. 30, President Nixon assigned first priority to energy. Listing "key areas in which landmark accomplishments" were possible in 1974, Nixon started with energy: "We will break the back of the energy crisis," he pledged. "We will lay the foundation for our future capacity to meet America's energy needs from America's own resources."

Text of the message's energy section:

Let me begin by reporting a new development which I know will be welcome news to every American. As you know, we have committed ourselves to an active role in helping to achieve a just and durable peace in the Middle East on the basis of full implementation of Security Council Resolutions 242 and 338. The first step in the process is the disengagement of Egyptian and Israeli forces, which is now taking place.

Because of this hopeful development, I can announce tonight that I have been assured through my personal contacts with friendly leaders in the Middle Eastern area that an urgent meeting will be called in the immediate future to discuss the lifting of the oil embargo.

Now this—this is an encouraging sign. However, it should be clearly understood by our friends in the Middle East that the United States will not be coerced on this issue.

Regardless of the outcome of this meeting, the cooperation of the American people in our energy conservation program has already gone a long way toward achieving a goal to which I am deeply dedicated.

Let us do everything we can to avoid gasoline rationing in the United States of America.

Last week I sent to the Congress the comprehensive special message setting forth our energy situation recommending the legislative measures which are necessary to a program for meeting our needs. If the embargo is lifted, this will ease the crisis. But it will not mean an end to the energy shortage in America.

Voluntary conservation will continue to be necessary. . . . The new legislation I have requested will also remain necessary. Therefore, I urge again that the energy measures that I have proposed be made the first priority of this session of the Congress.

These measures will require the oil companies and other energy producers to provide the public with the necessary information on their supplies. They will prevent the injustice of windfall profits for a few as a result of the sacrifices of the millions of Americans.

And they will give us the organization, the incentives, the authorities needed to deal with the short-term emergency and the move toward meeting our long-term needs. . . .

Let this be our national goal. At the end of this decade, in the year 1980, the United States will not be dependent on any other country for the energy we need to provide our jobs, to heat our homes and to keep our transportation moving.

To indicate the size of the government commitment to spur energy research and development, we plan to spend $10 billion in federal funds over the next five years. That is an enormous amount. But during the same five years, private enterprise will be investing as much as $200 billion, and in 10 years $500 billion to develop the new resources, the new technology, the new capacity America will require for its energy needs in the 1980s.

Economic report to Congress. In his annual economic message to Congress and a companion report prepared by the Council of Economic Advisers (CEA), President Nixon revealed the Administration's strategies for dealing with a "highly uncertain" outlook for 1974 when inflation and recession both threatened to overtake the economy.

The forecasts, which were submitted to Congress Feb. 1, stressed the need for policy flexibility in dealing with the possibility of economic stagnation, prolonged inflation and the domestic consequences of international economic upheavals.

Although the unpredictable course of the worldwide energy crisis was a major factor in setting the tentative tone of the report, the papers also reflected the Administration's economic and political reverses of 1973.

On the oil crisis, the council said the situation "does not require any decline in the level of economic activity in the industrialized world," but added there was a danger of a world recession if nations reacted to the high oil prices by "squeezing down the economy at home or by checking imports and spurring exports."

Budget's energy proposals. President Nixon submitted his fiscal 1975 budget to Congress Feb. 4, 1974.

Nixon budgeted federal spending on energy research and development at about $1.6 billion, an increase from an estimated $942 million in fiscal 1974. In addition to focusing on coal and nuclear fission and fusion, he sought $48 million for development of geothermal energy and $50 million for solar energy projects. Large sums—$179 million—were allotted to environmental control, with emphasis on increasing supplies of low-pollution coal, and research on energy conservation, $129 million.

Development of a coal liquefaction plant on a cost-sharing basis with industry was being planned. Projects also were being funded for augmenting coal production.

The Atomic Energy Commission budget reflected the energy crisis. For the first time in its history, the agency would spend more (58%) on civilian activities than on its military aspect. Funding for nuclear weapons testing, for the first time in 10 years, fell below $100 million, to $97.5 million.

The energy situation also affected the National Science Foundation, whose budget rose $32 million to $630 million, not all of the increase devoted to special energy projects.

Nixon sees 'crisis' at end. President Nixon, in an opening statement at his press conference Feb. 25, gave a "brief report on the energy situation—the progress we have made to date and the problems that we will have in the future."

His assessments proved controversial, especially his assertion that "while the crisis has passed, the problem still remains." Nixon also said that because of the "cooperation of the American people" and Federal Energy Office Administrator William Simon's management, "we have now passed through that [heating oil] crisis. The home fuel oil, as far as it's concerned, as we know, has been furnished. No one has suffered as a result. And as far as our [industrial] plants are concerned, all have the fuel that is required to keep the plants going."

Nixon said he had become familiar with the "major problem" that remained, the gasoline shortage, as he had "driven around the Miami area and also in the Washington area." But, he added on an optimistic note:

"I am able to report tonight that as a result of the cooperation of the American people, as a result, too, of our own energy conservation program within the government, that I now believe confidently that there is much better than an even chance that there will be no need for gas rationing in the United States."

Nixon's remarks renewed a policy conflict tinged with political overtones between Simon and Roy Ash, director of the Office of Management and Budget. Ash had told reporters Feb. 12 that the energy situation was "manageable, one-time and short-term." Simon, who had consistently emphasized the severity and long-term nature of the crisis, objected to Ash's assessment. Simon responded in a television broadcast Feb. 13 that Ash should "keep his cotton pick'n hands off energy policy. Perhaps I should call a press briefing on the budget," he added.

In an appearance before a Senate subcommittee Feb. 26, Simon, who testified as the Administration's top energy spokesman, was chided by legislators about the President's energy analysis. They asked how citizens could be expected to practice voluntary conservation if they "are told the crisis is over."

Vice President Gerald Ford, in an address Feb. 20, had anticipated Nixon's optimistic view. "We're almost over the hump of the temporary energy crisis and if we're able to get a break on the oil embargo there's no question we'll be out of the woods as far as the short term is concerned," Ford said.

House and Senate Majority Leaders Thomas P. O'Neill (Mass.) and Mike Mansfield (Mont.) also expressed disagreement with Nixon. Mansfield, who said he had waited 35 minutes to buy gasoline, said, "The shortage remains and so does the crisis." O'Neill concurred. "It is all very well for President Nixon to say the energy crisis is over. He doesn't have to wait in gas lines," he said.

Nixon's other statements during the press conference:

- He said he "will have to veto" an emergency energy bill passed by Congress because "what it does is simply to manage the shortage rather than to deal with the real problem." A rollback in oil prices, Nixon warned, would lead to "more long gas lines" requiring nationwide gas rationing. "That would mean 17,000–20,000 more federal bureaucrats to run the system at a cost of a billion and a half dollars." He renewed his call for Congress to "act responsibly and pass the Administration's energy proposals."
- Nixon said "the lines for gasoline will become shorter in the spring and summer months," but he added, "the price of gasoline is not going to go down until more supplies of gasoline come into the country and also until other fuels come on stream."

Nixon discussed the embargo issue briefly, noting that negotiations were under way to secure a military disengagement on the Syrian-Israeli front. "That following on the disengagement on the Egyptian front, I think will have a positive effect, although it is not linked to the problem of the embargo directly."

A reporter asked Nixon to comment on the shah of Iran's remark that "the U.S. is getting as much oil now as it did before the embargo."

Nixon refused to become embroiled in the controversy, saying only that the shah's "information is different from ours. We are getting substantially less from the oil producing countries in the Mideast than we were before the embargo."

In testimony before Congress Feb. 25, William Simon had accused the shah of making remarks that were "irresponsible, reckless and just plain ridiculous," a charge that Nixon disavowed.

Government & private roles. Speaking before the Executive Club of Chicago March 15, Nixon again took up the energy problem.

The President said the federal role in energy was "minimal" compared to the role of the private sector, that the government would spend probably $15 billion in the next five years to reach for self-sufficiency while private industry would spend, according to estimates, at least $500 billion. "The way to get this country moving is to energize private industry," he said.

There was an energy shortage in America, Nixon said, and it "has been dealt with very effectively by this Administration." The dire predictions of unemployment had not been realized, he said, and "we have, I believe, broken the back" of the energy crisis "although it will still be a continual nagging problem."

Rather than blame the big oil companies, which could do something about it "if they had the oil," he said, "the thing to do is to develop the resources of this country" to make it self-sufficient in energy.

To attain this, he advocated deregulation of natural gas, relaxing inhibitions in the environmental field for development of coal resources, development of shale oil and naval reserves and development of nuclear power, which he considered clean fuel and safe.

Nixon said Congress should be pressed to act on his energy proposals. The way to deal with the shortage, he said, "is not to demagogue about it but do something about it and it's time for the Congress to get off its something and do something about it right now."

In a related development, independent oil producers said drilling of new oil wells in the U.S. was being delayed by the shortage of tubular steel. Government officials told the New York Times Feb. 5 that "excessive stockpiling" a "maldistribution" of existing supplies made the effects of the shortage more severe. According to the Federal Energy Office, pipe inventories maintained by the 22 largest U.S. oil companies were 30% higher in December 1973 than their monthly averages had been since early 1972. Eight of those companies held 74% of the available pipe supplies, according to the FEO.

Legislation Rejected

Energy bill vetoed, veto upheld. President Nixon March 6, 1974 vetoed an emergency energy bill that had been passed by 67-32 Senate vote Feb. 19 and 258-151 House vote Feb. 27. The veto was sustained March 6 by a 58-40 Senate vote that fell eight votes short of the two-thirds majority necessary to override.

Despite the veto, there was no indication that the issue was dead. Sen. Henry Jackson (D, Wash.), a principal supporter of the legislation,which included a provision for reducing oil prices, warned that "the rollback provision is going to be an ongoing fight." "We can always put a rollback in a bill they [the Administration] must have," he said.

After the Senate's March 6 vote, the House voted 218-175 to insert a nearly identical price rollback measure into a bill that would establish a Federal Energy Administration. (Under the price amendment, producers of less than 30,000 barrels a day of "new" crude oil would be exempt from the rollback provision.)

In his veto message, Nixon said the bill would "set domestic crude oil prices at such low levels that the oil industry would be unable to sustain its present production of petroleum products, including gasoline. It would result in reduced energy supplies, longer lines at the gas pump" and increased unemployment.

Jackson termed the veto a "flagrant show of contempt for the impact of fuel shortages and soaring fuel prices." By his action, Nixon "defends and advocates higher oil prices," Jackson declared.

Nixon said March 6 that he wanted "to congratulate" those senators voting to sustain the veto. With their vote, he said, they "vetoed longer gas lines and vetoed nationwide rationing."

The President repeated his calls for speedy Congressional action on the Administration's other energy bills, including one that would open the Elk Hills, Calif. naval oil reserve to private development.

Rep. John E. Moss (D, Calif.) March 6 released a secret Justice Department memorandum written in 1970 by Richard G. Kleindienst, then deputy attorney general, to Robert P. Mayo, who was then director of the Bureau of the Budget. The paper warned that if the reserve were opened to private commercial use, military supplies would be "considerably depleted," the government would have difficulty getting a fair price for its oil and Standard Oil Co. (California), which operated the Elk Hills fields for the Navy, could begin pumping oil from the pools without restriction.

The emergency energy bill, which included a proposal to give the President statutory authority to order gasoline rationing, had been in trouble for months. The Senate had voted by 57-37 Jan. 29 to send it back to a House-Senate conference committee. A coalition of oil-state senators, Nixon supporters, southerners, and Democratic liberals led the move to recommit the bill in a joint effort to kill a provision for taxing the oil industry's windfall profits.

President Nixon had voiced strong opposition to the tax measure in a letter to Senate Minority Leader Hugh Scott (Penn.). Nixon urged deletion of the "unworkable" tax provision in conference.

Although the Jan. 29 vote was regarded as a victory for Nixon, Sen. Gaylord Nelson (D, Wis.), author of the motion to recommit, represented those who wished a stronger excess profits tax on the oil industry rewritten in the Senate Finance Committee and the House Ways and Means Committee as well as those objecting to features in the bill allowing a relaxation of clean air standards.

Also included in the emergency legislation was authority for the FEO to order further energy conservation measures subject to a six-month time limit and veto by either house of Congress; authority to make fuel-saving cutbacks in transportation; authority for the FEO to make generating plants convert from oil to coal burning; authority for the FEO to order increased domestic oil production; authorization for increased unemployment compensation for those laid off as a result of the fuel crisis; grants of limited immunity from antitrust prosecution for retailers and the oil industry, under close government supervision; requirements for energy producers to file product information reports with the FEO; authorization for a delay in certain 1976 auto emission requirements; authority for issuing loans to homeowners and small businessmen to make energy conservation improvements.

After deleting the windfall-profits provision, the conference committee reported the bill Feb. 6. But no action was taken immediately because Republicans and senators from oil producing states opposed another provision in the bill that would mandate a rollback in the price of crude oil from "new" wells and "stripper" wells producing less than 10 barrels a day to $5.25 a barrel. The current price of "new" oil, which was not under price controls, was $10.35 a barrel. The provision would have allowed the President subsequently to raise the ceiling price of oil to $7.09 a barrel, unless Congress disapproved within 30 days.

There were three Senate efforts to recommit the bill Feb. 19 (for deletion of the price rollback provision) but each vote failed despite a warning issued that day by Federal Energy Office Administrator William Simon. He said Nixon would veto the bill because the price reduction measure was "inflexible."

This warning was repeated as the House cast votes on the bill Feb. 27.

During his press conference Feb. 25, President Nixon had stated his intention to veto the bill if the House approved it.

He repeated that pledge Feb. 28 in a Washington appearance before the Young Republican Leadership Conference.

Nixon again called on Congress to enact pending Administration proposals designed to increase fuel supplies over the long range.

"By increasing the supply, the price will go down, the gas lines will certainly disappear and we can move forward as a country with the energy that we need," Nixon said.

After the Embargo

The lifting of the Arab oil embargo in March 1974 was greeted in the U.S. with expressions of relief coupled with warnings that the nation must continue efforts to make itself invulnerable to future economic warfare of this sort.

Nixon bars rationing, pledges more gasoline allocations. Addressing the National Association of Broadcasters at a meeting in Washington March 19, President Nixon announced four basic energy policy decisions made as a result of the lifting of the embargo. He flatly ruled out the need for compulsory gasoline rationing and rescinded an order banning Sunday sales of gasoline, effective March 24. Nixon said he had directed the FEO to increase fuel allocations to industry and agriculture "so that they can have the necessary energy to operate at full capacity."

He also pledged to increase gasoline allocations to the states "with the purpose of diminishing . . . and eventually eliminating" lengthy lines at gasoline stations, but he warned consumers that the upward "pressure on prices will continue because the oil we import from abroad—from, for example, the Arab oil producing countries—costs approximately twice as much as the oil we purchase in the U.S."

Nixon urged motorists to continue their voluntary fuel conservation efforts, such as car pooling, and to maintain reduced driving speeds because, despite the embargo's end, "we still have an anticipated shortage of perhaps 5%-8% in the U.S."

Nixon concluded his statement with an attack on Congress, which he said exhibited the "greatest shortage of energy" in its failure to act on "17 bills" proposed by the Administration to increase energy supplies.

Nixon referred to several measures awaiting Congressional passage that he said could close the energy gap—deregulation of natural gas prices; development of federally owned energy supplies, "particularly in Elk Hills," Calif. where the Navy owned property containing petroleum reserves; relaxation of environmental restrictions limiting development of alternative fuel supplies, such as coal; construction of deepwater ports for supertankers; and a speedup in nuclear plant construction.

Project Independence, Nixon's plan calling for national self-sufficiency in energy by 1980, could be achieved if "Congress quits dragging its feet," Nixon said.

(An FEO task force studying the requirements for a "maximum" program of energy development by 1980 put the investment costs of such a program at $255 billion, excluding the costs of building refineries abroad. Despite the high price, the study concluded that maximum energy resource development and conservation would not eliminate the U.S. need for imported petroleum supplies but would reduce the nation's reliance on foreign supplies to 4.4 million barrels a day in 1980 and, by 1985, to 1.5 million barrels a day. The study was published March 12.)

In remarks March 19, FEO Administrator William Simon reiterated President Nixon's warning that prices would not necessarily drop because supplies of Arab oil products were slated to increase with the lifting of the embargo. Domestic prices of gasoline and heating oil would reflect the dramatic rise in the price of foreign supplies as additional imports were received in this country, Simon said.

He also warned that the country could not revert to its "business as usual" attitude on fuel consumption. A "temporary surplus of certain supplies or at least a temporary sufficiency" resulting from an end to the embargo did not diminish the need for continued conservation of energy sources that remained in essentially scarce supply, Simon said.

Gasoline supplies above 1973 levels. As news of the embargo's end reached Washington, the American Petroleum Institute (API) released data showing that more gasoline was in storage and pipelines in the week ending March 9 than was available during the same period of 1973. Heating oil stocks also were higher than 1973 levels but heavy fuel oil supplies fell below last year's totals. Imports of crude oil totaled 2.12 million barrels, down slightly from the previous week, according to the API report issued March 14. (An FEO study for the same period showed a small rise in crude imports to 2.56 million barrels. The report was published March 17.)

The American Automobile Association's (AAA) statistics appeared to corroborate the API's report. A spot check of more than 6,000 gasoline stations across the country showed only 3% of the dealers completely out of fuel, compared with 20% two weeks earlier. According to the AAA March 19, 16 states and six turnpike systems were using versions of the odd-even gasoline rationing plan first established by Oregon.

In spite of official warnings that prices would climb, Gulf Oil Corp. March 20 announced a 1.5¢ a gallon cut in the price of distillate fuels, a 5¢ a gallon cut in jet fuel price and a $1 a barrel reduction in the price of residual fuel oil. The lifting of the embargo was cited as part of the reason for the price rollback.

(Officials said price competition within the industry within the past weeks also had prompted the price reduction.)

Federal Energy Administration created. President Nixon May 7, 1974 signed the first of his energy proposals to pass Congress. It was a bill establishing the Federal Energy Administration (FEA), which would replace the Federal Energy Office created by executive authority Dec. 4, 1973.

The legislation authorized the FEA to administer an emergency energy rationing plan and gasoline allocation programs for the states, but powers granted to the new agency were more limited than Nixon had requested. The FEA administrator was specifically empowered to impose energy conservation measures, prohibit unreasonable profits and develop export/import policies for energy resources. Nixon had asked Congress to give the agency the authority "to take all actions needed" to deal with energy shortages.

The law gave the FEA a two-year existence and required the President to submit the names of those nominated for top FEA posts to the Senate for consideration.

The measure, which had passed the Senate Dec. 19, 1973, was approved by the

House March 7. Final action came May 2 when the Senate adopted a conference report on the bill, already approved in the House April 29.

In its March 7 consideration of the measure, the House had reversed an earlier decision and had voted, 216–163, to reject a provision setting a ceiling on the price of "new" oil.

The provision was killed after Rep. John Anderson (R, Ill.) cited an apparent loophole—refiners producing less than 30,000 barrels a day would have been exempt from the price ceiling, according to the bill. Anderson said Standard Oil Co. (Ohio) produced only 25,000 barrels a day and would qualify for an exemption. Other supporters of the rollback were persuaded that an excess profits tax, currently under consideration in committee, was a preferable method for dealing with the oil industry's windfall profits.

After deleting the price provision, the House passed the Federal Energy Administration bill 353–29.

Gasoline supply situation. In the aftermath of the lifting of the Arab oil embargo and President Nixon's statement that the energy crisis had eased, there were widespread reports that energy conservation efforts were being abandoned by consumers across the country.

New Jersey dropped its odd-even gasoline rationing plan April 1, and Washington D. C. officials announced March 23 that the area's mandatory odd-even system of gasoline distribution would be ended. Hawaii, the first state to impose mandatory distribution rules, suspended them April 30. New York State's plan was abandoned May 1. Connecticut June 12 lifted sales curbs on gasoline stations along state highways.

The Federal Energy Office (which gained statutory authority as the Federal Energy Administration July 1) released its first analysis of the post-embargo fuel supply situation April 4. As a feature of the report, Deputy FEO Director John C. Sawhill assured motorists that "there will be no reoccurrence of gasoline lines" but he tied the pledge to his expectation that "the American people [would continue] to practice conservation." Sawhill repeated that assertion April 20 shortly after being named head of the FEO, succeeding William Simon, who was confirmed as Treasury secretary. (Sawhill was later designated administrator of the FEA.)

With demand for gasoline rising and conservation efforts waning, the FEA remained cautious in its administration of the allocation program for states. It was initially reported that May allocations would be boosted to at least 95% of the consumption level in the corresponding month of 1972, but Sawhill announced May 8 that gasoline distribution would remain at the April level, which was 90% of consumption.

The FEA justified its decision to limit gasoline supplies by releasing figures showing figures for the week ending April 26 that indicated gasoline demand had climbed 9% in one week. The weekly consumption level of 6.7 million barrels a day was a record high for the year.

The American Automobile Association May 14 confirmed reports that gasoline supplies were gradually becoming more plentiful, but also noted that rising prices had not yet stabilized. Its weekly survey of more than 5,600 gasoline stations across the country showed that average gasoline prices were up 1¢ to 55¢ a gallon for regular and 59¢ a gallon for premium.

The FEO issued June gasoline quotas for the states June 10. Each state was allocated a minimum 90% of its June 1972 consumption level, adjusted for population growth. The nationwide total of gasoline supplies for the month was estimated at 9.3 billion gallons, up 4.3% from the May average.

More fuel directives. The FEO-FEA continued to issue new fuel directives and revise past decisions. A new priority list for fuel uses was issued March 28. The agricultural production category was tightened, but most other changes were of a technical nature designed to give the program greater uniformity and clarity.

The FEO authorized price increases for a wide variety of petroleum products, according to regulations published April 2 in the Federal Register but not announced publicly. Wholesale and retail increases were permitted to cover marketing costs for gasoline, middle distillates, residual fuel oil and propane.

In March, three refiners—Ashland Oil Inc., Continental Oil Co. and Koch Industries Inc., accepted the FEO's finding that they had prematurely passed on increases in Canadian crude oil. Price rollbacks totaling $44 million were ordered March 26.

Under federal regulations, oil companies were allowed to pass crude oil costs directly on to purchasers of refined petroleum products.

Revisions proposed by the FEO in March for its controversial crude oil allocation program were finalized May 15 by the FEO, effective from June 1 through Aug. 31. Disincentives were removed in the new system, which no longer penalized large refiners for replacing oil that was allocated away to small and independent refiners. (The replacement was not subject to sharing orders.)

Congress OKs air rule changes. By voice votes in the House June 11 and in the Senate June 12, 1974, Congress approved a compromise bill setting rules for conversion to coal by oil- and gas-burning plants and extending to the 1977 model year the Environmental Protection Agency's (EPA) strict standards for auto exhaust emissions. The provisions were similar to those requested by the Nixon Administration March 22.

The principal differences between House and Senate versions had been in the coal-conversion standards. The conference committee measure, which leaned towards the original Senate bill, provided that the administrator of the Federal Energy Administration (FEA) could order immediate conversion to coal if a plant could comply with the emission standards of the Clean Air Act. If the plant could not comply immediately, conversion could be ordered after acquisition of control equipment or low-sulfur coal.

Plants which converted to coal voluntarily between Sept. 15, 1973 and March 15, 1974, or converted afterwards under FEA order, would be exempt from strict compliance with state emission rules, if emissions did not exceed EPA standards set for the affected air quality region.

Two other provisions in the Senate bill—EPA exemption from environmental impact statement requirements, and extension of funding for enforcement of the Clean Air Act—were included in the compromise measure.

Sawhill critical of oil industry. FEO Administrator John Sawhill June 29 accused major oil companies of "foot-dragging and calculated resistance" to the government's regulations requiring the majors to share their crude oil supplies with independent refiners.

"We have reason to believe," another FEO spokesman added, "that a half dozen of the 15 companies [required to sell crude oil] have yet to make any sales" to independents for the three-month period that began June 1. (After July 8, the government could order firms not in compliance to make specified sales to refiners with short supplies.)

Sawhill noted that the allocation program had been mandated by Congress. If the FEO were to "scrap" the program, he said, "crude-short refiners could be thrown into the world market where the prices are higher—and their competitive position would deteriorate." That prospect, he warned, "would conflict with our statutory mandate to protect the viability of the small and independent sectors of the petroleum industry."

(Sawhill's criticism of major oil companies marked the first break in a previously close, cooperative relationship between the industry and the energy agency.)

Regulation of crude oil supplies had been required because major oil companies had greater access than independent refiners to price controlled, domestic "old" oil (that which was pumped at pre-1972 production levels). Independent refiners had been forced to rely on imported crude supplies, which cost about twice as much as domestic oil.

Recently revised allocation regulations required the majors to sell domestic crude at prices that were averages of their costs for foreign and domestic oil.

A major legal challenge to the program was rejected July 1 when Judge Aubrey E. Robinson Jr. of U.S. district court in Washington refused to issue a preliminary injunction suspending the government regulations. Robinson ruled that the plaintiff, Exxon Corp., failed to demonstrate that it was "suffering irreparable injury" because of the allocation requirements.

Trade deficit widens. The U.S. suffered its third worst trade deficit on record in July 1974 when imports exceeded exports by a seasonally adjusted $728.4 million, according to the Commerce Department Aug. 26.

Worsening trade figures were directly related to rising fuel costs, spokesmen said. A 15% surge in fuel imports from June to July, combined with a record price for foreign oil—$11.69 a barrel, added $300 million to the nation's oil bill for the month. Total oil costs in July were $2.3 billion. Since January, the cost of oil had soared to $13.2 billion, which was far in excess of the $3.6 billion paid during the same period of 1973.

The monthly trade deficit grew to a record $1.13 billion in August, when oil imports rose to 206.6 million barrels, the price to a record $11.73 a barrel and the total cost of imported oil to a one-month record of $2.52 billion. But the monthly trade deficit finally began to decline in September, when it dropped to a seasonally adjusted $233.3 million, largely because of a 14% decline in oil imports that was due to a large buildup of petroleum inventories during preceding months.

Utility cutbacks. Inflation, high interest rates and generally poor economic conditions resulted in decisions of several utility companies to cancel or postpone the construction of new plants and other planned expansion moves.

A major utility, Consumers Power Co. of Michigan, was forced to pay 11.375% interest on its new bonds, the Wall Street Journal reported July 18. The Michigan utility had announced June 4 that it was retrenching because of rising costs and declining sales resulting from power conservation efforts. A 5% layoff in its 11,500 work force was planned, and in a later announcement, the firm canceled construction of two nuclear units. Arizona Public Service Co. also announced June 4 that it would cut spending by $63 million through 1976 by deferring nuclear expansion projects.

Four other major utilities joined the move toward spending cutbacks: Boston Edison Co. announced June 28 it would defer construction of a third nuclear unit; General Public Utilities Corp. July 24 announced a 25% cutback in its budget through 1976; Virginia Electric & Power Co. said July 24 that it would trim $100 million from its $579 million budget for 1974; and the Public Service Co. of Colorado announced the same day that its 1974 budget would be reduced from $210 million to about $145 million. Nuclear power plant deferrals had also been announced by the Carolina Power & Light Co., Detroit Edison, Duke Power and Potomac Electric Power Co., according to the New York Times July 29.

The Atomic Industrial Forum, the Times added, listed only 48 "operable nuclear power plants in the U.S." In its report, the association declared, "A deepening financial gloom is gripping the electric utility industry generally and especially their ability to raise large blocks of capital. Coupled with uncertain load-growth patterns for many companies, the capital squeeze has, not unexpectedly, put pressure on the largest, most capital intensive generating projects. These tend to be nuclear plants."

"Critical conditions in the financial markets" prompted Duke Power Co. officials to authorize a "significant cutback" in construction, it was reported Aug. 21. For the period 1974–79, total capital expenditures would be reduced to $3.2 billion from the originally projected $4.7 billion.

Cleveland Electric Illuminating Co. announced Sept. 9 that five utilities in Ohio and Pennsylvania participating in construction of electricity generating plants had agreed to delay construction of six units, including five nuclear units, and cancel another in an effort to cut costs by $700 million during 1975–79. Detroit Edison Co., Philadelphia Electric Co. and New York State Electric & Gas Corp. announced major capital spending reductions Sept. 4. The Detroit cutback, estimated at 37%, was the third made recently in its 1975 budget. Philadelphia planned an 18% reduction in construction for 1974–1978 and the New York cutback was estimated at 9%.

Southern Co. announced Sept. 12 that it was cutting its 1975–77 construction budget "by approximately one-third—a reduction totaling about $1.7 billion," with most of the cutback affecting nuclear plants in Alabama and Georgia.

One of the hardest hit utilities, Detroit Edison Co., announced Nov. 7 that it was essentially halting all major construction projects and cutting its 1975 construction budget for the fourth time this year. The new 1975 spending allocation was $230 million, down from the original projection of $558 million. A total of 3,500 jobs would be lost because of the cutbacks.

A private study, made by National Economic Research Associates Inc. and published Oct. 11, showed that the nation's utilities had reduced their construction budgets by 18% through 1978 because of rising costs and a slowdown in the growth of energy consumption.

The spending cutbacks, some of which reflected cancellations but most of them visible as postponements, totaled $16.1 billion as of Oct. 1 and represented 132,-490,000 kilowatts of capacity. Utilities' projected construction budgets through 1978 now totaled $88.1 billion.

According to the report, "a little more than half of the 175,918,000 kilowatt total of all nuclear capacity being planned" had been postponed or canceled—an amount that was "equal in magnitude to a half dozen Tennessee Valley Authorities and is enough capacity to serve a dozen New York Cities."

The Atomic Energy Commission earlier had anticipated that nuclear generating plants could provide 15% of the nation's energy supplies by 1980, compared with the current rate of 7.4%.

The study warned that unless "cutbacks in electric utility construction budgets are not soon restored, then our only options five or six years from now will be voluntary conservation or involuntary curtailment of services." Approximately 8–10 years were required to bring a nuclear power plant into operation from the proposal stage. A conventional generation plant required about five years to become operational.

According to the report, only four of the nation's 15 largest utilities had not yet announced cutbacks; 39 utilities across the country had announced cuts in their construction budgets.

According to the utilities industry, the cutbacks resulted from new, scaled-down estimates of energy consumption, the nation's generally gloomy economic trends, the high cost of borrowed money, and a widespread slump in the securities market, in which equity securities of many utilities were selling at less than book value.

Because of these financial considerations, utilities had coupled spending cutbacks with sharply higher rate increases. According to the National Utility Service Inc. Oct. 14, electricity rates at the nation's 50 largest utilities had climbed an average 55.4% during the first half of 1974. For all of 1973, the average rate increase was 12.3%.

In the first six months of 1974, 46 utilities obtained rate increases valued at $1.38 billion, and 41 were awaiting approval of $1.02 billion in increases. During all of 1973, 128 rate increases were approved totaling $1.08 billion.

(The Edison Electric Institute reported that in 1974, "the electric power systems of the contiguous United States placed 36.4 million kilowatts of additional electric generating capacity into commercial operation. . . . 15 units aggregating 1.5 million kilowatts were hydraulic, and 114 units aggregating 34.9 million kilowatts were thermal. Twelve units included in the thermal capacity are powered by nuclear fuel. 49 are conventional steam turbine-generators. 44 are gas turbine-generators, and 9 are diesel engine-generators. This newly added capacity increased the total capability . . . of the U.S. electric power systems to 460.0 million kilowatts at the year end—an increase of 6.4% over the total at the end of 1973.")

Embargo effects analyzed. As the direct impact of the recent crisis in energy supplies lessened and the indirect price consequences became widespread, analysts began to examine the effects of the Arab oil embargo and warn that further shortages could be expected.

Because of the embargo, the nation's economic output declined by $10–$20 billion during the first quarter of 1974, according to a report issued Sept. 2 by the Federal Energy Administration.

The curb on oil imports also caused the civilian labor force to shrink by about 500,000 persons. An estimated 80% of the industrial layoffs could be "traced to the decline in demand for automotive or recreational vehicles." (Of those laid off, about 85% were semi-skilled workers, 5% clerical and 3% professional.) According to Labor Department figures cited in the FEA report, 150,000–225,000 jobs were lost between November 1973 and March 1974 "as a direct result of employers' inability to acquire sufficient supplies of petroleum, principally in gasoline stations and airlines. A decline of approximately 310,000 jobs also occurred indirectly in industries whose products or processes were subject to reduced demand from either real or anticipated fuel shortages," the Labor Department said. Most of the energy crisis-related unemployment was concentrated in the Midwest, the FEA said, with Michigan accounting for 70% of that total.

The embargo also caused a loss in state and federal gasoline taxes, used to finance road maintenance, education and other budget areas. In February, 17 states reported a total drop of 8.3% in gallons of petroleum products sold, compared with the previous year, according to Department of Transportation statistics. The FEA estimated that this reduction in gasoline consumption resulted in a $700 million loss in tax money.

Increased energy prices also caused the Consumer Price Index to move sharply higher, the FEA added. The report concluded that "energy shortages are as potentially damaging as failures of the economic system to fully employ labor and capital."

FUEL END-USERS
(In Percent)

Petroleum

	1974 [1]	1970	1960	1950	Percent energy wasted
Transportation	55	53	53	52	76
Residential/commercial	15	18	21	20	26
Industrial	10	11	14	16	24
Nonenergy (petrochemicals, etc.)	12	11	9	7	(2)
Electrical generation	8	7	3	5	65

Natural Gas

	1974	1970	1960	1950
Industrial	43	43	48	53
Residential/commercial	33	33	33	27
Electrical generation	21	18	13	10
Nonenergy (petrochemicals, etc.)	2	3	3	7
Transportation	2	3	3	3

Coal

	1974 [1]	1970	1960	1950
Electrical generation	59	57	40	18
Industrial	38	38	46	45
Residential/commercial	2	3	10	22
Nonenergy (chemicals etc.)	1	2	2	2
Transportation	---	---	2	13

[1] Estimate.
[2] Used as product.

20,000 gas stations closed. According to a nationwide survey of retail gasoline stations published Sept. 26, nearly 20,000 service stations went out of business during the 12-month period ending June 30. According to Audits and Surveys Inc., a marketing research firm, the number of gas stations dropped from 215,880 to 196,130. Retail outlets specializing in automobiles, parts, fuel or service declined by about 6% to 302,640.

Consumption of petroleum products down. The American Petroleum Institute (API) announced Nov. 10 that U.S. consumption of petroleum products during the first 10 months of 1974 averaged 16.5 million barrels a day, down 3.7% from the same period of 1973. Gasoline use declined 2.4%; distillate fuel consumption fell 4.9%; heavy fuel oil 4.5%; kerosene 24.4% and jet fuel, 6%–8.2%.

Domestic oil production during the same 10-month period also declined. Production at a rate of 8.5 million barrels a day was off 3.5% from the previous year. Imports of crude oil and petroleum products averaged slightly more than 6 million barrels a day, a decline of 2.2% from 1973 levels. 1974 imports represented 36.5% of domestic demand and would have been higher if the oil embargo had not limited shipments.

Despite the falloff in production and imports, sizable stockpiles of oil had been accumulated, according to API. Gasoline supplies were 4.6% higher than the previous year's level. Distillate fuel stockpiles were up 3.3%.

The figures indicated that consumption was increasing toward the end of 1974. During the first quarter, when the embargo was still in effect, consumption was down 7.1% from the first three months of 1973; in the second quarter, consumption was down 2.4% and by the third period, consumption was off only 1.5%.

Prices & Profits

Rising costs. Sen. Philip A. Hart (D, Mich.) charged that price increases for gasoline and petroleum products had cost consumers more than $30 billion on an annual basis since January 1973—or more than the cost of the Vietnam war, which was put at $26–$30 billion annually.

Hart's remarks before the Senate Commerce Committee were reported April 29, 1974. In a report May 8, it was noted that skyrocketing fuel costs had raised the price of producing three metals used in the military aircraft and space shuttle programs. The inflationary spiral was expected to add almost $1.5 billion to the costs of aluminum, titanium and magnesium because the process used to produce these metals required heavy outlays of electricity.

The Federal Energy Office proposed changes May 16 for its petroleum price regulations. One of the revisions, which would disallow some of the costs oil companies had claimed for crude and refined products purchased abroad from their affiliates, could result in a price rollback by a number of oil companies because these overseas costs had been used to compute domestic prices under government guidelines.

Use of these "transfer prices" to the "disadvantage of American consumers" "may explain in part the significant increases in international profits reported by the major oil companies," FEO Administrator John Sawhill said.

The proposed rule changes resulted from an audit of 30 large refiners undertaken since January by the FEO and the Internal Revenue Service. The change in regulations governing calculation of domestic U.S. oil prices, under which the FEO asserted its authority to intervene in companies' cost accounting procedures, could establish an important precedent.

Sawhill revealed April 25 that the joint FEO-IRS investigation was considering 47,000 cases of alleged price gouging by the energy industry at the wholesale and retail level. The evidence indicated that U.S. consumers may have been overcharged by as much as $100 million, Sawhill said. He added that $14.2 million in wholesale and retail refunds had been ordered as a result of price violations uncovered during the continuing investigations.

'73 oil price increase disputed. Consumer advocate Ralph Nader made public June 7 internal documents from the Cost of Living Council (CLC) indicating that the CLC had authorized a $1 a barrel increase in the price of "old" oil despite the objections of its staff and an independent consultant hired by the CLC.

The new information was obtained from the CLC when Nader invoked the Freedom of Information Act.

The documents pertained to the CLC's decision, announced Dec. 19, 1973, to allow a $1 increase to $5.25 a barrel in the price of crude oil that was produced in the U.S. at pre-1972 production levels.

According to the newly released documents, the CLC staff felt that the increase was "arbitrary" and unjustified. Since production of domestic oil already was near maximum levels, the staff paper declared, no additional increase in oil supplies could be expected as a result of the price hike. Secondary and tertiary recovery methods of oil could be stimulated by a price increase, the staff added, but these supplies were likely to increase regardless of any pricing decision, and in any case, would produce no immediate rise in supply.

The CLC's independent consultant, the Stanford Research Institute, asserted that the $4.25 price ceiling was not holding down production—the bottleneck was occurring because of an equipment shortage, and a price increase would not alleviate the problem. The Stanford Research Institute also said that consumer demand for oil would not show a significant decline if a moderate increase in petroleum prices were implemented.

The effort to reduce soaring demand by making oil products more costly had been a basic tenet of Administration policy since the imposition of the Arab oil embargo.

Other CLC documents obtained by Nader showed that Herbert Stein, chairman of the President's Council of Economic Advisers, favored a $12 a barrel ceiling, coupled with a windfall profits tax limiting the oil companies' yield to $6.50. William Simon, who then was head of the Federal Energy Office, and other high Administration officials were said to have supported a $7–$8 price level.

CLC Director John T. Dunlop, who supported a price ceiling of $5 a barrel, successfully opposed this upper limit, documents showed. In a memo to George Shultz, then Treasury secretary, Dunlop wrote, "If a price of $7–$8 is now established, I recommend full decontrol of the economy because even orderly phasing out of a controls system cannot be viable under such a shock." (Controls expired automatically April 30.)

New oil (that which was produced at levels exceeding those of 1972) was not subject to price controls. The free play of market forces had caused the price of new oil to rise to more than $10 a barrel.

The Stanford group, which favored a tax on oil products to curb consumption, said increases in new oil had provided the oil industry with sufficient incentives to increase production.

The CLC staff and the Stanford group also warned that higher domestic prices could trigger a new round of price increases by oil exporting countries.

John Sawhill, administrator of the Federal Energy Office until the agency was superseded by the Federal Energy Administration July 1, defended the $1 a barrel price increase June 7 at Senate hearings considering his nomination as administrator of the FEA.

Sawhill opposed an oil price rollback and said "circumstances have changed since" December 1973. "The [$5.25] price looks much smaller now relative to the world price of oil," Sawhill said.

'Double-dip' pricing, conflict of interest controversies. Controversy developed over an FEA regulation written in January 1974 at the height of the Arab oil embargo. John Sawhill said the agency rule permitted oil companies to overcharge consumers by $100 million in 1974.

In a letter Sept. 13 to Sen. Henry Jackson (D, Wash.), chairman of the Permanent Investigations Subcommittee, Sawhill said a continuing investigation of about 10 major oil companies revealed that they had profited from the agency's "double recovery" rule which, until it was revised in May, allowed refiners to double their costs on the volume of crude oil they were required to sell to other refiners under the government's oil-allocation rules. Unwarranted profits based on this practice, known in the oil industry as "double-dipping," could total up to $300 million, Sawhill said.

The controversy had its origins in the agency's oil swap rule, which required large refiners to share their supplies of crude oil with independent refiners, whose supplies were limited and higher priced, in order to increase domestic production of petroleum products.

Sawhill also said the Federal Bureau of Investigation (FBI) was examining a possible conflict of interest involving the role of Robert C. Bowen, a Phillips Petroleum Co. engineer who had served with the Federal Energy Office for a year on leave from Phillips. Bowen had returned to work for Phillips in June after he was dismissed by Sawhill because an independent investigation of Bowen's possible conflict of interest was forwarded to the Justice Department for review. The probe, conducted by the General Accounting Office, had been requested by Rep. Charles Vanick (D, Ohio), because of reports that Bowen had been responsible for drafting price regulations.

The Continental Oil Co. said Sept. 19 that it had been told by the FEA that it was permissible to use the double dipping loophole to make double recovery on costs.

The Washington Post reported Sept. 22 that Mobil Oil Corp. had warned Treasury Secretary William E. Simon, then head of the FEO, that the loophole would permit sellers of crude oil to make an excess profit of $4–$5 a barrel. The warning went unheeded, according to the Post.

The Los Angeles Times also reported Sept. 22 that the Internal Revenue Service warned the FEO in January that its regulation allowing two price markups could result in "unwarranted profits."

A subcommittee of the House Small Business Committee held hearings during September and October on whether the FEO had written the loophole intentionally or inadvertently.

Bowen testified Sept. 25 that his intention was only to provide oil refiners, forced to share their crude oil supplies, with a financial incentive to continue importing a maximum amount of oil during the embargo. Bowen also denied that he wrote the actual double recovery portion of the regulation, saying he did not know who inserted the loophole provision.

Attorneys for the FEA testified Sept. 25 that Bowen and his superior at the agency, William A. Johnson, then FEO assistant administrator for policy analysis, had ignored their warnings that the double recovery regulation would result in unwarranted profits to some oil companies. Phillips netted profits of about $52 million from the double dipping practice, according to Rep. John D. Dingell (D, Mich.), chairman of the subcommittee.

William N. Walker, former FEO-FEA general counsel, identified Bowen as the "key party" in drafting the crude oil allocation rule. Simon personally approved the Bowen draft after hearing legal objections to the rule, Walker testified.

In testimony Sept. 26, Sawhill admitted that he had been wrong when he said Bowen's work had been restricted to technical matters, not policy making decisions, while at the FEO.

Bowen served at the FEO under a presidential leave plan, the Executive Interchange Program. Sawhill told the subcommittee that he had "no intention" of using the program because "it doesn't make sense to bring people from industry into a regulatory agency like ours where potential for conflict of interest exists." Later in the hearing, Sawhill said he had been reminded that three persons on loan from Dow Chemical, Mobil Oil and General Components Co. currently worked for the FEA.

Sawhill also testified that Gulf Oil Corp., the nation's largest seller of crude, continued to use the double dip practice, but Sawhill added that their profits would be disallowed. The agency would try to recover $40 million in windfall earnings made by the six oil companies which profited from the loophole, Sawhill said, adding that another $292 million in double dip credits, on five oil companies' books but not yet passed along to consumers, also would be disallowed.

Simon, who had refused to appear before the subcommittee, bowed before a threat of subpoena and testified Oct. 2 that he had not approved the double recovery regulation, but instead, had delegated all authority for reviewing regulations to John Sawhill, then deputy administrator, and William Johnson. (They had both testified earlier that they had not approved the controversial regulation.)

Simon admitted that he had approved Bowen's hiring, but only after being assured by Johnson and Edward Schmults, then FEO general counsel, that Bowen's role as technical consultant did not violate conflict of interest standards. Simon also conceded that William Walker had warned him in April that Bowen was involved in making policy recommendations, but that he had relied on Johnson's view that Bowen's activities remained "generally consistent" with his original terms of employment that were restricted to technical matters.

Although he was shown an FEO memorandum of recommendations prepared by Bowen for Simon and Johnson in September 1973, Simon continued to insist that Bowen "didn't participate in policy decisions" while serving at the agency.

Assistant Attorney General Henry Petersen was questioned Oct. 3 about the Justice Department's decision, made public Aug. 30 in a letter to Rep. Vanick, that prosecution of Bowen for conflict of interest violations was "not warranted." Petersen told Vanick the criminal investigation "had been completed" and showed only that a "violation, if any [committed] by Mr. Bowen was inadvertent and technical."

However, it was revealed during subcommittee hearings that the Justice Department "investigation" consisted of one request for an interview with an FEA attorney. Bowen and other key FEO-FEA officials were never interviewed. On the same day that the interview was requested by the FBI, the Justice Department decided that there was no evidence to warrant Bowen's prosecution because he had made a full disclosure of his Phillips' ties to Johnson, his superior at the agency.

Record oil quarterly & 6-month profits. Oil companies reported continued high earnings levels in the first quarter of 1974. When measured in percentage gains over the first quarter of 1973, profits rose as much as 748% at Occidental and as little as 29% at Standard (Ohio). According to the New York Times May 1, the 20 major oil companies showed a 79% gain in first quarter profits,which totaled $3.3 billion.

Among the reports:

Occidental Petroleum Corp., up 748% (over the first period of 1973) to $67.8 million (net income), announced April 23; Royal Dutch/Shell Group of Companies, up 178% to $776.1 million on revenue of $7.61 billion (up 94%), reported May 9; Phillips Petroleum Co., up 150% to $108.6 million (operating earnings) on gross revenue of $1.15 billion (up 69%), announced April 30; Texaco Inc., up 123.2% to $589.4 million on revenue of $2.49 billion, announced April 23; Atlantic Richfield Corp., up 86.7% to $93.9 million on revenue of $1.56 billion (up 56%), announced April 30; Standard Oil Co. (California), up 92% to $292.9 million on revenue of $3.91 billion (up 110%), announced April 25; Cities Service Co., up 87% to $68.8 million on revenue of $703.2 million (up 34%), announced April 30; Sun Oil Co., up 85% to $90.8 million, announced April 16; Standard Oil Co. (Indiana), up 81% to $219 million on 55% gain in revenue, announced April 22; Gulf Oil Corp., up 76% to $290 million on revenue of $4.16 billion, announced April 22; Mobil Oil Corp., up 66% to $258.6 million on revenue of $4.4 billion (up 57%), announced April 29: Shell Oil Co., up 52% to $121.8 million on revenue of $1.89 billion (up 46%), announced April 25; Exxon Corp., up 39% to $705 million, announced April 23; Standard Oil Co. (Ohio), up 29% to $22.6 million on revenue of $482.9 million (up 27%), announced April 18.

Oil firms continued to make money during the next quarter. The profits of 30 of the largest international oil companies rose 93% during the first half of 1974, according to Chase Manhattan Bank Sept. 10. Net worldwide income for the six-month period totaled $8.9 billion; net income in the U.S. was $2.9 billion, up 42% from the same period in 1973. The oil companies' capital development also increased over the period, rising 122% in the U.S. to $6.5 billion. Nearly $4 billion was spent in foreign development, according to the bank.

Some individual profit reports, with percentage changes over a one-year period:

British Petroleum Co.—second quarter earnings, $226.7 million, up 71%, revenue up 204%; first half earnings $909.6 million, up 277%, revenue up 179%; announced Sept. 5.

Exxon Corp.—second quarter earnings, $850 million, up 67%; first half earnings, $1.56 billion, up 53%; announced July 19. (First quarter earnings were up 39%. Earnings for all of 1973 rose 5% over the previous year's level.)

Gulf Oil Corp.—second quarter earnings, $250 million, up 28%, revenue up 103%; first half earnings, $540 million, up 50%, revenue up 109%; announced July 22.

Standard Oil Co. (Indiana)—second quarter earnings, $280 million, up 131%, revenue up 62%; first half earnings, $499 million, up 106%, revenue up 59%; announced July 23.

Mobil Oil Corp.—second quarter earnings, $367.4 million, up 99%, revenue up 72%; first half earnings, $626 million, up 84%, revenue up 65%; announced July 24.

Cities Service Co.—second quarter operating profits, $53.8 million, up 76%, revenue up 45%; first half operating earnings, $122.6 million, up 82%, revenue up 38%; announced July 23.

Continental Oil Co.—second quarter earnings, $100.4 million, up 94%, revenue up 75%; first half earnings, $209.6 million, up 111%, revenue up 73%; announced July 24.

Phillips Petroleum Co.—second quarter earnings, $123.8 million, up 167%, revenue up 90%; first half earnings, $204.7 million, up 128%, revenue up 80%; announced July 24.

Occidental Petroleum Co.—second quarter earnings, $92.6 million, up 292%, revenue up 100%; first half earnings, $160.4 million, up 403%, revenue up 97%; announced July 24.

Standard Oil Co. (Ohio)—second quarter earnings, $50.3 million, up 19%, revenue up 40%; first half earnings, $72.9 million, up 22%, revenue up 34%; announced July 24.

Ashland Oil Co.—fiscal third quarter earnings, $32 million, up 45%, revenue up 77%; fiscal nine months earnings, $85.7 million, up 42%, revenue up 55%; announced July 24.

Texaco Inc.—second quarter earnings, $460.4 million, up 72%, revenue up 123%; first half earnings, $1.05 billion, up 97%, revenue up 111%; announced July 25.

Sun Oil Co.—second quarter earnings, $127 million, up 163%, revenue up 80%; first half earnings, $218.2 million, up 124%, revenue up 72%; announced July 25.

Royal-Dutch Shell Group—second quarter earnings, $588.2 million, up 74%, revenue up 103%; first half earnings, $1.34 billion, up 121%, revenue up 98%; announced Aug. 8.

Shell Oil Co.—second quarter earnings $124.5 million, up 39%, revenue up 25%; first half earnings, $246.4 million, up 45%, revenue up 50%; announced July 23.

Standard Oil Co. (California)—second quarter earnings, $285.3 million, up 57%; revenue up 123%; first half earnings, $578.2 million, up 73%, revenue up 118%; announced July 31.

Atlantic Richfield Corp.—second quarter earnings, $139.7 million, up 104%, revenue up 63%; first half earnings, $233.7 million, up 96%, revenue up 57%; announced July 26.

Getty Oil Co.—second quarter earnings, $62.2 million, up 167%; first half earnings, $138.8 million, up 141%; announced July 26.

Antitrust actions brought by states. An antitrust suit filed by the State of Florida against 15 of the nation's major oil companies was dismissed by U.S. district court in Tallahassee, the Wall Street Journal reported July 24.

The court ruled that the state attorney general lacked the authority to bring such a suit on behalf of such a broad group—the state, its counties, cities and citizens.

The attorney general of Kansas filed suit Oct. 8 against 12 large oil companies. They were charged in federal district court with controlling prices and eliminating competition from smaller, independent rivals. "These companies are organized in a manner which enables them to control the flow of petroleum and its derivatives from the well-head through a multilevel system to the eventual consumer at the pump," the suit charged.

In New York, seven major oil firms pleaded not guilty Sept. 5 to criminal charges that they had conspired to fix gasoline prices during 1972 with the aim of driving independent gasoline dealers out of business. Named as defendants were Exxon, Mobil, Gulf, Texaco, Amoco, Shell and Sun.

It was alleged that they "refrained from competing among themselves in terms of price in areas where independent marketers were not a significant competitive factor, while concentrating discriminatory rebates in areas where independent marketers were a significant competitive factor."

Three of the oil firms—Exxon, Gulf and Mobil—were also accused of agreeing to thwart "competition in public bidding for contracts for the sale of gasoline to" the state and New York City.

A civil complaint had been filed by the state against the same seven defendants in 1973.

New natural gas price. The Federal Power Commission (FPC) June 27, 1974 authorized a single national price for "new" natural gas pumped from wells that began operation after Jan. 1, 1973. The ruling, which the FPC termed a "landmark decision," affected new gas sold in interstate commerce and replaced the FPC's current policy of setting ceiling prices for separate geographic areas under limited term and emergency sales rate procedures.

The new rate—42¢ per 1,000 cubic feet—eventually would mean sharply higher prices for gas wholesalers and consumers. Area gas prices established in previous FPC rulings had ranged from 19.9¢ to 34¢ per 1,000 cubic feet. The average price of gas sold nationwide, including gas that the FPC had said could be sold at prices above the regulated ceiling level, was 27¢ per 1,000 cubic feet in March.

The new rate also allowed for an automatic annual price increase of 1¢ per 1,000 cubic feet, a "gathering allowance" for certain producing areas of 1¢–2.5¢ per 1,000 cubic feet, and an additional 1¢ per 1,000 cubic feet increase for producers who transported gas from offshore to onshore. Rate reviews were planned every two years, the FPC said. The agency also noted that its "optional" pricing procedure adopted in 1972, allowing producers to apply for rates above the area ceilings, was not abrogated by the new national price policy. (In an earlier decision, the FPC had authorized a 55¢ rate for one area producer.)

The pricing action was necessitated, according to the FPC, by a "national emergency" in which consumer demand for natural gas was expected to exceed supplies into the 1980s while at the same time national reserves of the fuel continued to decline. All five FPC commissioners agreed that higher wellhead prices were justified by costs and were needed to spur exploration and production in order to increase natural gas supplies. (The new rate was expected to provide producers with a 15% rate of return on capital and thus stimulate drilling and production capacity.)

Four commissioners voted to establish the uniform price at 42¢ but one member, Rush Moody Jr., was in partial dissent, claiming that the 42¢ rate was too low. The natural gas industry was critical of the FPC action for similar reasons.

Moody's view was echoed June 25 by Federal Energy Office Administrator John Sawhill, who said the FPC's new rate was a "step in the right direction" but was still too low "to stimulate exploration needed to provide adequate supplies of natural gas." Sawhill urged Congress to remove all federal price controls from natural gas and allow market conditions to determine prices. (Gas sold intrastate and hence not subject to federal regulations had been selling as high as $1.30 per 1,000 cubic feet on the free market. Deregulation of natural gas prices had been a cornerstone of the Administration's energy policy, but observers believed that Congressional action was unlikely because the FPC had, in effect, preempted the issue.)

Gas shortage seen—It was feared that the U.S. would suffer a shortage of natural gas supplies during the 1974–75 winter that would be at least as severe as the previous year's shortage of oil. The FPC had warned June 12 that the chronic gas shortage had worsened in the last 12 months, and within a few months, supplies would fall short of demand by 10%, equivalent to the loss of about 330 million barrels of oil. The shortage could become a "severe crisis" over the next five years, according to the FPC. (Natural gas accounted for more than 31% of the nation's energy supplies.)

In a report published June 8 surveying national gas estimates for the next 25 years, the FPC said that to overcome the expected shortage, the nation must utilize nuclear explosives or massive injections of water to break up gas reserves locked in underground reservoirs.

Unless new recovery methods were used, the FPC said, the U.S. would be producing less than 19 trillion cubic feet of gas in 1980, an amount which was 3–4 trillion cubic feet less than was now being consumed.

FEA Administrator John Sawhill also warned of severe dislocations in natural gas supplies. Sawhill said Aug. 19 that supplies would likely be curtailed in the Northeast during the winter, resulting in plant closings and loss of jobs.

According to the New York Times Aug. 20, there had been an 8.3% average rate of curtailment of non-interruptible gas supplies by the 20 major pipeline firms in the Northeast between April 1973 and March 1974. Those same suppliers were projecting a 12.2% curtailment for the April 1974–March 1975 period, the Times said.

In the Northwest, wholesale deliveries to the area's major pipeline company had been cut 15%, the Wall Street Journal reported July 12, prompting Oregon to set a ceiling on new natural gas customers enrolled by gas distribution firms.

GAO cites FPC improprieties—The General Accounting Office (GAO) charged Sept. 14 that the FPC used "improper" procedures in granting price increases for natural gas. After completing its 10-month investigation of the agency, the GAO also said the FPC had failed to safeguard against conflict of interest by employing officials who held stock in companies regulated by the agency.

Rep. John E. Moss (D, Calif.), who had requested the investigation, termed the report "one of the most powerful indictments of a federal regulatory agency within memory." The study focused on the FPC's performance under its present chairman, John Nassikas, who had been confirmed in August 1969.

In questioning the propriety of several price rulings, the GAO noted that the FPC had granted 96 unauthorized extensions through the end of 1973 to its 60-day emergency period, when producers were allowed to charge prices above the federally regulated ceiling in hopes that the higher prices would stimulate sales during a supply shortage period. Because of the extensions, some gas was sold at uncontrolled prices for up to 300 days, in violation of the FPC's mandate to regulate interstate gas sales, the GAO charged.

The FPC also failed to adhere to its own price setting rules, according to the GAO, when it failed to obtain accurate information on the price and volume of gas sold under emergency conditions, making it difficult to determine the "benefits and costs of these emergency sale procedures." The GAO also charged that the FPC failed to make speedy reviews of other contracts, which permitted gas to be sold at prices higher than the prevailing ceiling. Because of the delay, excessive prices were permitted and no refunds were ordered, the GAO declared.

The FPC "needs to improve its procedures or revise its regulations to provide effective protection against excessive rates and charges," the report stated.

The GAO also charged that there had been a "total breakdown" in FPC enforcement of conflict of interest rules. There was "widespread noncompliance" with agency regulations requiring officials to file reports of their financial holdings, and when filed, the FPC failed to review them, according to the GAO.

As a consequence of the GAO report, the FPC ordered 19 officials, including seven administrative law judges, to dispose of their securities holdings. The FPC replied to the GAO charges in an appendix to the report, claiming that it had corrected the conflict of interest abuses. It either denied wrongdoing regarding pricing policies or said that certain deficient procedures had been revised.

Court rejects gas rate increase. A three judge panel of the U.S. Court of Appeals in Washington Oct. 7 set aside a price increase for natural gas approved in May 1973 by the Federal Power Commission (FPC). Acting on an appeal by Consumer Union, the court ordered the FPC to reconsider its decision authorizing an unprecedented 73% price increase under the FPC's optional pricing procedure.

The court did not question the validity of the procedure, which allowed producers and pipeliners to set charges above existing interstate ceiling prices in various geographic areas. But the court cited "serious weaknesses" in the FPC's justification of the record price increase related to the way the agency calculated the costs of producing the gas and the weight given non-production cost factors.

North Atlantic air fares rise. Representatives of scheduled airlines agreed Aug. 24 to raise passenger fares on North Atlantic flights an average 10%, effective Nov. 1. It was the fifth general fare increase of 1974 and meant that the passenger's cost of flying the transatlantic route was about 35% higher than at the end of 1973.

Participants at the International Air Transport Association (IATA) conference in Montreux, Switzerland agreed that the cheapest excursion fare would be increased 18%–20%. The price of a first class ticket would rise about 7%.

An IATA official said the rise was necessitated by soaring operating costs, chiefly higher priced fuel. Aviation fuel currently cost 40¢–50¢ a gallon compared with 11¢–12¢ one year earlier. The sharply higher fuel bills had caused airlines to lose $800 million over the past 16 months.

Ford Administration

Gerald R. Ford was inaugurated as U.S. President Aug. 9, 1974 to succeed Richard M. Nixon. Ford's energy policy seemed soon to be diverging from the Nixon policy.

Ford on environment & energy. In his first environmental policy message, President Ford said Aug. 15, 1974 that the "zero economic growth" approach to conserving natural resources "flies in the face of human nature" and must be compromised to meet economic and energy needs.

Ford's message was delivered at the Expo '74 fair in Spokane, Wash. by Interior Secretary Rogers C. B. Morton.

Ford said the previous winter's energy crisis had demonstrated the need for more coal mining, offshore oil exploration, oil shale development and nuclear power plant construction.

Rejecting the zero-growth argument that economic expansion and a clean environment were inconsistent, Ford said expansion was necessary to rebuild cities, improve transportation systems, generate employment and fund important environmental programs already under way.

Ford contended that meeting energy needs did not mean "that we are changing our unalterable course to improve the environment, and it doesn't mean that we are retreating or giving up the fight." He added, however, that "it does mean stretching the timetable in some cases," adjusting some long-range goals "to accommodate the needs of the immediate present," and some "trade-offs."

Ford noted that in the last five years the U.S. had launched "the greatest environmental clean-up in history," but that the energy situation had added a "critical new element in the environmental equation." Full energy production would "entail environmental costs or risks of one kind or another," Ford said, "but we must all be prepared to bear those costs."

FEA proposes propane price increase. The Federal Energy Administration (FEA) Sept. 6 proposed that propane prices be increased 3¢–5¢ a gallon, bringing the total average price of bottled gas to 14¢–15¢ a gallon and raising the cost to consumers by up to $450 million.

The FEA conceded that propane producers would reap higher profits from such an increase, but the agency contended that the action was needed to stimulate production in order to insure adequate supply during the coming winter and avert future shortages. Officials claimed that the proposed increase was "in consonance with" a law requiring that prices be based on a May 15, 1973 level, because only production-related costs were being added to the base rate.

Current supplies of propane were double the previous winter's level, but officials warned that a cold winter could rapidly deplete those reserves.

An estimated 70% of the nation's propane supply, about 9 billion gallons annually, was produced from natural gas, the remainder deriving from crude oil production.

FPC eases gas price restraints. The FPC proposed Sept. 9 that small gas producers be allowed to sell gas in the interstate market at prices 50% above the ceiling rates set by the FPC for large gas producers. The price differential was needed, according to the agency, because small producers, whose annual output was less than 10 billion cubic feet, faced greater drilling risks, higher costs and more problems in obtaining capital. (An estimated 12% of the nation's gas supply was provided by small producers.)

In a separate ruling issued the same day, the FPC also reinstituted its 60-day emergency and limited terms contracts (abolished in June). The effect was to decontrol prices paid by pipeliners to gas producers for a specific period of time. The action was taken to alleviate the expected winter gas shortage, the agency said.

An FPC staff report issued Sept. 12 recommended that a single nationwide price of 24.5¢ per 1,000 cubic feet be set for "old" gas sold by producers to interstate pipeliners. ("Old" gas was that which was produced at wells operating prior to 1973. It constituted 90%–95% of gas sold in the interstate market.)

It was recommended that this proposed national pricing system replace the FPC's practice of setting different regional prices. The average price of old gas currently was 21¢ per 1,000 cubic feet, but area rates ranged from 4.5¢ to 35¢ per 1,000 feet.

The FPC Dec. 4 raised its new uniform gas price to 50¢ per 1,000 cubic feet, an 8¢ increase over the previous national ceiling price of 42¢.

The new rate, which was made retroactive to June 21, applied to gas sold on an interstate basis to pipeline companies by producers from "new" wells or that were switched from the intrastate market after Jan. 1, 1973.

The FPC conceded that the new rate could result in up to a 16% rate increase for consumers, but argued that the rate increase was "more than counterbalanced by a more probable assurance of continued service."

In justifying its action, the FPC said the new rate would enable producers to recover all costs and make a 15% return on investment, thus stimulating exploration and development of the nation's gas supplies.

One commissioner, Rush Moody Jr., argued that the rate still was not high enough and he dissented from the FPC decision. (The FPC also modified another part of its June ruling and increased the gathering allowance for gas that was expensive to produce from 1¢ per 1,000 cubic feet to 1.5¢.)

FPC Chairman John Nassikas joined Interior Secretary Rogers C. B. Morton Dec. 4 in urging Congress to decontrol all "new" natural gas supplies to spur increased production. They coupled that proposal with a call for a tax on the gas industry's excess profits.

Court reverses FPC order—A federal appeals court in New Orleans Nov. 10

ordered the FPC to allow the delivery of greater amounts of cheap natural gas to public utilities. The FPC earlier had ruled that United Gas Pipe Line Co. must reduce its deliveries to utilities, forcing them to switch to more expensive fuels, and thereby conserving gas for use by home owners. The court ruled that the FPC exceeded its statutory authority.

Utility rate increases urged. Federal energy and financial officials met in Washington Sept. 11 with state utility regulators to discuss the liquidity crisis facing electric utilities.

Treasury Secretary William P. Simon urged state officials to alleviate the cash squeeze facing most utilities because of soaring fuel costs and high interest rates by allowing quick rate increases. The action was needed, Simon said, to avert "blackouts, brownouts, and, worse, economic stagnation" caused by widespread deferrals of capital expansion projects during a period of tight money.

Under Simon's plan, which also had the backing of Federal Energy Administrator John Sawhill and Federal Power Commission Chairman John Nassikas, regulators would permit utilities to include in their rate calculations the costs of ongoing construction work. Power companies' costs also would be passed through automatically to consumers under the rate plan favored by federal officials. (They did not comment on an industry proposal that the government guarantee debt financing for utilities.)

Sawhill justified the need for rate increases over the long term, saying utilities "must be permitted to earn a sufficient rate of return so that their stocks will again be attractive to investors."

While agreeing that many utilities were caught in a money bind, state officials were not receptive to the federal proposal. Some regulators accused the Administration of making their state agencies the scapegoats for what they called the economic mistakes of the Nixon and Ford Administrations. They charged that federal efforts should be aimed at aiding the utilities by lowering high interest rates and reducing the cost of foreign oil.

A Michigan regulator said much of the rate increases granted utilities were used to pay federal income taxes, and he suggested tax law changes for this revenue having the effect of bringing a utility's equity earnings up to the generally authorized 12%.

Richard Morgan, a spokesman for several consumer organizations, objected to the assumption about the need for industry growth that was shared by the utilities and by federal officials. Consumers should not be required "to raise capital for expansion consumers don't want or need," he said. As an alternative proposal to across-the-board rate increases, Morgan offered a plan for "lifeline" rates that would guarantee to all residential consumers a "basic amount of electricity" at prices lower than those now charged. In contrast to the current rate system, which permitted increased consumption at lower costs, Morgan's plan would force consumers using more than the "lifeline" minimum to pay higher prices per kilowatt hour.

Natural resources meeting. Interior Secretary Rogers C. B. Morton and other Administration officials discussed the inflationary impact of terminating oil price controls during a meeting in Dallas Sept. 16 with 45 representatives of the oil, natural gas, coal, metals mining and electric power industries.

Although most industry executives charged that federal price controls caused shortages and aggravated inflation, Morton warned that phasing out controls might take "two years or more" to prevent an international oil cartel from driving up the unregulated price of fuel and adding to the nation's inflationary problems.

(The price of "old" oil, that which was produced at pre-1972 output levels, was subject to a federal price ceiling of $5.25 a barrel. The remaining "new" oil sold at free market prices. According to the Wall Street Journal Sept. 17, new oil was selling for about $10 a barrel, while imported oil cost an average $9.50 a barrel.)

Morton and Treasury Secretary William P. Simon had earlier indicated strong backing for an end to oil price controls although neither proposed a timetable for the decontrol action. Their remarks were made before a Sept. 10

meeting of the National Petroleum Council, an industry group that advised the Interior Department.

Consumer representatives at the Dallas meeting called for a rollback in the price of crude oil. As an alternative to price decontrol, Rep. John Anderson (R, Ill.) urged enactment of a bill pending in Congress imposing a temporary windfall profits tax on oil producers. Anderson also called for the repeal of the oil depletion allowance to reduce inflationary pressures. Anderson challenged charges by industry spokesmen that environmental and health safeguards spurred inflation by increasing the costs of energy production. The short-term costs of these regulations should be viewed against their long-term benefits, Anderson said. It was also suggested that investment in energy conservation, as opposed to investments in new supplies, should be emphasized to alleviate shortages.

There was also support for a change in U.S. foreign policy that aimed at confronting the oil producing nations to force a reduction in oil prices.

Sen. Floyd K. Haskell (D, Colo.), who attended the Dallas meeting, and Sen. Henry M. Jackson (D, Wash.) Sept. 16 attacked the Administration's energy policies, which they said "had not focused" on alternatives to the two-tiered price system for old and new oil. The legislators, both members of the Senate Interior Committee, charged that new oil could be brought under federal price control at a saving of $5–$10 billion in consumers' costs without significantly hurting industry incentives for domestic oil production.

Project Independence, they said, would be "massively inflationary, bidding up the prices of those commodities, services and capital goods that are in the shortest supply," in an effort to make the nation self-sufficient in energy supplies.

FEA acts on crude oil transfer costs. The Federal Energy Administration Oct. 28, 1974 issued two regulations dealing with the transfer costs paid by some international oil companies for crude oil imported from foreign affiliates. Under current price controls, these transfer costs could be passed on to consumers. The new regulations could result in an oil price rollback if evidence of overcharging were found. Of the 4 million barrels of oil imported to the U.S. daily, an estimated 3 million barrels of crude oil were bought by refiners from foreign affiliates, the FEA said.

FEA Administrator John Sawhill said the agency was suspicious that some foreign affiliates were inflating the transfer costs because some of the international oil companies had been paying "significantly higher" landing prices than independent refiners.

In one regulation adopted Oct. 28, the FEA disallowed some of the costs claimed by oil companies in purchases from overseas affiliates. Their transfer costs would have to reflect competitive market prices, according to the new rule, which was made retroactive to October 1973. If excess costs were used to justify price increases in the U.S., the agency would require a price rollback.

Sawhill said the FEA would collect price data on foreign crude oil transactions from all oil companies on a monthly basis in order to enforce the regulation.

The other FEA regulation dealing with transfer costs was issued as a proposal setting a company-by-company ceiling price on landed costs (reflecting transfer charges) of imported crude. The rule was designed to limit an oil company's profit margin to its May 1973 levels plus 25¢ a barrel.

Price equalization plan adopted. Regulations that would equalize the cost of crude oil among refiners and the cost of fuel oil among distributors were adopted by the Federal Energy Administration (FEA) Dec. 2. The regulations would allocate low-priced domestic oil proportionately among all refiners with the aim of eliminating the price disparities that existed between refiners with greater access to cheaper domestic oil and those which depended on the higher-priced domestic oil and imported supplies.

Although the rules were not expected to cause any net change in the nationwide price of oil, areas of the East, which relied heavily on imported supplies, were expected to benefit from the rules because the burden of using high price oil would be shared by all parts of the nation.

Savings to New England consumers could total $360 million annually, FEA officials said. In other areas, which had easy access to cheaper domestic fuel, prices were expected to rise slightly.

The new regulations, which would take effect gradually over the next two months, were designed to replace the two-tiered price control system for domestic "old" and "new" oil. (Old oil came from wells drilled prior to 1972 and was subject to federal price control; new oil came from wells developed after that point and was not controlled.)

Under the new system, refiners with access to the cheaper old oil were required to sell the rights to that oil to other refiners, reducing their competitive edge and equalizing the cost of oil production. Distributors of home heating oil and residual fuel oil would also be assigned rights to old domestic crude in order to equalize their production costs. (This feature had not been part of the original equalization program proposed in August. The measure was added after Interior Secretary Rogers C. B. Morton yielded to Congressional pressure from representatives of east coast states.)

The new rules were aimed at protecting independent refiners and distributors who had relied on major oil companies for their supplies.

Court rejects Nader suit. Judge Gerhard Gesell of U.S. district court in Washington Dec. 13 rejected a suit brought by consumer advocate Ralph Nader seeking to roll back a $1 a barrel increase in the federally controlled price of "old" oil.

There was "clearly a rational basis" for the Cost of Living Council's decision in December 1973 authorizing the increase, Gesell ruled, adding that the emergency resulting from the Arab oil embargo was justification for the lack of public hearings and comment on the council's action.

15 oil firms roll back prices. Fifteen oil companies agreed to price rollbacks totaling about $77 million to correct alleged overcharges to customers, the Federal Energy Administration (FEA) announced Dec. 17. The FEA also disallowed about $375 million in "banked" costs claimed under a controversial agency ruling, since revised, permitting the oil industry to make a double recovery on costs. These expenses could have been passed on to consumers.

The FEA ordered price rollbacks of $10 million from Standard Oil Co. (Ohio), $6.9 million from Exxon Corp. and $600,000 from Mobil Oil Corp.

The remaining adjustments were settled voluntarily or through consent agreements:

Kerr-McGee Corp., $12.8 million in rollbacks, $2.9 million in cost reductions; Skelly Oil Co., $6.2 million cost reductions; Murphy Oil Corp., $1.4 million cost reductions; Phillips Petroleum Co., $31.7 million cost reductions, $215,000 rollbacks; Amerada-Hess Corp., $6.7 million cost reductions; Texaco Inc., $25.5 million cost reductions; Delta Refining Co., $157,000 cost reductions; Sun Oil Co., $700,000 refund to identifiable customers; Getty Oil Co., $300,000 rollbacks and $940,000 cost reductions; Shell Oil Co., $1 million cost reductions; Charter Oil Co., $4 million rollbacks; and Atlantic-Richfield Co., $1.5 million rollbacks and $8.1 million cost reductions.

The FEA revised its price regulations Nov. 6 to limit the amount of "banked" costs that oil companies could pass on to consumers in any single month to 10% of the total costs that they had delayed passing through by Oct. 31. The action was designed to protect the public from sudden large price increases in any one month. For competitive reasons, oil companies had been accumulating banked costs and postponing recovery of these expenses.

Policies & Programs

Anti-inflation program. The development of a new national energy policy was proposed by President Ford Oct. 8, 1974 in an anti-inflation program presented before a televised joint session of Congress.

The U.S. had "a real energy problem," Ford said. "One third of our oil, 17% of America's total energy, now comes from foreign sources we cannot control—at high cartel prices costing you and me $16 billion more than just a year ago. The primary solution has to be at home."

Ford named Interior Secretary Rogers C.B. Morton to head a newly established National Energy Board charged with developing a national energy policy. "His marching orders are to reduce imports of foreign oil by 1 million barrels per day by the end of 1975, whether by savings here at home or by increasing our own sources," Ford said. Morton was also directed to increase domestic energy supplies by "promptly utilizing coal resources and expanding the recovery of domestic oil still in the ground in old wells," the President added.

In his legislative requests, Ford asked Congress to give priority to four areas involving the deregulation of natural gas, increased use of the naval oil reserves in California and Alaska, amendments to the Clean Air Act, and passage of strip mine legislation that would "insure adequate supply with common sense environmental protection."

Ford also said he would seek new legislation requiring the use of cleaner coal processes and nuclear fuel in new electric plants and "the quick conversion of existing oil plants." Ford proposed setting a target date of 1980 for "eliminating oil-fired plants from the nation's base loaded electrical capacity."

The President said he would use the existing Defense Production Act to allocate scarce materials, but amendments to the law might be required. He also promised to meet with automobile industry officials "to assure, either by agreement or by law," a program that could achieve a 40% increase in gasoline mileage within a four-year development deadline.

The development of coal gasification programs and efforts to make new use of nonfossil fuels were long-term programs needed to increase domestic energy supplies, Ford said.

New energy agency established. A bill establishing a new Energy Research and Development Administration was approved by the House Oct. 9, Senate Oct. 10 and signed by President Ford Oct. 11. The new agency, which would consolidate federal energy research, replaced the Atomic Energy Commission (AEC), which was abolished under the legislation. The AEC, with a $1.7 billion fiscal 1975 budget, oversaw a $10 billion complex of national laboratories.

The AEC's licensing and regulation activities would be handled by a new Nuclear Regulatory Commission. The bill specified that the heads of the three regulatory offices—reactors, materials safeguard and research—would have direct access to the commissioners of the new agency. The commissioners were required to make public disclosure of any safety-related "abnormal occurrences" at nuclear plants within 15 days. Manufacturers and nuclear facility officials were required to report equipment or operating defects threatening public safety. A maximum penalty of $5,000 was set for each individual violation.

The nuclear weapons development aspect of the AEC operation was tentatively put under the aegis of the new agency, which, along with the Defense Department, would prepare recommendations for the future disposition of the responsibility.

An Energy Resources Council (formerly the National Energy Board) authorized under the legislation was established by President Ford by executive order Oct. 11, with Interior Secretary Rogers C. B. Morton as chairman.

Policy making group revamped. President Ford Oct. 29 announced the appointment of a "new team that will be in charge of the energy problem," under the direction of Interior Secretary Rogers C. B. Morton as head of the Energy Resources Council.

Federal Energy Administration Director John Sawhill was the principal victim of the administrative shakeup. Ford, announcing Sawhill's resignation, said he would be replaced by Andrew E. Gibson, who had previously served President Nixon as maritime administrator and as assistant secretary in the Commerce Department. But Gibson's nomination ran into trouble and was later withdrawn.

Ford named Robert C. Seamans Jr. to head the new Energy Research and Development Administration; Dixy Lee Ray to the new post of assistant secretary of state for oceans and international, environmental and scientific matters; and William A. Anders to serve as chairman of the new Nuclear Regulatory Commission.

Two reporters questioned Ford about Sawhill's abrupt departure: "Why have you dumped John Sawhill? Was his advice too blunt and politically unattractive at this time? What policy differences did you and [Interior Secretary] Morton have with Mr. Sawhill which precipitated his resignation?"

Ford denied that he had any "major policy differences" with Sawhill but said that Morton, who had overall charge for determining national energy policy, had the "right" to "make recommendations for those that will work with him on the council." Ford also promised that Sawhill "will be offered a first class assignment in this Administration." Morton later said Sawhill "lacked a sense of working together."

Sawhill had been outspoken in his call for increased energy conservation. He had publicly advocated the use of mandatory energy saving measures, such as a sharp increase in the gasoline excise tax, a controversial subject in Congress. Morton's efforts to solve the nation's energy problem emphasized increased fuel production.

Gibson's nomination to succeed Sawhill also had a controversial aspect. After leaving the Commerce Department in 1973, Gibson had been president of Interstate Oil Transport Co. of Philadelphia, which was owned by a shipping firm with major international tanker operations. Interstate barges carried petroleum for oil companies and utilities on the East Coast and Gulf of Mexico. According to the White House, Gibson severed his ties to Interstate in May.

Sawhill said Nov. 19 that the oil industry had lobbied for his ouster. Sawhill added, however, that he did not know if its influence had played any part in President Ford's decision to dismiss him. The remarks were made in testimony before the Joint Economic Committee of Congress.

Zarb confirmed as FEA administrator. Frank G. Zarb, 39, associate director of the Office of Management and Budget specializing in energy matters, was confirmed Dec. 11 as administrator of the FEA succeeding John C. Sawhill. (The nomination had been submitted to the Senate Dec. 2.)

President Ford had announced his selection of Zarb, a former investment banker, Nov. 25 following the withdrawal of Ford's controversial nomination of Andrew J. Gibson to replace Sawhill at the FEA.

Charges of possible conflict of interest had been leveled against Gibson, the former president of an oil hauling transportation firm. Zarb also had past ties to the oil industry, according to a revised personal biography issued by the FEA. The FEA stated that Cities Service Co. hired Zarb as a trainee in 1957 and that he left the firm in 1961, the Wall Street Journal reported Dec. 2. He became a partner in the Wall Street investment and securities firm of Hayden, Stone and joined the Nixon Administration in 1971 as an assistant secretary of labor.

Zarb was expected to continue to hold the post of executive director of the Energy Resources Council.

Melvin A. Conant, formerly of Exxon Corp., was confirmed Dec. 13 as assistant FEA administrator for international energy affairs.

Conant had left Exxon after serving 10 years as its senior government relations counselor in Europe, the Middle East and Asia. Upon his joining the FEA in January as director of its international trade office, Exxon had paid Conant the lump sum of $90,000. The payment raised conflict of interest charges and delayed Senate consideration of his nomination to the top FEA post by President Nixon. President Ford had resubmitted the nomination Nov. 18.

Report calls for stress on conservation. The Ford Foundation's Energy Policy Project concluded after a three-year study that too much stress was being placed on increasing national energy supplies and not enough attention was being focused on conservation efforts.

According to the project report issued Oct. 17, the nation's energy consumption growth rate could be cut by more than half without harming the economy, allowing "massive new commitments" to offshore oil drilling, oil imports, nuclear power and development of coal and shale deposits to be postponed for 10 years.

The Ford study concluded that national energy consumption, which had been expanding at an average annual rate of 4.5% from 1965 through 1973, could be cut to 2% without affecting the U.S. stan-

Western lands under 1970 legislation, but dard of living. "Energy growth and economic growth can be uncoupled," project director S. David Freeman said.

However, that view was not unanimously held. Eight of the project's 21 advisory board members said in a separate statement that "even the most austere self-discipline will fail to resolve the real and ultimate problem of [energy] supply." A ninth member, Mobil Oil Corp. President William P. Tavoulareas, characterized the majority's report as a "formula for economic stagnation" and one that would intensify "government interference in the lives of the American people."

Among the project's recommendations for government imposed conservation rules: higher energy prices, elimination of the oil depletion allowance, enactment of pollution taxes, charging consumers for the cost of stockpiling oil, federal assistance (perhaps as energy stamps) to low income families hard hit by energy shortages or price increases, immediate gasoline rationing and enactment of minimum gasoline economy standards for cars in order to raise average fuel economy to 20 miles a gallon by 1980, federal loans for installation of better building insulation and upgrading construction codes to require improved insulation, and redesign of electricity rates "to eliminate promotional discounts and to reflect peak load costs."

The project also called for the immediate termination of the government's commitment to build a nuclear breeder reactor (one that makes more fuel than it uses), reassessment of federal plans to increase offshore oil leases, and increased research spending on energy derived from unconventional sources, such as solar, and solid and organic wastes.

The report also called for regulation of electric power rates by regional commissions to "assure that utility expansion plans were integrated into regional grids."

In other findings, the report concluded that the U.S. "energy supplying industry does not constitute a monopoly by economic standards, although there are indications of diminished competition in some areas." To overcome this trend, the report urged increased antitrust enforcement and an end to the practice of joint bidding on offshore oil leases.

Project Independence blueprint. The Federal Energy Administration formally presented its 800-page "Blueprint for Project Independence" to the Energy Resources Council Nov. 13, 1974. The study, in preparation for seven months at a cost of $5 million, originally had been intended to serve as a master plan for the nation's energy policy, but differences within the Nixon and Ford Administrations over energy affairs had resulted in a lack of continuity on policy matters and had diluted the report's impact as a policy setting document.

The plan was presented by Federal Energy Administrator John Sawhill, who had been dismissed by President Ford because of differences over the fuel conservation issue. Eric R. Zausner, who had been Sawhill's assistant at the FEA, was the report's principal author.

At a news briefing Nov. 13, an official of the Office of Management and Budget said other groups within the executive branch would be able to present alternative policy suggestions. All options would then be analyzed by the White House and presented by President Ford in a speech early in 1975 setting forth a definitive national policy on energy.

Sawhill urged quick action rather than continued debate. "The time for writing reports is over. The longer we wait, the fewer options we have," Sawhill said.

Although President Nixon had set national self-sufficiency in energy needs as the major goal of Project Independence when it was announced in November 1973, designers of the FEA blueprint rejected the idea, opting instead for a goal of independence from "insecure" foreign oil by 1985 and emphasis on strict conservation measures to reduce the volume of energy imports. "If we want to reduce our dependence on imports, we must reduce our energy demands. This is an inescapable conclusion," Sawhill declared.

Highlights of the FEA report:

Conservation—Adoption of a 15¢ a gallon federal tax on gasoline; a mandatory fuel economy standard requiring that cars get at least 20 miles to the gallon; a 25% tax credit for insulating existing housing and a 15% tax credit for improving the energy efficiency of commercial buildings; national heating, cooling and lighting standards for all commercial buildings; efficiency standards for electrical appliances. Greater use of public transportation and redesign of electricity rates could reduce national energy growth by more than half to 2% a year by 1985.

There were two alternatives to strict conservation programs, according to the study. One was intensive exploration and drilling for oil and natural gas, but the FEA noted that environmental hazards, as well as the social and financial costs of the operation, posed strong drawbacks to this approach.

Another choice was establishment of a 1 billion barrel stockpile of oil in underground salt caverns. Maintenance costs for the stockpile were estimated at $1.4 billion a year.

U.S. ENERGY USE

CONSUMPTION IN ENERGY EQUIVALENTS OF BILLION BARRELS OF OIL PER YEAR

	1974	1970	1960	1950
Petroleum	5.90	5.00	3.45	2.35
Natural gas	4.30	3.62	2.15	1.07
Coal	2.45	2.36	1.78	2.22
Hydroelectric	.52	.39	.21	.12
Nuclear	.14	.10		
Total	13.30	11.47	7.55	5.76

Note: The 13,300,000,000 barrels of oil equivalent consumed by the United States also equals 78 quadrillion Btu's . . . or an average of 37,500,000 barrels of oil a day . . . or 215 trillion Btu's a day.

CONSUMPTION BY ACTUAL FUEL TYPE

[In percent]

	1974	1970	1960	1950
Petroleum	44	44	44	39
Natural gas	32	31	27	18
Coal	18	21	25	38
Hydroelectric	4	4	4	5
Nuclear	1			

Note: Energy imports amounted to 15 percent of total needs and are projected to reach 25 percent by 1985.

THE ENERGY USERS, BY SECTOR

[In percent]

	1974	1970	1960	1950
Industrial	32	31	34	36
Transportation	25	24	25	27
Residential/commercial	22	24	24	23
Electrical generation energy loss	15	15	11	9
Nonenergy (petrochemicals, etc.)	6	6	5	5

Note: 56 percent of all transportation energy use is by automobiles (also 14 percent of total energy use).

Regarding development of U.S. energy sources:

Oil—Domestic production peaked in 1970 and would continue to decline for several years, despite price increases awarded oil producers as incentives, because the development of new fields required several years. By 1985, domestic production was estimated at 12–20 million barrels a day, with output rising as government-set prices were raised. "Of the estimated 200–400 billion barrels of undiscovered oil in the U.S., one third is in offshore regions." Oil spills from offshore drillings were not regarded as a major threat.

Natural gas—Reserves had been falling since 1967 and consumption currently was 2–3 times greater than annual discoveries of new supplies. "The outlook for increased gas supplies is not promising" if prices remained at their current, federally-regulated level. Higher prices could stimulate recovery from offshore areas and Alaska.

Coal—Coal, which had provided 90% of the nation's energy supply in 1900, now comprised 17% of all energy consumption; two-thirds of the current use was in electric utility consumption. At the 1974 production rate of 599 million tons, the U.S. had another 800 years' reserves of coal, most of it high in sulfur. Water shortages were possible consequences of some coal gasification plans.

Electricity—Nuclear power plants, which had fallen behind schedule for becoming operational, nonetheless were expected to provide 30% of all electricity by 1985. Rate redesigns could diminish peak power loads and reduce the need for new generating capacity.

Oil shale—Production could reach 250,000 barrels a day by 1985 at $11 a barrel, but a strong push to recover oil shale from the Colorado Rockies could result in a severe water shortage in the area. Development also might require federal price support.

Solar and geothermal energy—Solar energy would become a significant energy source after 1985 because of technological advances and the high recovery and storage costs of fossil fuels. Geothermal energy development had been hampered by delays in awarding federal leases on it might prove a significant energy source by 2000, with relatively small environmental costs.

Ford, GM officials back gas tax—Proponents of the gasoline tax increase won support from unlikely sources—officials of the nation's two largest auto makers. Henry Ford 2nd, chairman of the Ford Motor Co., said Nov. 22 that he favored cutting fuel consumption by imposing a 10¢ a gallon increase on the federal gasoline tax.

The revenues could be used in three ways, Ford said: to extend unemployment compensation for auto workers laid off during the current industry-wide recession; to finance a tax break for families earning less than $17,500 a year; and to provide gasoline stamps for the unemployed and those living under the federally defined poverty level.

Thomas A. Murphy, chairman of General Motors Corp., said Dec. 2 that he supported voluntary fuel conservation measures, but added that he believed the Administration "ought to consider" a gasoline tax increase if it were necessary to reduce oil imports.

Treasury Secretary William Simon also indicated some support for mandatory conservation measures. Simon said Dec. 3 that if voluntary efforts proved inadequate in reducing oil imports by the end of January 1975, the Administration would be forced to resort to stronger measures. An increase in gasoline taxes was only one such approach, Simon said, noting that mandatory oil allocations could be imposed if necessary.

Burns asks austerity policy. Arthur Burns, chairman of the Federal Reserve Board, Nov. 27 criticized President Ford's fuel saving program, which relied on voluntary measures, and urged the Administration to adopt an "austerity policy" on energy conservation.

"I hope we won't waste precious time if the voluntary program proves to be inadequate, as many people, including myself, believe it will," Burns told the Joint Economic Committee of Congress. If stronger fuel saving measures were not adopted, he warned, the "alternative drift may lead to a permanent decline in our nation's economic and political power in a very troubled world."

Among the mandatory conservation efforts Burns supported was a "sizable" gasoline tax increase. "We have been lecturing the rest of the world [on cutting fuel consumption] while our own practices leave much to be desired," Burns said. He noted that although Britain, Italy, France, Japan, West Germany and other nations had raised gasoline taxes, "we haven't touched the tax on gasoline."

Burns also said he favored a tax on imported oil and a tax on autos based on horsepower or weight. He called for faster development of domestic energy sources and an increase in U.S. oil storage capacity to mitigate the impact of any new embargo on oil supplies.

Burns called on the U.S. to exert leadership in devising a strategy that would lead to a reduction in oil prices and a diminution of the power of the oil producing nations' cartel. He gave only limited support to the $25 billion oil facility proposed by Secretary of State Henry Kissinger as a means for dealing with the severe balance of payments problems confronting oil importing nations. Burns told the committee he would support the proposal only if the mutual aid program were limited to one or two years.

"Financial cooperation alone is not enough. Even with an orderly financing of [balance of payments] deficits, the immense burden of carrying and ultimately repaying the debts will still remain. The fundamental problem is the huge oil bill of the importing countries, and a fundamental solution requires that the price of oil be reduced," Burns said.

Coalition asks tough policy. A coalition of more than 100 prominent citizens, headed by former Commerce Secretary Peter G. Peterson, urged President Ford and the Congress to develop a "tough" energy policy to deal with the nation's continuing energy problems.

"America must curtail its need for imported oil and go full speed ahead to develop alternative energy sources," the group, Citizens for a Strong Energy Program, said in an open letter published Dec. 26, 1974.

The group's chief goal was a reduction in domestic energy consumption by 1 million barrels a day by July 4, 1976. To achieve that end, they supported stand-by gasoline rationing plans, tariffs on imported oil, ceilings on oil imports, gasoline or fuel oil taxes with rebates for the poor, energy rates that penalized wasteful use of energy and major research programs for nuclear and non-nuclear fuel development.

Multinational panel submits report. After a two-year study of the operations of U.S.-based multinational oil companies, the Senate Foreign Relations Committee's multinationals subcommittee, in a report Jan. 11, 1975, called on the U.S. to break the oil producing nations' cartel by making drastic cutbacks in oil consumption and by restricting the power of major international oil firms to negotiate long-term price and supply arrangements with foreign oil exporting states.

Among the recommendations in the majority report, submitted by the five member panel and outlined by subcommittee chairman Frank Church (D, Ida.):

■ A 15% cutback in oil consumption as part of a larger plan to reduce consumption in the oil consuming nations by a total of 7 million barrels a day. The gradual, mandatory program would aim at reducing U.S. consumption by 2.5 million barrels a day through use of import quotas or ceilings and gasoline rationing. Church was unable to say what effect a 15% fuel cutback would have on the economy, but he favored a tax on auto horsepower or weight, improved mass transit systems, rationing and government imposed controls on gasoline to mitigate the impact of the drastic curtailment in gasoline supplies.

Sen. Clifford Case (R, N.J.) dissented from the majority report in favoring a ban on all foreign oil imports.

■ Legislation requiring that oil companies obtain government approval on foreign oil contracts, thereby making the U.S. a party to oil and price negotiations. Church urged Congress to consider establishing a federal corporation that supplanted private oil firms in purchasing oil directly from foreign oil producers and then allocated the supplies among domestic refiners.

■ Congress was urged to restructure taxes paid by oil firms to provide incentives for their investments in nations not belonging to the Organization of Petroleum Exporting Countries (OPEC) or in OPEC nations that were unlikely to restrict production. The subcommittee also favored ending the oil depletion allowance and other tax credits that favored foreign over domestic production.

■ The government was urged to launch a "crash program" to develop alternative energy sources "without abandoning suitable environmental safeguards."

■ Oil buying arrangements between oil producing states and major oil firms raised "significant antitrust issues," the report charged, and should be "tested" by the Justice Department.

"This is the overriding lesson of the petroleum crisis," the subcommittee said: "In a democracy, important questions of policy with respect to a vital commodity like oil, the life blood of an industrial society, cannot be left to private companies acting in accord with private interests and a closed circle of government officials. They must be surfaced for public debate and education so that a coherent policy can be evolved with a firm basis of public support."

Legislative Action

Strip-mine bills vetoed. A bill to regulate the strip-mining of coal on federal lands was cleared by Congress Dec. 16, 1974. It was vetoed by President Ford Dec. 30 on the ground that it would slow the production of coal at a time "when the nation can ill afford significant losses from this critical energy source."

The bill would have required coal strip-miners to prevent environmental damage and restore mine land to its "approximate original contour."

A major review of national energy policies was under way, Ford said, and "unnecessary restrictions on coal production would limit our nation's freedom to adopt the best energy options." He said the bill would increase U.S. dependence on foreign oil.

There was a flurry of publicity about Ford's rental of a ski chalet in Vail, Colo., from Dallas millionaire Richard D. Bass, who held a 20,700-acre federal coal lease in Wyoming that reportedly would cost him more than $100 million for reclaiming land if Ford signed the bill. White House Press Secretary Ron Nessen had told reporters Dec. 19 that Ford was unaware of the Bass lease and did not make decisions on legislation in any event "on the basis of whose house he rents at Christmas."

Ford did not sign the bill within 10 days of receiving it from Congress, which had adjourned, so the bill was killed by pocket veto.

The passage of the bill Dec. 16 had been preceded by a long struggle over the measure in committee. A two-month deadlock had developed in the House-Senate conference over the rights of surface-rights owners. The final version was resolved Dec. 3 and the bill's House backers then sought, to avoid a pocket veto, immediate House consideration by suspension of the rules. This failed Dec. 9, by 212–150, or 30 short of the required two-thirds majority. The requisite rule from the Rules Committee was sought, and obtained after three days, by a 9–4 vote Dec. 12 for floor consideration by a straight up-and-down vote, parliamentary challenges ruled out. One of the day's delay was attributed to a report Dec. 11 that outgoing Federal Energy Administration Chief John C. Sawhill had dropped the agency's opposition and concluded "that this bill is one we could live with."

The agency's estimates of potential tonnage loss of production resulting from regulation reportedly had been drastically revised downward; that an early estimate of a 100-million-ton annual loss in output was attributed to misleading data supplied by the coal industry. Frank Zarb, who succeeded Sawhill Dec. 12 as FEA administrator, gave the agency's estimate Dec. 13 as an output loss from the bill of 48 million-141 million tons of coal in 1977. The bill included a moratorium on federal coal leasing until Feb. 1, 1976. A moratorium was already in effect pending passage of legislation.

There were conflicting views of the bill's impact on production. Bethlehem Steel Corp. broke ranks with industry opposition to the bill by urging Ford Dec. 17 to sign the bill. Board Chairman Lewis W. Foy said the bill would have little effect on output or production costs.

Some 260,000 acres of federal lands estimated to contain about six billion tons of coal were under lease to coal companies, who also were seeking to obtain the consent from surface owners for future leasing of underlying coal by the Interior Department. The bill would require the coal companies to obtain the consent of farmers and ranchers with surface rights prior to any mining. A fair market price would be determined by three appraisers, which the land owner could accept or reject. Side agreement between coal companies and surface-rights owners would be barred. The companies were permitted to purchase the surface rights outright, if the surface owner agreed.

Another controversial feature of the bill was a fee to be applied against all coal mined in the U.S.—strip and underground mining. The fee would be 10% of coal's value at the mine mouth, or 35¢ a ton, whichever was smaller. On the current production of more than 600 million tons a year, the fee would raise $200 million, the revenue to be used to reclaim the more than a million acres of abandoned strip-mined land, to reimburse local communities for adverse mining impact and to build miners' facilities.

1975 strip-mine bill vetoed—A similar strip-mine regulation bill was passed by Senate voice vote May 5, 1975 and by 293–115 House vote May 7. President Ford vetoed it May 20, and his veto was upheld June 10 when the House voted by only 278–143 to override the veto. The vote was three short of the two-thirds note necessary to overturn a veto.

Before the bill's passage, a protest against it had been staged in Washington April 8. About 2,000 strip mine workers from Appalachia, fearful that a production cutback would lead to loss of jobs, participated. A convoy of 600 giant coal trucks was driven through downtown Washington for three hours, circling the Capital and the White House with air horns blasting. Several hundred strip mine workers massed at both locations. The protest was sponsored by strip-mine companies and related firms, which were said to have contributed $400,000 for hotel rooms and gasoline plus $50 in expenses for each protester.

Rep. Morris K. Udall (D, Ariz.), manager of the measure in the House, issued a statement deploring the protest. It was, he said April 8, "a mischievous and purposeful effort by a misguided segment of the coal industry to mislead and foster fear among workers and their families."

President Ford urged Congress April 22 to formulate a bill that would "strike a balance" between the environmental and production needs. Because of the energy problem, the President said in a letter to House and Senate leaders, the nation needed to double coal production in the next decade and open 250 major new mines, the majority of them surface mines. "I believe we can achieve these goals and still meet reasonable environmental protection standards," Ford said.

The President's decision to veto the bill had been announced May 19 and justified by Federal Energy Administrator Frank G. Zarb. He said up to 36,000 jobs would be lost if the legislation became law. "We can't visit that kind of thing on the American people even if the objective is as noble as this one," he said. He said the President "still believes a federal strip-mine bill is in order." and obtainable in 1975.

In his veto message May 20, Ford warned that Congress' controversial measure would reduce coal production and increase unemployment and electric bills to the consumer. The President said coal was "needed more than ever" in light of the need to reduce the nations' dependence on foreign oil as an energy source. He estimated the loss of coal production as a result of the bill at from 40 to 162 million tons a year, up to a quarter of the annual tonnage.

A principal sponsor of the measure, Sen. Henry M. Jackson (D, Wash.), deplored later May 20 "the President's lack of sensitivity to the need to balance energy production with protection of our land."

Jackson disputed Ford on the production issue, saying the President's estimates of production loss, supplied by the Federal Energy Administration, were "vague" and without "any rational basis." He claimed the environmental benefits of the bill could be gained "without a loss in coal production."

Another principal sponsor of the bill, Rep. Morris K. Udall (D, Ariz.), sent a letter protesting the veto to members of the House May 20. In it, he countered Ford's point about higher electric bills by citing support for the strip mining bill by consumer groups.

The tonnage-loss estimate of the Environmental Protection Agency, which made its own study of the issue, was 89.7 million tons, the New York Times reported June 8, saying it had come into possession of an unpublished study by the EPA recently completed and limited in circulation to top federal officials. The study also gave the tonnage-loss estimate—86 million tons a year—of the National Economic Research Associates, a private consulting firm whose estimates had been used in the past by coal-consuming electric utilities, allies with the coal companies in opposing passage of strip-mine legislation. Udall credited the utilities with "a very thorough job" of lobbying against the bill through advertisements. "Many people," he said June 10, "were frightened about their utility bills."

Oil tanker bill vetoed—President Ford Dec. 30, 1974 had withheld his signature and pocket vetoed a bill that would have required at least 20% of imported oil to arrive on U.S. tankers. The requirement would have risen to 25% in mid-1975 and 30% in mid-1977. Currently, about 6% of imported oil came in on U.S. ships.

Ford said in his veto statement Dec. 30 the bill "would have an adverse impact" on the country's economy and its foreign relations. "It would create serious inflationary pressures" by increasing the price of oil, he said, and "it would further stimulate inflation in the ship construction industry." He also objected that the bill "would serve as a precedent for other countries to increase protection of their industries."

Deepwater ports bill passed. A bill to regulate the construction and operation of deepwater ports for receiving oil from huge supertankers was cleared by both houses of Congress Dec. 17, 1974 and was signed by President Ford Jan. 4, 1975.

Petroleum would be pumped to land from the "superports" through underwater pipelines. Handling of the oil in this manner was considered less susceptible to spills, although environmentalists contended that potential spills 20 miles from the shore would be harder to remedy.

The bill provided for liability against spill damage claims of up to $50 million. Claims of more than that were to be paid from a $100 million fund financed by a user charge of 2¢ a barrel on oil unloaded at the ports.

Governors of states adjacent to the proposed ports were given veto power over proposed projects.

93rd Congress' record. The Library of Congress provided a summary of the 93rd Congress' (1973–74) legislative actions on energy. The 93rd Congress had passed 48 energy-related bills, of which 43 were signed into law and 5 were vetoed. According to the summary:

> The 93rd Congress has involved itself in energy-related legislation on an unprecedented scale. More than 2000 bills have been introduced, and more than 30 standing congressional committees have collectively held over 1000 days of hearings on nearly every aspect of energy policy programs and problems. . . .
>
> *Major Accomplishments:* Although several important issues are still pending, the record shows major accomplishments in many key energy areas:
>
> *Measures which encourage conservation of supply and use of energy:* Emergency Petroleum Allocation Act (P.L. 93–159 and P.L. 93–511); encouragement of increased railroad and other mass transit facilities to conserve gasoline (P.L. 93–87, P.L. 93–146, P.L. 93–236 and P.L. 93–503); lowered speed limits (P.L. 93–239 and P.L. 93–643); and extension of the Defense Production Act through June 30, 1975. (P.L. 93–426).
>
> *Measures to manage and expand production of energy supplies:* Trans-Alaska Pipeline (P.L. 93–153); Coal Conversion and Clean Air Act Amendments (P.L. 93–319); a consolidated energy research and development appropriations act providing $2.2 billion for FY 1975 (P.L. 93–322); and a new ten-year research and development program, administered by ERDA, mandating higher priorities for non nuclear sources (P.L. 93–577).
>
> *New initiatives to develop commercial application of solar and geothermal energy technology:* Housing and Community Development Act, promotes energy conservation and use of solar energy in building (P.L. 93–383); Geothermal Energy Act (P.L. 93–410); Solar Heating and Cooling Demonstration Act (P.L. 93–409), and a Solar Energy Research and Development Program (P.L. 93–473).
>
> *Several organizational changes* which provide a beginning for reducing the fragmented pattern of U.S. energy policy formulation and program administration: Federal Energy Administration (P.L. 93–275) and legislation to create a more permanent Energy Research and Development Administration and a Nuclear Regulatory Commission (P.L. 93–438). An Energy Resources Council to coordinate existing programs and recommended new policies was also authorized by P.L. 93–438.
>
> MAJOR ISSUES STILL PENDING
>
> End use rationing and mandatory conservation. The President's veto of the Energy Emergency Act (S. 2589) left the Executive Branch without specific authority to institute mandatory end use rationing and other energy conservation measures. A revised version of S. 3267, reported in the Senate on December 5, 1974, would provide standby emergency authority for these purposes.
>
> Revisions in oil and gas depletion allowances and other long standing petroleum production incentives. The issue is complicated by the fact that domestic production of oil and gas has been plateauing so the question of whether continued incentives would produce more than marginal increases in output is an open one.
>
> Deregulation of natural gas. Similar problems arise as between proponents who wish deregulation as an incentive to increased production and/or to encourage further conservation of use.
>
> Petroleum pricing problems. Current issues center on the effect of escalating prices on domestic inflation, but higher prices are also being promoted as incentives to increased production from marginal fields.
>
> Petroleum import policy. To what degree should we depend on foreign imports to meet current demand, even if prices should become lower in the future? If imports continue, how should the country protect itself against interruptions in supply? What form should strategic reserves take and in what volume?
>
> The crisis atmosphere and accompanying legislative activity of the fall and winter of 1973/74 appear to have been replaced by growing Congressional recognition that the basic energy issues have no easy short term solutions. Intermediate and long term solutions require and are receiving widening Congressional involvement, continuing legislative action and increasingly informed concern.
>
> STATUS OF MAJOR ENERGY LEGISLATION ENACTED BY THE 93D CONGRESS: 1973–74
>
> I. ENACTED AND PUBLIC LAW
>
> *Public Law and bill number*
>
> 1. Economic Stabilization Act Amendments: Public Law 93–28, S. 398.
>
> 2. Rural Electrification Act: Public Law 93–32, S. 394.

3. Water Resources Planning Act: Public Law 93–55, S. 1501.

4. Federal-Aid Highway Act: Public Law 93–87, S. 502.

5. To Amend the Euratom Corporation Act of 1958, as amended: Public Law 93–88, S. 1993.

6. Oil Pollution Act Amendments: Public Law 93–119, H.R. 5451.

7. Rail Passenger Corporation: Public Law 93–146, S. 2016.

8. Mineral Leasing Act of 1920, Amendments and Trans-Alaska Oil Pipeline Authorization: Public Law 93–153, S. 2081.

9. Emergency Petroleum Allocation Act of 1973: Public Law 93–159, S. 1570.

10. Emergency Daylight Saving Time Energy Conservation Act of 1973: Public Law 93–182, H.R. 11324.

11. Northeast Rail Service Act: Public Law 93–236, H.R. 9142.

12. Emergency Highway Energy Conservation Act: Public Law 93–239, H.R. 11372.

13. Supplemental Appropriation Act to Explore and Develop Naval Petroleum Reserves: Public Law 93–245, H.R. 11576.

14. Intervention on the High Seas Act: Public Law 93–248, S. 1070.

15. Fuel Cost Pass-Through for Truckers: Public Law 93–249, S.J. Res. 185.

16. Federal Energy Administration: Public Law 93–275, S. 2776.

17. AEC Authorization Act: Public Law 93–276, S. 3292.

18. NASA Authorization Act: Public Law 93–316, H.R. 13998.

19. Energy Supply and Environmental Coordination Act of 1974: Public Law 93–319, H.R. 14368.

20. Special Energy Research and Development Appropriations for 1975: Public Law 93–322, H.R. 14434.

21. AEC Omnibus Legislation of 1974: Public Law 93–377, S. 3669.

22. Housing and Commodity Development Act: Public Law 93–383, S. 3066.

23. Aid Energy Affected Small Business: Public Law 93–386, S. 3331.

24. Natural Gas Pipeline Safety Act, as amended, additional appropriations: Public Law 93–403, H.R. 15205.

25. The Solar Heating and Cooling Demonstration Act of 1974: Public Law 93–409, H.R. 11864.

26. Geothermal Energy Research, Development and Demonstration Act: Public Law 93–410, H.R. 14920.

27. Public Works and Economic Development Act: Public Law 93–423, H.R. 14883.

28. Defense Production Act of 1950, extended for 2 years: Public Law 93–426, S. 3270.

29. Emergency Daylight Saving Time Energy Conservation Act of 1973, amendments: Public Law 93–434, H.R. 16102.

30. Energy Reorganization Act of 1974: Public Law 93–438, H.R. 11510.

31. Federal Columbia River Transmission System: Public Law 93–454; S. 3362.

32. Solar Energy Research, Development and Demonstration Act of 1974: Public Law 93–473, S. 3234.

33. Foreign Investment Study Act: Public Law 93–479, S. 2840.

34. To amend the tariff schedule of the United States to provide for the duty free entry of methanol imported for use as fuel: Public Law 93–482, H.R. 11251.

35. International Nuclear Cooperation: Public Law 93–485, S. 3698.

36. National Railroad Passenger Corporation: Public Law 93–496, H.R. 15427.

37. Export Administration Amendments: Public Law 93–500, S. 3792.

38. Urban Mass Transportation Act: Public Law 93–503, S. 386.

39. Emergency Petroleum Allocation Extension Act of 1974: Public Law 93–511, H.R. 16757.

40. Federal Nonnuclear Energy Research and Development Act of 1974: Public Law 93–577, S. 1283.

41. Deep Water Port Act: Public Law 93–627, H.R. 10701.

42. Federal-Aid Highway Act Amendments of 1975: Public Law 93–643, S. 3934.

43. Export-Import Bank Act Amendments (requires semi-annual report on impact of domestic energy resources of each loan involving export of any energy-related product or service): Public Law 93–646, H.R. 15977.

II. VETOED

1. Energy Emergency Act (S. 2589, H.R. 11450), March 6, 1974.

2. Atomic Energy Act, Price-Anderson provisions (H.R. 15323), October 12, 1974.

3. TVA Pollution Control Cost Credit (H.R. 11929, S. 3057), December 21, 1974.

4. Energy Transportation Security Act (H.R. 8193, December 30, 1974.

5. Surface Mining Control and Reclamation Act (S. 425, H.R. 11500), December 30.

III. SELECTED ENERGY BILLS PENDING ON THE CALENDARS AT ADJOURNMENT

Council on Energy Policy (S. 70).

National Land Use Policy (S. 268, H.R. 10294).

National Resource Lands Management (S. 424).

Oil Shale Revenues (S. 3009).

Outer Continental Shelf Supply Act (S. 3221).

Elk Hills and Naval Petroleum Reserves (S.J. Res. 176).

Federal Coal Leasing Policy (S. 3528).

Construct Coal Slurry Pipelines (S. 3879).

Aid for Independent Oil Refiners (S. 2743).

Protect Franchised Dealers in Petroleum Products (S. 1694).

Aid to Utility Cooperatives Serving Food Producers (S. 2150).

Oil and Gas Energy Tax Act (H.R. 14462).

Energy Tax and Individual Relief Act (H.R. 17488).

Standby Energy Emergency Authorities (S. 3267, H.R. 13834).

Energy Conservation and Greater Efficiencies in Electrical Energy (S. 2532).

Resource Conservation and Energy Recovery Act (S. 3954).

Energy Conservation in Federal Facilities (H.R. 1565).
Promote Efficient Use of Energy, Establish Architectural Guidelines (H.R. 11714).
Federal Mass Transportation Act (H.R. 12859).

Ford's Proposals & Actions

Ford's State of the Union message. President Ford delivered a somber State of the Union message to Congress Jan. 15. He proposed immediate revision of economic policy to combat recession and a massive effort to achieve energy independence for the nation.

The President had revealed the outlines of his program in a broadcast to the nation Jan. 13.

In his message Jan. 15, addressed to a joint session of Congress and televised nationally, Ford asked Congress not to legislate restrictions on his conduct of foreign policy, which he said was the responsibility of the president. He deplored the growth of the federal budget to "shocking" proportions and cautioned Congress he would veto any new spending programs it adopted this year. He would propose no new spending programs, he said, except in energy and would present legislation "to restrain the growth of a number of existing programs."

"I must say to you that the state of the union is not good," he told Congress. The economy was beset with unemployment, recession and inflation. Federal deficits of $30 billion in fiscal 1975 and $45 billion in fiscal 1976 were anticipated. The national debt was expected to rise to over $500 billion. Plant capacity and productivity were not increasing fast enough. "We depend on others for essential energy," Ford said, and "some people question their government's ability to make hard decisions and stick with them."

Ford said he was sure that his program would begin to restore the nation's status of "surplus capacity in total energy" that it had in the 1960s. But he cautioned that it would impose burdens and require sacrifices. He assured that "the burdens will not fall more harshly on those less able to bear them."

The goals were to reduce oil imports to "end vulnerability to economic disruption by foreign suppliers by 1985" and to develop energy technology and resources.

Immediate action was needed to cut imports, Ford said, and he urged quick action on legislation to allow commercial production at the Elk Hills, Calif. Naval Petroleum Reserve.

He would submit legislation to enable more power plants to convert to coal.

Using presidential power, he would raise the fee on all imported crude oil and petroleum products. Crude oil fee levels would be increased $1 per barrel on Feb. 1, by $2 per barrel on March 1 and by $3 per barrel on April 1.

These were "interim" actions, he said, pending a broader program he requested that Congress enact within 90 days. This would include:

- Excise taxes and import fees totaling $2 per barrel on product imports and on all crude oil.
- Deregulation of new natural gas and a natural gas excise tax.
- A windfall profits tax. Ford asked for the windfall profits tax by April 1 since he planned to act to decontrol the price of domestic crude oil on that date.

The President said he was prepared to use presidential power to limit imports as necessary to guarantee success of his energy conservation program.

Before deciding on the program, he said, he considered rationing and higher gasoline taxes as alternatives but rejected them as ineffective and inequitable.

Ford spoke of "a massive program" that must be launched to attain energy independence within a decade. The largest part of increased oil production, he said, must come from "new frontier areas" on the outer continental shelf and from the Naval Petroleum Reserve in Alaska. He planned "to move ahead" with development of the outer-shelf areas "where the environmental risks are acceptable."

On coal, Ford recommended "a reasonable compromise on environmental concerns." He said he would submit legislation to allow "greater coal use without sacrificing clean air goals." Referring to his veto of the 1964 strip-mining bill, Ford said he would sign a version "with appropriate changes."

As for nuclear power, he would submit legislation to expedite leasing and rapid selection of sites.

For utilities, whose financial problems were worsening, Ford said, he proposed that the one-year investment tax credit of 12% be extended by an additional two years "to specifically speed the construction of power plants that do not use natural gas or oil." He also would submit proposals for selective reform of state utility commission regulations.

To provide stability for production, Ford said he planned to request legislation to authorize and require tariffs, import quotas or price floors to protect domestic energy prices.

To cut long-term consumption, Ford proposed:

- Mandatory thermal efficiency standards for new buildings.
- A tax credit of up to $150 for homeowners installing insulation.
- Aid to low-income families to buy insulation.
- Revision and deferment of automotive pollution standards for five years "which will enable us to improve new automobile gas mileage by 40% by 1980."

As a foil against foreign disruption, Ford requested stand-by emergency legislation and a strategic storage program of one billion barrels of oil for domestic needs and 300 million barrels for defense.

The President listed these goals to be attained by 1985: one million barrels of synthetic fuels and shale oil production per day; 200 major nuclear power plants; 250 major new coal mines; 150 major coal-fired power plants; 30 major new refineries; 20 major new synthetic fuel plants; thousands of new oil wells; insulation of 18 million homes; and sale of millions of new automobiles, trucks and buses using "much less fuel."

Text of the energy section of the message:

The economic disruptions we and others are experiencing stems in part from the fact that the world price of petroleum has quadrupled in the last year. But, in all honesty, we cannot put all of the blame on the oil exporting nations. We, the United States, are not blameless. Our growing dependence upon foreign sources has been adding to our vulnerability for years and years. And we did nothing to prepare ourselves for such an event as the embargo of 1973.

During the 1960s, this country had a surplus capacity of crude oil, which we were able to make available to our trading partners whenever there was a disruption of supply. This surplus capacity enabled us to influence both supplies and prices of crude oil throughout the world. Our excess capacity neutralized any effort at establishing an effective cartel, and thus the rest of the world was assured of adequate supplies of oil at reasonable prices.

By 1970 our capacity, our surplus capacity had vanished, and as a consequence the latent power of the oil cartel could emerge in full force. Europe and Japan, both heavily dependent on imported oil, now struggle to keep their economies in balance. Even the United States, our country, which is far more self-sufficient than most other industrial countries, has been put under serious pressure.

I am proposing a program which will begin to restore our country's surplus capacity in total energy. In this way, we will be able to assure ourselves reliable and adequate energy and help foster a new world energy stability for other major consuming nations.

But this nation and, in fact, the world must face the prospect of energy difficulties between now and 1985. This program will impose burdens on all of us with the aim of reducing our consumption of energy and increasing our production. Great attention has been paid to the considerations of fairness, and I can assure you that the burdens will not fall more harshly on those less able to bear them. I am recommending a plan to make us invulnerable to cutoffs of foreign oil. It will require sacrifices. But it—and this is most important—it will work.

I have set the following national energy goals to assure that our future is as secure and as productive as our past:

First, we must reduce oil imports by one million barrels per day by the end of this year and by two million barrels per day by the end of 1977.

Second, we must end vulnerability to economic disruption by foreign suppliers by 1985.

Third, we must develop our energy technology and resources so that the United States has the ability to supply a significant share of the energy needs of the free world by the end of this century.

To attain these objectives, we need immediate action to cut imports. Unfortunately, in the short-term there are only a limited number of actions which can increase domestic supply. I will press for all of them.

I urge quick action on the necessary legislation to allow commercial production at the Elk Hills, Calif., Naval Petroleum Reserve. In order that we make greater use of domestic coal resources, I am submitting amendments to the Energy Supply and Environmental Coordination Act which will greatly increase the number of power plants that can be promptly converted to coal.

Obviously, voluntary conservation continues to be essential, but tougher programs are needed—and needed now. Therefore, I am using Presidential powers to raise the fee on all imported crude oil and petroleum products.

Crude oil fee levels will be increased $1 per barrel on Feb. 1, by $2 per barrel on March 1 and by $3 per barrel on April 1. I will take action to reduce undue hardship on any geographical region. The foregoing are interim administrative actions. They will be rescinded when the broader but necessary legislation is enacted.

To that end, I am requesting the Congress to act within 90 days on a more comprehensive energy tax program. It includes:

- Excise taxes and import fees totaling $2 per barrel on product imports and on all crude oil.
- Deregulation of new natural gas and enactment of a natural gas excise tax.
- I plan to take Presidential initiative to decontrol the price of domestic crude oil on April 1.
- I urge the Congress to enact a windfall profits tax by that date to ensure that oil producers do not profit unduly.

The sooner Congress acts, the more effective the oil conservation program will be and the quicker the federal revenues can be returned to our people. I am prepared to use Presidential authority to limit imports, as necessary, to guarantee success.

I want you to know that before deciding on my

energy conservation program, I considered rationing and higher gasoline taxes as alternatives. In my judgment, neither would achieve the desired results and both would produce unacceptable inequities.

A massive program must be initiated to increase energy supply, to cut demand and provide new stand-by emergency programs to achieve the independence we want by 1985.

The largest part of increased oil production must come from new frontier areas on the outer continental shelf and from the Naval Petroleum Reserve No. 4 in Alaska. It is the intent of this Administration to move ahead with exploration, leasing and production on those frontier areas of the outer continental shelf where the environmental risks are acceptable.

Use of our most abundant domestic resource—coal—is severely limited. We must strike a reasonable compromise on environmental concerns with coal. I am submitting clean air amendments which will allow greater coal use without sacrificing clean air goals.

I vetoed the strip-mining legislation passed by the last Congress. With appropriate changes, I will sign a revised version when it comes to the White House.

I am proposing a number of actions to energize our nuclear power program. I will submit legislation to expedite nuclear leasing and the rapid selection of sites.

In recent months, utilities have canceled or postponed over 60% of planned nuclear expansion and 30% of planned additions to nonnuclear capacity. Financing problems for that industry are worsening. I am therefore recommending that the one-year investment tax credit of 12% be extended an additional two years to specifically speed the construction of power plants that do not use natural gas or oil. I am also submitting proposals for selective reform of state utility commission regulations.

To provide the critical stability for our domestic energy production in the face of world price uncertainty, I will request legislation to authorize and require tariffs, import quotas or price floors to protect our energy prices at levels which will achieve energy independence.

Increasing energy supplies is not enough. We must take additional steps to cut long-term consumption. I therefore propose to the Congress:

- Legislation to make thermal efficiency standards mandatory for all new buildings in the United States.
- A new tax credit of up to $150 for those homeowners who install insulation equipment.
- The establishment of an energy conservation program to help low-income families purchase insulation supplies.
- Legislation to modify and defer automotive pollution standards for five years which will enable us to improve new automobile gas mileage by 40% by 1980.

These proposals and actions, cumulatively, can reduce our dependence on foreign energy supplies to three to five million barrels per day by 1985.

To make the United States invulnerable to foreign disruption, I propose stand-by emergency legislation and a strategic storage program of one billion barrels of oil for domestic needs and 300 million barrels for national defense purposes.

I will ask for the funds needed for energy research and development activities. I have established a goal of one million barrels of synthetic fuels and shale oil production per day by 1985 together with an incentive program to achieve it.

I have a very deep belief in America's capabilities. Within the next 10 years, my program envisions: 200 major nuclear power plants; 250 major new coal mines; 150 major coal-fired power plants; 30 major new refineries; 20 major new synthetic fuel plants; the drilling of many thousands of new oilwells; the insulation of 18 million homes; and the manufacturing and the sale of millions of new automobiles, trucks and buses that use much less fuel.

I happen to believe that we can do it. In another crisis—the one in 1942—President Franklin D. Roosevelt said this country would build 60,000 military aircraft. By 1943, production in that program had reached 125,000 aircraft annually. They did it then. We can do it now. If the Congress and the American people will work with me to attain these targets, they will be achieved and will be surpassed.

Press conference. Ford amplified his views at a press conference Jan. 21.

Ford was asked about his statements on the use of military force in an extreme situation. He said that "on any occasion where there was any commitment of U.S. military personnel to any engagement, we would use the complete constitutional process that is required of the President."

In this context, he was also asked about the United Nations charter requirement that members refrain from the threat of the use of force. Stressing the hypothetical aspect of the question to which Secretary of State Henry Kissinger first responded—that the U.S. would use force in an extreme case, Ford said he fully endorsed Kissinger's response. The extreme case, he said, was "if a country is being strangled and I use strangled in the sense of the hypothetical question, that in effect means that a country has the right to protect itself against death."

"Would a new oil embargo be considered strangulation?" he was asked. "Certainly none comparable to the one in 1973," Ford responded.

Ford said he would veto a mandatory gasoline rationing program if it were passed by Congress. In his opening statement, Ford dismissed rationing, which was favored by many Democratic congressmen as a means of conserving fuel, as short term, "ineffective" and "inequitable." To reach the Administration's goal of reducing oil imports by 1 million barrels a day by 1975, Ford said a rationing system would have to limit "each driver to less than nine gallons a week."

"To really curb demand," Ford said, "we would have to embark on a long-range rationing program of more than five years." Moreover, he continued, "rationing provides no stimulus to increase domestic petroleum supply or accelerate alternative energy sources. By concentrating exclusively on gasoline rationing, many other areas for energy conservation are overlooked."

Ford challenged Congress to act quickly on his proposals and to pass them substantially intact. "I will not sit by and watch the nation continue to talk about an energy crisis and do nothing about it," he said. In announcing his own effort to take the "first step toward regaining our energy freedom," Ford said he would use executive authority to raise tariffs on imported oil as a temporary measure until Congress passed legislation setting higher fees.

This step, Ford said, would "set in motion the most important and far-reaching energy conservation program in our nation's history. We must reverse our increasing dependence on imported oil. It seriously threatens our national security and the very existence of our freedom and leadership in the free world."

Ford was asked why the Administration had not included penalties against automobile horsepower and weight among its energy and tax proposals. Responding to a charge by some critics that his was a "made in Detroit" plan designed to "rescue or revive the auto industry," Ford said auto taxes, gas rationing and closure of gas stations on Sunday had been rejected as "little pieces" of a plan. His program was "comprehensive" and "well integrated," Ford said.

Ford would veto forced rationing. President Ford embarked on the first of a series of speaking engagements to sell his plan of action to the public and defend it against Democratic opponents in Congress.

In a Washington address Jan. 22, Ford told the Conference Board, a nonprofit institute for business and economic research, he would veto any "mandatory rationing program" involving petroleum products.

White House Press Secretary Ron Nessen Jan. 22 cited Federal Energy Administration estimates that a rationing system would cost $2 billion annually to administer; 15,000-20,000 full time federal workers would be needed to implement the program; and 3,000 offices would be required at the state and local level. In the rush to obtain rationing coupons, Nessen claimed that the number of licensed drivers would rise from 12 million to 140 million, and gasoline purchased legally on the "white market" through the sale of coupons would cost up to $1.25 a gallon.

All of the figures were based on the assumption that gasoline rationing would be the only measure used to attain a 1 million barrel a day reduction in oil imports, a premise most experts considered unrealistic since Congress was also considering a combination of tax, rationing and allocation measures.

According to a Gallup poll published in Newsweek magazine Jan. 20, 55% of those responding favored a nationwide rationing program, compared with 32% who preferred to cut consumption through use of higher gasoline taxes.

Ford OKs higher oil tariffs, rebuffs bids for delay. President Ford signed a proclamation Jan. 23 that was intended to reduce the nation's consumption of foreign oil by making imports more costly. Tariffs on imported crude oil would rise $1 a month for three months: increasing $1 a barrel on Feb. 1, to $2 on March 1 and to $3 on April 1. (The current levy was 18¢ a barrel.)

Tariffs on finished petroleum products also would be increased, but by varying amounts. According to the White House, the $1 increase in duties would cause the cost of petroleum products to rise an average 1¢ a gallon by March. The ultimate effect of a $3 boost in tariffs would be an increase of about 3.5¢ a gallon, according to the Federal Energy Administration.

Ford defended his order as a "firm" but "fair" way to reduce national dependence on foreign oil. "Each day without action increases the threat to our national security and welfare," he said in a prepared statement.

"We've diddled and dawdled long enough," Ford said at an impromptu press conference after signing the executive order. It was a remark aimed at his critics in Congress, where opposition to higher tariffs was strong and widespread. Ford acknowledged that Congress could nullify the tariff increase by repealing the law which authorized his order, but he warned that such an actior would constitute a "backward step." Ford had made a similar charge during his press conference Jan. 21 when asked if he would consider delaying the tariff increase.

Governors of 10 Northeastern states that were heavily dependent on supplies of costly imported oil vehemently opposed his action. They met with Ford just before he signed the proclamation in an unsuccessful effort to dissuade him from authorizing the tariff increase. All of the governors disagreed with Ford's assessment of the inflationary impact of higher import fees. Ford claimed that a rise in tariff's would add only $250 to annual household fuel bills and that electric rates would climb by only 1.5% in New England. According to governors of the six New England states, higher tariffs alone would cost their area $890 million.

Gov. Milton J. Shapp (D, Penn.) said he "told the President I thought his program represented a blueprint for economic disaster, that there would be a shock wave of inflation through the country greater than the one we had when the Arabs lifted their embargo."

Gov. Michael Dukakis (D, Mass.) accused Ford of "holding the Northeast hostage for his program." "I don't think he liked the characterization, but he didn't object when the other governors used terms like 'a lever' or 'leverage' to get your program through [Congress]," Dukakis said.

Gov. Hugh Carey (D, N.Y.) agreed with Dukakis, charging that Ford's strategy was to "coerce Congress into action" on the rest of his energy proposals by imposing higher tariffs. Carey, a former New York congressman and member of the House Ways and Means Committee, claimed that in ordering higher tariffs, Ford "was acting outside the powers granted him by the Trade Expansion Act," which the White House cited as authority for the executive action. Congress "never intended" the legislation to be used for "unilateral actions of this kind," Carey said. "We've suffered in the past from an excessive use of presidential power," he declared.

Ford Jan. 22 had rebuffed a direct appeal for delay from Rep. Al Ullman (D, Ore.), chairman of the House Ways and Means Committee. Ullman asked that action setting higher tariffs be postponed temporarily because the order raised "serious legal questions." Ford's proclamation was issued under a section of the trade law permitting him to raise tariffs when the secretary of the Treasury concluded that certain imports threatened national security. Public hearings usually preceded such a determination, but Secretary William Simon waived them.

Ford defends tariff action—Ford again took the offensive when he appeared on a nationally televised interview Jan. 23 defending his order raising oil import fees.

Failure to act on the tariff question, Ford said, "would have been a sign of weakness around the world."

In remarks on other aspects of his energy proposals before Congress, Ford said the aim of the plan was to reduce the "percentage of reliance we have or will have on foreign oil." The U.S. currently imported 38% of its total crude oil supplies, according to Ford. Passage of his plan would cause imports to drop to about 10%, the President said.

If Congress rejected his energy program, Ford said, he would accept arbitrary allocation of petroleum products as an alternative approach to cut demand and limit foreign oil imports, but he restated his opposition to fuel rationing as a measure of "last resort."

Albert, Humphrey give Democratic reply. Two prominent Democrats in Congress replied to President Ford's plan for reviving the economy and reducing dependence on foreign oil; both criticized his energy saving proposals.

House Speaker Carl Albert (D, Okla.) Jan. 20 said that the Democrats favored a "more moderate approach" which combined gasoline rationing, an excise tax on automobiles, gasless days and other measures, he said.

CBS asked Sen. Hubert H. Humphrey (D, Minn.) to deliver a reply Jan. 22 to the President's program. Humphrey rejected Ford's energy proposals as "the least desirable set of alternatives," charging that they would cost consumers $45 billion a year.

Budget proposals. President Ford Feb. 3, 1975 submitted to Congress a $349.4 billion budget ($51.9 billion in deficit) for fiscal 1976. The budget was predicated on Ford's anti-recession and energy program.

The energy section requested legislation to permit commercial production of oil from military petroleum reserves. The revenue from sale of the oil would finance exploration of further federal reserves in Alaska and the cost of setting up a national strategic oil stockpile. Plans for the stockpile were to be submitted to Congress within a year.

A stepped-up program for development of oil and gas from federal offshore areas was planned. The leasing revenues, which totaled $6.7 billion in fiscal 1974, were estimated at $5 billion in the current fiscal year and $8 billion the next. The areas designated for development were off the East and California coasts and in the Gulf of Alaska.

A supplemental appropriation of $3 million was requested for fiscal 1975 to help the states prepare for offshore oil and gas development.

Federal spending for civilian energy research and development was projected at $2.1 billion in fiscal 1976, a 33% increase. Of the $1.6 billion requested for nuclear energy development, $261 million was for a liquid-metal fast breeder reactor.

Other energy research received considerable boosts in the new budget over previous levels of funding. Solar-energy projects were allotted $57 million, up 551%; geothermal projects $28.3 million, up 105%; coal projects, principally in liquefaction, $279 million, up 60%; oil shale projects $311 million, up 135%. Research in energy conservation was budgeted at $32.1 million, up 93%.

The budget also incorporated the Administration's plans to increase energy taxes by $30 billion annually and return these monies to the economy. In addition to the temporary increase in the federal fee on imported oil already put into effect, the Ford plans included an excise tax of $2 a barrel on domestic and imported oil; an excise tax of 37 cents a thousand cubic feet on natural gas; ending federal price controls on domestic oil on April 1; a windfall-profits tax on oil firms, and ending price regulation of newly discovered natural gas.

Economic Report. "The economy is in a severe recession," President Ford told Congress in the opening sentence of his annual Economic Report Feb. 4. Unemployment, inflation and interest rates were all too high, he said, and the nation faced an energy problem demanding conservation and development of new energy sources.

Ford advocated passage of his tax and energy proposals to help stabilize the economy.

Ford emphasized the need for Americans to adjust to higher pricing for energy products and to reduce U.S. dependence on unreliable sources of oil. He reiterated his stand that long-term energy rationing would be "both intolerable and ineffective," and he called attention to his proposal for a program of permanent tax cuts to compensate consumers for the coming higher costs of energy.

Excerpts from the Economic Report:

The economy is in a severe recession. Unemployment is too high and will rise higher. The rate of inflation is also too high although some progress has been made in lowering it. . . .

Moreover, even as we seek solutions to these problems, we must also seek solutions to our energy problem. We must embark upon effective programs to conserve energy and develop new sources if we are to reduce the proportion of our oil imported from unreliable sources. Failure or delay in this endeavor will mean a continued increase in this nation's dependence on foreign sources of oil.

We therefore confront three problems: the immediate problem of recession and unemployment, the continuing problem of inflation, and the newer problem of reducing America's vulnerability to oil embargoes. . . .

The achievement of our independence in energy will be neither quick nor easy. No matter what programs are adopted, perseverence by the American people and a willingness to accept inconvenience will be required in order to reach this important goal. The American economy was built on the basis of low-cost energy. The design of our industrial plants and production processes reflect this central element in the American experience. Cheap energy freed the architects of our office buildings from the need to plan for energy efficiency. It made private homes cheaper because expensive insulation was not required when energy was more abundant. Cheap energy also made suburban life accessible to more citizens , and it has given the mobility of the automobile to rural and city dwellers alike.

The low cost of energy during most of the twentieth century was made possible by abundant resources of domestic oil, natural gas and coal. This era has now come to an end. We have held the price of natural gas below the levels required to encourage investment in exploration and development of new supplies, and below the price which would have encouraged more careful use.

By taking advantage of relatively inexpensive foreign supplies of oil, we improved the quality of life for Americans and saved our own oil for future use. By neglecting to prepare for the possibility of import disruptions, however, we left ourselves overly dependent upon unreliable foreign supplies.

Present circumstances and the future security of the American economy leave no choice but to adjust to a higher relative price of energy products. We have, in fact, already begun to do so, although I emphasize that there is a long way to go. Consumers have already become more conscious of energy efficiency in their purchases.

The higher cost of energy has already induced industry to save energy by introducing new production techniques and by investing in energy-conserving capital equipment. These efforts must be stimulated and maintained until our consumption patterns and our industrial structure adjust to the new relationship between the costs of energy, labor, and capital.

This process of adjustment has been slowed because U.S. energy costs have not been allowed to increase at an appropriate rate. Prices of about two-thirds of our domestic crude oil are still being held at less than half the cost of imported oil, and natural gas prices are being held at even lower levels. Such artificially low prices encourage the wasteful use of energy and inhibit future production.

If there is no change in our pricing policy for domestic energy and in our consumption habits, by 1985 one-half of our oil will have to be imported, much of it from unreliable sources. Since our economy depends so heavily on energy, it is imperative that we make ourselves less vulnerable to supply cutoffs and the monopolistic pricing of some foreign oil producers.

The need for reliable energy supplies for our economy is the foundation of my proposed energy program. The principal purpose is to permit and encourage our economy to adjust its consumption of energy to the new realities of the market-place during the last part of the twentieth century.

The reduction in our dependence on unreliable sources of oil will require Government action, but even in this vital area the role of Government in economic life should be limited to those functions that it can perform better than the private sector.

There are two courses open to us in resolving our energy problem: the first is administered rationing and allocation; the second is use of the price mechanism. An energy rationing program might be acceptable for a brief period, but an effective program will require us to hold down consumption for an extended period.

A rationing program for a period of five years or more would be both intolerable and ineffective. The costs in slower decision-making alone would be enormous. Rationing would mean that every new company would have to petition the government for a license to purchase or sell fuel. It would mean that any new plant expansion or any new industrial process would require approval. It would mean similar restrictions on homebuilders, who already find it impossible in much of the nation to obtain natural gas hookups. After five or 10 years such a rigid program would surely sap the vitality of the American economy by substituting bureaucratic decisions for those of the marketplace. It would be impossible to devise a fair long-term rationing system.

The only practical and effective way to achieve energy independence, therefore, is by allowing prices of oil and gas to move higher—high enough to discourage consumption and encourage the exploration and development of new energy sources.

I have, therefore, recommended an excise tax on domestic crude oil and natural gas and an import fee on imported oil, as well as decontrol of the price of crude oil. These actions will raise the price of all energy-intensive products and reduce oil consumption and imports. I have requested the Congress to enact a tax on producers of domestic crude oil to prevent windfall profits as a result of price decontrol.

Other aspects of my program will provide assurances that imports will not be allowed to disrupt the domestic energy market. Amendments to the Clean Air Act to allow more use of coal without major environmental damage, and incentives to speed the development of nuclear energy and synthetic fuels will simultaneously increase domestic energy production.

Taken as a whole, the energy package will reduce the damage from any future import disruption to manageable proportions. The energy program, however, will entail costs. The import fee and tax combination will raise approximately $30 billion from energy consumers.

However, I have also proposed a fair and equitable program of permanent tax reductions to compensate consumers for these higher costs. These will include income tax reductions of $16 billion for individuals, along with direct rebates of $2 billion to low-income citizens who pay little or no taxes, corporate tax reductions of $6 billion, a $2 billion increase in revenue-sharing payments to state and local governments, and a $3 billion increase in Federal expenditures.

Although appropriate fiscal and energy policies are central to restoring the balance of our economy, they will be supplemented by initiatives in a number of other areas. . . .

We sometimes discover when we seek to accomplish several objectives simultaneously that the goals are not always completely compatible. Action to achieve one goal sometimes works to the detriment of another. I recognize that the $16 billion antirecession tax cut, which adds to an already large Federal deficit, might delay achieving price stability. But a prompt tax cut is essential. My program will raise the price of energy to consumers; but when completed this necessary adjustment should not hamper our progress toward the goal of a much slower rate of increase in the general price level in the years ahead. . . .

The proposals I have made to deal with the problems of recession, inflation, and energy recognize that the American economy is more and more a part of the world economy. What we do affects the economies of other nations, and what happens abroad affects our economy. Close communication, coordination of policies, and consultations with the leaders of other nations will be essential. . . .

Ford seeks support on tours. President Ford sought support for his anti-recession and energy program in a series of regional visits, speeches, press conferences and meetings with groups of governors Feb.

In Houston Feb. 10, Ford promised independent oil producers more tax relief to permit new exploration. In Topeka Feb. 11, he said that he thought there was "some justification" to amend his "windfall profits" tax proposal to grant exemption for profit plowed back into exploration and development of additional sources of oil. This would reduce the proposed $12 billion of additional tax by some $3–$4 billion, he said, which was not "a serious change" but one that would have to be offset elsewhere in his energy program.

Ford said a revision of the oil proposal had been advocated "by individuals both in and out of government" whom he saw in Houston and Topeka. The President was flanked at a Houston dinner Feb. 10 by Govs. Dolph Briscoe (D, Tex.) and David Boren (D, Okla.). Boren said next day he and Briscoe had pleaded at length for easing the Administration's proposed tax increases for the oil industry.

Court upholds tariff increase—A federal district judge Feb. 21 refused to issue a temporary injunction sought by nine states against the $3 a barrel tariff increase planned by President Ford.

The states, eight in the Northeastern area and Minnesota, argued that the levy would cause undue economic damage to their regions because they relied heavily on imported oil. Judge John Pratt rejected their arguments, contending that "our continued dependence on foreign oil threatens our national security, the economy, the posture of our defense, and the conduct of our foreign affairs."

The states, which also claimed that Ford's action was illegal because no public hearings had been held on the issue, had sought to block the increase until public hearings could be held and an environmental impact statement could be considered.

Northeastern states represented in the suit were Connecticut, New York, New Jersey, Massachusetts, Vermont, Pennsylvania, Rhode Island and Maine.

Democrats present rival plans. Congressional Democrats offered two widely divergent energy plans as alternatives to the program presented by President Ford. Two Senior Administration spokesmen, Treasury Secretary William E. Simon and Federal Energy Administrator Frank G. Zarb, indicated March 3 that the White House preferred the plan presented by the Democrats on the House Ways and Means Committee to the plan drawn up by the House and Senate leadership.

The committee proposals were drafted by eight task forces of the committee's Democrats and outlined March 2 by committee chairman Al Ullman (Ore.). Features of the Ullman plan:

- A steep increase in federal gasoline taxes to cut consumption. The tax would be raised 5c a gallon either on July 1, or Jan. 1, 1976. The tax would then rise 5c a gallon every six months thereafter until it reached 40c a gallon. However, half of the tax would be rebated through use of gasoline coupons and a refundable tax credit every year.
- In another effort to reduce consumption, taxes would be imposed on automobiles and other engine-powered vehicles with poor fuel economy. It was suggested that cars getting less than 17 miles to a gallon of gasoline be taxed up to $1,000 each.
- Import quotas would be phased in gradually in order to reduce U.S. reliance on foreign oil by 25% by 1985. A special allocation system would ease the burden on regions heavily dependent on imported fuel. A new federal agency would be created to purchase foreign oil.
- A windfall profits tax would be imposed on oil and natural gas, with the revenue earmarked for exploration and conservation of fuel supplies.
- Oil and gas prices would be decontrolled gradually over a five–eight year period.

The Democratic leadership's plan, which was released Feb. 27, called for a 5c a gallon tax on gasoline. The revenue would be used to finance studies of alternative energy sources. Purchasers of fuel-efficient cars would be offered tax rebates and excise taxes on heavy cars with poor gasoline economy would be sharply increased. Statutory standards would be written setting stiff automobile fuel economy. The national 55 mile-an-hour speed limit would be strictly enforced. Gasoline stations would be closed on Sundays.

If these conservation steps proved inadequate, import quotas on foreign oil would be imposed. According to the Democratic leadership, its plan would reduce fuel consumption by 500,000 barrels a day in the first year, half the saving expected from the Administration proposals.

Rep. James C. Wright (D, Tex.), chairman of the House task force that devised the plan, said the Democrats differed "very profoundly" with Ford's "concept that the only way to achieve [energy] conservation is to make the stuff so God-awful expensive that the consumer can't afford it." Democratic spokesmen also said that steps designed to revive the economy must take precedence over energy-saving measures.

Ford vetoes oil tariff bill; delays implementation. President Ford March 4 vetoed a bill passed by Congress suspending his power to raise the tariff on imported oil by $3 a barrel over a three-month period, but as a conciliatory gesture to Congress, Ford voluntarily delayed for 60 days implementation of the rate increase scheduled to take effect April 1 and rescinded the previous increase he had authorized by administrative action March 1. The $1-a-barrel levy increase which was implemented by executive action Feb. 1 remained in effect. (The vetoed bill, passed by 309–114 House vote Feb. 5 and 66–28 Senate vote Feb. 19, provided for a 90-day suspension of the President's tariff-raising power.)

The White House claimed to have enough votes in the Senate to sustain the veto. However, a test of the issue was avoided when Senate Democratic leaders decided to bury the veto message in the Senate Finance Committee while efforts were made to reach a compromise with the White House on a mutually acceptable energy program.

In another effort to mollify opponents of his energy program, Ford also said he would postpone plans to decontrol the price of "old" domestic oil* by administrative action April 1.

*Oil production from a given well for each month of 1972 formed the base level. "New" oil was the volume of domestic crude that exceeded the base level for any given month. "Old" oil was the base level minus an amount of released oil equal to new production from the well.

If a well produced 1,000 barrels of crude in March 1972, this was the base level. If, in March 1975, that same well was producing 1,100 barrels, new oil totaled 100 barrels, released oil totaled 100 barrels, and old oil totaled 900 barrels.

Released oil, which was not subject to federal price controls, provided an incentive that allowed producers to double the amount of oil production at decontrolled prices. The prices of new oil and oil from stripper wells producing less than 10 barrels a day also were uncontrolled. The price of old oil was set by Congress at $5.25 a barrel.

Ford May 1 again postponed a $1-a-barrel increase in the tax on imported oil, a move that was seen as a conciliatory gesture toward the Democratic-controlled Congress, which was working on alternative energy-saving approaches to the nation's fuel crisis.

At the same time, however, Ford also directed the Federal Energy Administration (FEA) to develop a plan for phasing out all remaining price controls on domestically-produced oil over a two year period, an action that could provoke a confrontation with the Congressional leadership, which opposed a fast-paced decontrol of U.S. oil.

In a compromise of its own, the House leadership decided to postpone a vote to override Ford's veto while his voluntary suspension of the tax increase remained in effect. The House voted March 11 to send the bill suspending the President's authority to raise imported oil tariffs to the Ways and Means Committee; however, the bill could be returned to the House floor at any time for a vote to override.

In announcing the White House action May 1, President Ford's spokesman, FEA Administrator Frank Zarb, said that another 30-day suspension of the May 1 increase would give the House an opportunity to act on energy legislation being drawn up by the Ways and Means Committee. If the House-passed bill met with White House approval and the Senate seemed likely to pass a similar measure, Zarb said, Ford might withdraw altogether his plan to impose the second and third tariff increases on imported oil.

Ford originally had proposed that all price controls on oil be lifted April 1. The new proposal of two years was seen as a concession to Congress. According to the plan outlined by Zarb, the Administration would free a portion of each oil producer's controlled oil at a rate of about 4% a month, allowing the price to rise slowly to world price levels, currently above $12 a barrel. Zarb contended that decontrol would raise gasoline prices an average 5¢ a gallon.

Depletion repeal. A tax-cut bill that also repealed the 22% oil-and-gas depletion allowance was passed by 287–125 House vote March 25, 1975 and by 45–16 Senate vote the same day. President Ford signed it reluctantly March 29.

The tax allowance on oil and gas output was repealed retroactive to Jan. 1, 1975. However, the allowance was retained until July 1, 1976 for natural gas sold under federal price regulations (or until the controlled price was raised to take account of repeal of depletion). The allowance also was retained for natural gas sold under fixed price contracts until the price was raised.

Small producers were provided a permanent exemption that allowed independent oil companies to continue taking the depletion allowance on a basic daily output of oil and natural gas, averaging 2,000 barrels of oil or 12 million cubic feet of natural gas or an equivalent quantity of both oil and gas.

The daily production eligible for depletion would be reduced by 200 barrels a day for each year between 1976 and 1980, leaving the small-producer exemption at a permanent level of 1,000 barrels of oil per day or 6 million cubic feet of natural gas.

The depletion rate available for the small producers would be reduced to 20% in 1981, 18% in 1982, 16% in 1983, and to a permanent 15% rate in 1984.

The 22% depletion rate would be kept until 1984 for production of up to 1,000 barrels a day through costly secondary or tertiary recovery methods used to extract remaining oil and gas from wells that were mostly pumped out.

The deduction taken under the small-producer exemption would be limited to 65% of the taxpayer's income from all sources. The small-producer exemption would be denied to any taxpayer who sold oil or gas through retail outlets or operated a refinery processing more than 50,000 barrels of oil a day.

Under the measure, the amount of foreign tax payments on oil-related income that an oil company could take as a credit against U.S. taxes was limited to 52.8% of its 1975 income from foreign oil operations. The limit would be reduced to 50.4% in 1976 and 50% thereafter. Use of excess credits within those limits was allowed only to offset U.S. taxes on foreign oil-related income, not on income from other foreign sources.

After 1975, oil companies were denied the use of the per-country limitation option that allowed a company to compute its maximum foreign tax credits on a country-by-country basis.

It was required that oil companies recapture foreign oil-related losses that were deducted from income subject to U.S. taxes by taxing an equivalent amount of subsequent foreign oil-related profits as if earned in the U.S. (and therefore not eligible for deferral until transferred to the U.S.). The credit for foreign taxes on the subsequent profits also would be reduced in proportion to the amount treated as U.S. profits.

The foreign tax credit was denied for any taxes paid to a foreign country in buying or selling oil or gas from property that the nation had expropriated. Domestic international sales corporations (DISCs) were denied the deferral of taxes on half of the profits from exports of natural resources and energy products. Effective in 1976, certain existing exemptions were repealed from a 1962 law requiring current U.S. taxation of profits earned by subsidiaries set up by a U.S. corporation in tax haven countries that imposed little or no taxes. Allowance was made for deferral of U.S. taxes on all earnings by a foreign subsidiary if less than 10% of its income was defined as tax haven income.

Tariff raised $1 a barrel. In a nationwide TV broadcast May 27, 1975, President Ford said he would impose a second $1-a-barrel increase in the tariff on imported oil and set a 60¢-a-barrel fee on imports of refined petroleum products, effective June 1. Ford said he also would send Congress a plan for phasing out price controls on old oil—oil produced domestically at pre-1972 production levels.

Ford said he acted because, since February, Congress "has done nothing positive to end our energy dependence." He cited Congressional approval of a bill suspending for 90 days the President's power to increase imported oil tariffs, legislation that was passed in reaction to the first $1-a-barrel tariff increase set by Ford in February.

In addition to its "attempt to prevent the President from doing anything on his own" regarding energy, Ford said, Congress passed an "anti-energy" bill regulating strip mining. That bill, he said, "would reduce domestic coal production instead of increasing it; put thousands of people out of work; needlessly increase the cost of domestic coal production instead of increasing it; put thousands of people out of work; needlessly increase the cost of energy to consumers; raise electric bills for many; and compel us to import more foreign oil, not less."

Ford also accused Congress of doing "little or nothing to stimulate production of new energy sources," such as those at the Elk Hills (Calif.) naval petroleum reserve and reserves in Alaska and offshore under the continental shelf.

The U.S. "could save 30,000 barrels a day if only the Congress would allow more electric power plants to substitute American coal for foreign oil. Peaceful atomic power, which we pioneered, is advancing faster abroad than at home," Ford said.

In discussing the gravity of the energy crisis facing the country, the President emphasized the direct link between energy and unemployment. "Our American economy runs on energy," he said. "No energy, no jobs. It's as simple as that."

The "sudden fourfold increase in foreign oil prices and the 1973 embargo helped to throw us into this recession," Ford said, warning that another embargo could result in another recession. (He also noted that in five years, the nation's fuel bill had skyrocketed from $5 billion to $25 billion annually.)

While Congress continued to "drift, dawdle, and debate" the energy crisis, Ford said, the nation's dependence on foreign oil and its vulnerability to pressure from oil-producing states increased. "In 10 years, if we do nothing," he said, "we will be importing more than half our oil, at prices fixed by others, if they choose to sell to us at all." (The U.S. currently imported 37% of its oil supplies, which totaled an estimated 17 million barrels a day.)

"In two and a half years, we will be twice as vulnerable to a foreign oil embargo as we were two winters ago. Domestic oil production is going down; natural gas production is starting to dwindle and many areas face severe shortages next winter; coal production is still at the level of the 1940s," Ford said.

"This country needs to regain its independence from foreign sources of energy and the sooner the better," Ford declared.

The President's tariff program was designed to discourage energy consumption by making fuel more costly. According to White House estimates, higher import fees would result in a saving of 100,000 barrels a day by the end of 1975 and 350,000 barrels a day by the end of 1977. The additional taxes on imported oil were expected to cost the consumer a total of 2.9¢ per gallon at retail—1.5¢ of that from the import levies announced May 27. (Oil industry sources said the cost of the latest tariff increase could total 2.5¢ a gallon.)

According to the White House, decontrol of old oil prices would save 518,000 barrels a day by the end of 1977 and cost the consumer an extra 5¢ a gallon. (60% of all oil produced domestically was subject to federal price controls, which set a ceiling of $5.25 a barrel).

Ford also said he would ask Congress to enact a windfall profits tax with a plowback provision so that oil producers would use the excess revenues generated by higher prices in the exploration and development of new energy supplies.

The oil industry generally applauded Ford's move. Congressional Democrats, however, were critical of the Administration program and rejected his "do-nothing" charges.

In a rebuttal to Ford's speech, broadcast by CBS immediately following the President's address, Rep. Al Ullman (D, Ore.), chairman of the House Ways and Means Committee, said Ford's action would have "disastrous effects" on inflation and might cause the recession to worsen. Ford was not "going to add any new jobs by raising prices," Ullman declared.

Sen. Henry M. Jackson (D, Wash.), chairman of the Senate Interior Committee, charged that Ford "proposes to do to us in 1975 what the Arab oil cartel did in 1974" in raising prices and spurring inflation.

U.S. plans coal-burning utilities. The Federal Energy Administration (FEA) sent letters of intent May 6 and May 17 to a total of 13 public utilities advising them that they might be ordered to substitute coal for oil or natural gas as burner fuel. Authority to require the changeover, expected to affect 79 public utilities, stemmed from the Energy Supply and Environmental Coordination Act of 1974, which took effect June 22 of that year.

In a May 9 interview cited in the Washington Post the following day, EPA Administrator Russell E. Train emphasized the government's concern for the environmental aspects of the changeover plan. "Our statute prohibits any fuel conversions that result in any change to health standards. If we determine there is any risk to public health then we can pull the plug on this. In fact, we're directed to do so by law."

Long-range energy plan. The Energy Research and Development Administration June 30, 1975 a program for energy development that envisioned gradual transition from oil and natural gas as the primary sources of energy for the nation. Unless there was development of major new sources of domestic supply, the country faced continued dependence on imported oil, the report said, and dependence on imports made the U.S. "vulnerable to undesirable external influence on U.S. foreign and domestic policy."

Coal and uranium, or atomic power, were given top priority in the report for reducing the nation's dependence on oil and gas, although development of "a large number of technologies" was seen as essential to reduce oil imports to a low level or to nothing.

The report urged rapid development of "enhanced recovery" from U.S. oil fields, which in the past normally were abandoned with only one-third of the oil recovered. "Enhanced recovery will buy roughly 10 years of time," the report said, to develop new energy sources such as synthetic fuels, oil from shale and oil and gas from coal.

Conservation also was stressed in the report, which cited target areas—transportation, industrial processes, commercial buildings, houses and conversion of waste materials to energy.

The hoped-for transition would have oil and gas, currently comprising 75% of U.S. energy supplies, reduced to a 30% share by the end of the century. The goal for coal was to increase its share from 20% to the 30% level by the year 2000. For uranium, its 2% share, it was hoped, would rise to 30%.

In the more distant future, the report looked to the sun, the nuclear breeder reactor and nuclear fusion as vast energy sources, with solar electricity allotted equal ranking with the others.

Prices & Profits

Electric costs. Two Democratic Senators charged March 24, 1975 that automatic price increases granted to electric and gas utilities because of increased fuel prices cost consumers about $6.5 billion in 1974. The fuel adjustment increases represented the bulk of the total $9.6 billion rise in utility bills over the year, according to Sens. Edmund S. Muskie (Me.) and Lee Metcalf (Mont.).

The total increase to consumers was 1½ times the size of rate increases granted from 1948 through 1973. The survey, which was based on data from 37 state public service commissions and the District of Columbia, also contrasted the sharp rise in utility bills with an increase in electric power consumption of less than 1% during 1974 and a 4.2% decline in natural gas consumption. Automatic adjustments for higher fuel costs removed most of the incentives for efficient and cost control operations, the Senators charged. (Twenty-nine states required no hearings before added fuel costs were passed on to consumers.)

Electricity rates of the nation's 15 largest electric utilities soared 61.3% between June 1973 and December 1974, according to a private utility consulting firm March 24. If these trends continued, National Utility Service Inc. said, U.S. utility rates would be the highest in the world by the end of the decade. Over the 18-month period, only two countries recorded higher rate increases. In Belgium, utilities bills rose 85.4% and in Italy, charges were up 83.3%.

Within the U.S., electric rates charged by Consolidated Edison of New York were the highest in the nation—an average 5.93¢ per killowatt hour. Jersey Central Power and Light Co. was granted the largest rate increase over the 18-month period—105.6%.

The report called for a redesign of rate structures that would provide consumers with an incentive for savings. Under current rate structures, consumers paid a base price for a certain amount of electricity used and rates were reduced when consumption exceeded that amount. The report stated: "Load factors and time when energy is used must be given more prominent consideration in rate structures. Just as the utility industry requires incentives to invest in plants and equipment to meet future demand, so does the consumer need a stimulus to help maximize the utilization of utility plant facilities and control unnecessary expansion."

The Federal Energy Administration announced March 24 that the agency would fund several experimental rate structures in an effort to redesign the nation's electricity pricing policy. Under the new plan, consumers using electricity during peak load hours would be charged premium rates. Reduced rates would be in effect during off peak times.

Total electricity usage and peak demand had grown at a rate of 7% a year throughout the postwar period, according to the FEA. The aim of the experiment was to reduce the total growth rate to 5% and cut peak demand growth to 4% by encouraging more efficient use of electricity. (During off-peak hours, the FEA noted, utilities operated at an average 49% of capacity.)

Overcharges to utilities probed. The Federal Energy Administration (FEA) disclosed April 1 that fuel suppliers may have overcharged 72 of the nation's electrical utilities by as much as $19.7 million during the Arab oil embargo. Since utilities were regulated and allowed to pass on to consumers increases in fuel prices, the alleged price gouging eventually affected homeowners and business persons.

As a result of the FEA's continuing investigation, two suppliers signed consent agreements March 12 to repay overcharges: the Bessemer Oil Co. agreed to restore $478,803.60 to the Alabama Power Co., the Mississippi Power Co., and the Gulf Power Co.; and Accent Petroleum Co. of Colorado agreed to repay $14,113 to the Colorado Springs Department of Public Utilities.

The FEA also was negotiating for repayment of $710,000 to the Georgia Power Co. by an unidentified supplier.

The FEA investigation, which was being conducted in cooperation with the Customs Bureau and the Justice Department, involved charges that suppliers misrepresented relatively low-priced domestic fuel as higher-priced imported fuel in sales to utilities, that unnecessary handling of oil was used to increase prices or that paper sales of fuel were used to raise prices. The price-gouging probe, dubbed Project Escalator, involved investigations of more than 200 of the 407 suppliers of fuel oil to utilities. Similar investigation also had been launched by the FEA to determine whether prices of butane, propane and domestic crude oil had been illegally increased

Utilities' profits, rates increase. The earnings of electric utility companies during 1975's first quarter were 13%–15% higher than in the same period of 1974, the Edison Electric Institute, a trade association for investor-owned utilities, announced June 1.

Utilities were in better financial shape than a year earlier, the institute, said, because of rate increases granted by state regulatory agencies, a decline in interest rates on short-term loans and generally easier money conditions, a slowdown in inflation, higher sales, and stabilized fuel costs.

According to the trade association, rate increases totaling $2.2 billion were authorized during 1974 by state officials—a 105% rise from 1973. (Consumers' costs rose about 25% on national average as a result of the rate hike, the group added.) In a later report June 2, the institute said that $1 billion in electric rate increases were granted during the first three months of 1975.

Analysts also pointed to an easing of the utilities' financing problems. During the tight money conditions of 1974, utilities were forced to seek high-cost short-term loans to meet their capital needs. According to the New York Times June 2, utilities borrowed $6.5 billion during 1974 at rates from 10%–12%. Short-term rates currently were about 7%.

Sales also appeared to have rebounded after the cutback in customer demand following the Arab oil embargo. Sales during the first four months of 1975 were 2% higher than in the same period of 1974. Uncertainties clouded the growth picture, however. Utilities' spokesmen said that until the government devised an explicit energy policy providing for reduced consumption of foreign fuel and increased production of domestic supplies, utilities would experience difficulties meeting legal requirements that they provide service to customers upon demand.

New FEA price rules set. The Federal Energy Administration (FEA) announced Jan. 3, 1975 that with the beginning of the current year it was adopting new price regulations for propane, butane and natural gasoline in order to keep the price of these fuels "as low as possible without affecting dwindling supplies." (Natural gasoline was a liquid extracted from natural gas and blended with products refined from crude oil to produce motor gasoline.)

Among the highlights of the new regulations, as reported by the FEA, was retention of the May 15, 1973 basic price level but with the added stipulation that "sellers use for the purpose of computing current lawful prices" a price of 8.5¢ per gallon for propane, 9¢ for butane and 10¢ for natural gasoline. (After an FEA-authorized price increase for propane in September 1974, under a system which allowed sellers to add production-related costs to the basic price, the gas had been selling for an average price of 14¢–15¢ a gallon.) The new rules would allow "natural gas liquid processors" to add .5¢ a gallon to their prices "to reflect increased non-product costs" including "rents, labor costs, utility rates, and interest charges" incurred "as part of the processing operations." Natural gas "consumed in producing natural gas liquids" could be added to the 1973 basic price. Refiners, gas processors and resellers could apply increased propane costs selectively by charging, for example, a higher price to industrial users than to homeowners. According to the FEA news release, the previous rule had required suppliers to apply increases in the base price for propane "to all classes of purchasers by precisely the same amounts." As "an incentive to construct new facilities," a new rule permitted unspecified higher prices for natural gas liquids (natural gasoline) extracted in gas processing facilities built after January 1975.

Propane overcharge found—An investigation conducted by the FEA showed that rural users of bottled propane gas had been overcharged by about $80 million during the winter of the 1973 Arab oil embargo, it was reported March 17.

An FEA spokesman said the investigation, code-named Project Speculator, had gone on for about nine months and would continue and might lead to the disclosure of overcharges of as much as $200 million. He said the FEA had threatened legal action to force propane suppliers to roll back prices by nearly $50 million.

The Washington Post March 18 said Project Speculator was generating charges that propane suppliers in farming states of the Midwest had set up dummy corporations among themselves to pass along overcharges to consumers. The newspaper quoted an unnamed FEA source as saying there had been "an inordinate amount of monkeying around with propane prices where suppliers would literally gather around a table, selling bottled gas to each other with everybody taking his historic profit margin." The Post reported that propane was currently selling for 26¢–30¢ a gallon.

U.S. oil pricing held illegal. The Temporary Emergency Court of Appeals in Washington ruled Feb. 19, 1975, by 2–1, that the government had acted illegally when it decontrolled "new" crude oil prices in January 1974.

The principal finding of the court was that the Emergency Petroleum Allocation Act of 1973 had required the FEA to set prices for all types of crude oil and that the FEA was therefore in violation of this act when in January 1974 it allowed "new" oil to sell at prices determined by the market.

According to the majority opinion, "While the [Allocation] Act directs that the prices for all categories of crude oil be regulated, it does not specify a method for doing so. Consequently, the FEA has discretion in devising a regulatory scheme, but cannot adopt measures which contravene a statutory mandate.

"It isn't the function of this court to determine what the equitable price is, or should be," the decision said, "We merely hold that the President, through the FEA, by permitting the price of new crude oil to float at free market levels, hasn't struck any balance and, as a result, has failed to satisfy the requirement that prices be set at an equitable level."

A nation of 'gasoline consumers.' The Labor Department issued a preliminary report April 16 of a two-year survey of consumer spending and earnings. The study confirmed that the U.S. was a nation of "beef-eaters and gasoline consumers," according to Julius Shiskin, commissioner of labor statistics.

Gasoline expenditures accounted for 50% of a family's direct energy purchases; the average family spent $30.32 for its entire weekly food budget and $6.36 on gasoline per week.

MAJOR OIL COMPANIES' NET INCOMES

(Dollars in millions, after tax)

Company	1973	1974	Percent increase
Exxon	$2,443.0	$3,140.0	28.5
Texaco	1,292.4	1,588.4	22.8
Mobil	849.3	1,040.1	22.5
Standard, California	843.6	970.0	15.0
Standard, Indiana	511.2	970.3	89.8
Shell	332.7	620.5	86.5
Phillips	230.4	402.1	74.5
Union	180.2	188.0	59.8
Standard, Ohio	74.1	147.5	99.1
Gulf	800.0	1,065.0	33.1
Sun	230.0	378.0	64.3
Continental	242.7	327.6	35.0
Amerada Hess	245.8	201.9	(−)17.9
Arco	270.2	474.6	75.6
Getty	142.2	281.0	97.6
Occidental	71.9	280.7	290.4

The data were gathered from diaries kept by more than 10,000 families for two years beginning in mid-1972.

Fortune ranks 500 largest industrials. "Oil, inflation and recession had—not surprisingly—a lot to do with the record turned in by the 500 largest industrials in 1974," Fortune magazine reported in its May 1975 issue.

The performance of the petroleum-refining industry dominated the list, just as higher oil prices had a dominant impact on the national economy during 1974. Exxon Corp., the group's standard-bearer, was ranked No. 1 in sales, surpassing General Motors Corp., the nation's sales leader for the previous 40 years. Exxon also was the 1974 leader in assets, net income and stockholders' equity.

Despite the huge profits reported by oil companies in 1974, the petroleum refining industry ranked only fourth in terms of an annual earnings increase. The group's profits for 1974 were only 39.6% higher than in 1973, compared with a record 100.8% median increase in the mining industry's earnings. Metal manufacturers ranked second with a 78.6% median increase in profits and the chemicals group posted a 40.5% median gain.

Profits for the 500 increased 12.8%, but non-oil companies showed only 4.1% gain.

As a group, the petroleum refining industry reported an 80.4% median increase in sales during 1974—the largest gain registered by any industry since records were first compiled in 1955. Five of the seven largest companies on the list were oil companies. Among the top 50 industrials, 16 were oil companies; three of them, Sun Oil Co., Amerada Hess Corp., and Ashland Oil Corp., were new to the top-50 group, but each of the 16 oil companies moved up in rank.

Higher oil prices had a double effect on the sales ranking, Fortune noted, swelling the oil firms' sales while depressing sales in other key industries, chiefly the automotive group.

Offshore Oil

Offshore oil leases. The Interior Department announced Jan. 11, 1974 that it had accepted bids totaling almost $1.5 billion for drilling rights on 485,000 acres in the Gulf of Mexico off Florida, Alabama and Mississippi.

The bids included the highest offer in history for an offshore tract: $211,977,600 for a 5,760 acre tract off Florida bid by a consortium of Exxon Corp., Mobil Oil Corp. and Champlin Petroleum Co.

The department had opened the bidding over the protests of environmentalist groups and the State of Florida, who contended that well blowouts or tanker spills would threaten the numerous wildlife refuges along the coastline, as well as the area's extensive commercial fishing industry.

An environmental impact statement prepared by the department's Bureau of Land Management said shipment by tanker of some of the area's production would be necessary because of the distance from refineries. Any major spill, the statement said, could endanger some animal species already considered near extinction.

Under terms of the leasing, companies would be required to provide equipment to contain and remove pollution, and drilling platforms would have to be erected in locations that would least interfere with fishing boats.

The department April 9 announced its acceptance of high bids totaling $2.1 billion for drilling rights on 91 tracts in the Gulf of Mexico off Louisiana. High bids of $82.5 million for 23 other tracts were rejected as too low.

Total bids of $6.46 billion for 520,275 acres—a record for a single lease sale—had been submitted March 28. The $2.2 billion in high bids was also a record. Exxon Corp., bidding alone, submitted high bids of $247 million for six tracts. Most other bids were entered by consortiums.

The department Nov. 18 accepted high royalty bids for the development of eight tracts of offshore oil and gas leases in the Gulf of Mexico. It was the first time that tracts had been offered under the royalty bidding system, in which the winner agreed to pay a percentage of total gross value of the oil and gas extracted from the wells.

Under the bonus system that was currently in use, winners paid a fixed production royalty of 16.67% plus an advance sum for exploration rights.

The bidding change, which involved making a smaller initial capital outlay, was made to provide smaller companies with an incentive to seek federal leases, according to officials. The highest bid accepted was an 82.165% royalty plus a cash bonus of $125,000.

California bids sought—The Interior Department said Jan. 2 that it had asked oil companies to propose tracts for possible federal leasing in 1975 in a 7.7-million acre area off Southern California. The Santa Barbara channel, scene of a well blowout in 1969, was excluded. A federal ban on drilling on 35 tracts already under lease in the channel remained in effect.

The State of California had lifted its ban on operations in state-controlled offshore areas Dec. 11, 1973. The state emphasized that the action applied only to tracts already under lease, and would not involve new drilling platforms or the granting of new leases.

Five oil companies involved in the Santa Barbara blowout agreed July 23 to pay almost $9.5 million in actual damages to a group of plaintiffs.

The payments included $4.5 million to the state, $4 million to Santa Barbara, $775,000 to Santa Barbara County and $200,000 to the city of Carpenteria. The companies were Union Oil Co. of California, Mobil Oil Corp., Gulf Oil Corp., Texaco Inc. and Peter Bawden Drilling Co.

The settlement did not include punitive damages sought by the plaintiffs.

Caution on offshore development urged. The Council on Environmental Quality April 18, 1974 indirectly approved new development of offshore oil and gas reserves in parts of the Atlantic region, but declined to make direct recommendations on drilling in environmentally vulnerable areas.

Instead of specific recommendations, the final version of the council's report to President Nixon assessed the relative environmental dangers to various areas under consideration and stated that federal leasing could proceed "with caution" under carefully controlled conditions.

Draft versions of some sections of the report, leaked by various sources earlier in the month, recommended specific sites for possible drilling in the Gulf of Alaska, in the Georges Bank area off New England, off the mid-Atlantic states and in the Atlantic region off South Carolina and northern Florida.

Council Chairman Russell W. Peterson referred to the less-specific final version as a clarification of the council's true intentions, and said he hoped the government's leasing policies would give first priority to areas with the lowest environmental risks.

The council's rankings were based on the relative danger of oil from spills and blowout leaks reaching shore. The risks, the report stated, would be related to seismic conditions, storm dangers and tidal conditions. Under those criteria, the greatest dangers were posed in the Gulf of Alaska, the "Baltimore Canyon trough" off Long Island and in the South Carolina-northern Florida region. Drilling would be more feasible, the report suggested, off New England and the Maryland-Delaware-New Jersey area.

Offshore leasing plan detailed. John C. Whitaker, undersecretary of the interior, ordered preparation of a "firm plan" setting up leasing schedules for the federal government to sell offshore oil drilling rights to areas in the Atlantic Ocean and Alaskan waters, according to an internal department memo published by the Washington Post Oct. 4, 1974.

Whitaker called for leasing a total of 10 million acres in 1975, "not just 10 million acres offered" for sale; "a sale in '75 in both Alaska and the Atlantic"; and "a lease sale in the Atlantic in 1975 that would include some leases in all of the promising areas of the Atlantic, rather than just the sale limited to the Baltimore Canyon" [a trench in the Atlantic floor extending along the New Jersey and Delaware coasts about 60 miles offshore].

Whitaker confirmed the authenticity of the memo and said that leases to as many as 13 million offshore acres would have to be offered for sale in order to meet his goal of leasing 10 million acres.

Environmentalists and state officials denounced the government policy. A spokesman for the Environmental Policy Center charged that the department had made the decision to lease offshore rights without waiting for the results of state studies to determine the environmental impact of such drilling.

Thirteen Atlantic coastal states had filed suits claiming that they already controlled the Outer Continental Shelf beyond the three mile boundary as part of their territorial grants from the king of England. States currently derived no direct benefit from offshore drilling—revenue from the sale of leases went directly to the U.S. Treasury.

In a letter to President Ford Oct. 7, 20 senators criticized the department plan as "hasty and ill-conceived." They asked the Administration to revise its leasing schedules until results could be obtained from the states' impact studies.

According to a draft of the Interior Department's environmental impact study published Oct. 24, "past performance" indicated that "sooner or later a major spill will occur wherever there is significant development and production in potential areas. We are certain that thousands of minor spills will occur."

The forecasts were based on data from spills in the Gulf of Mexico between 1954 and 1974. In the 20-year period, the department said, there had been 44 oil pollution incidents of greater than 50 barrels of oil. A total of 320,000 barrels was spilled—more than half in one 1967 incident.

Based on the Gulf production and spill rate, the department calculated a spill rate for other offshore areas of up to 8,500 barrels a year.

In another effect on the environment, the study said that 8,000 acres of sea floor would be removed from fishing activities with the construction and operation of 1,600 drilling platforms on the Outer Continental Shelf.

Whitaker told a Congressional subcommittee Oct. 8 that the government planned to offer 19.1 million acres, including 3.5 million acres off New Jersey and Delaware, for offshore drilling bids by oil companies.

Interior Secretary Rogers C. B. Morton, chairman of the Energy Resources Council, Oct. 9 confirmed that stepped up offshore leasing was planned for 1975. "We will without question sell 10 million acres next year against 3 million this year, which was up three times from what we historically made available in the past," Morton said.

Ford seeks support for leasing plan— President Ford met with 23 governors of coastal states Nov. 13 to seek support for the government's accelerated leasing schedule. Citing the lack of domestic petroleum supplies as one of the nation's chief concerns, Ford said, "I believe that the outer continental shelf oil and gas deposits can provide the largest single source of increased domestic energy during the years when we need it most."

Ford sought to allay the governors' fears about possible environmental damage to the coastline. "The greatest danger to our coasts from oil spills is not from offshore production but instead from greatly expanded tanker traffic that would result from increasing imports," he said.

One of Ford's few supporters in the group was Louisiana Gov. Edwin W. Edwards (D), who said offshore drilling was "essential."

Leasing program curtailed–Morton announced Nov. 25 that the government would scale down plans to lease oil drilling rights to 10 million offshore acres as opposition to the project mounted from officials in coastal states. Morton also said that hearings on a draft environmental impact statement related to the leasing project would be postponed.

Morton made the disclosures in a meeting with Sens. John V. Tunney and Alan Cranston, both California Democrats, Burt Pines, attorney for the city of Los Angeles, and Richard Maullin, energy adviser to California's Gov.-elect Edmund G. Brown Jr. Pines had said Nov. 22 that if the Administration refused to order a delay in the hearings, he would seek a restraining order to allow state officials sufficient time to study the government's draft environmental impact statement. Distribution of the 1,300-word paper had not begun until Oct. 31 and some state officials had not yet received copies, according to Pines. Massachusetts, Oregon and Washington were prepared to join in such a suit, he added.

Tunney criticized the Administration's plan to lease the acreage under a "bonus bidding" system, which would permit the government to sell the leases "without knowledge of their worth." Under a "royalty bidding" system, the government would be entitled to a fair price for any oil discovered in the area.

A Senate study group also was critical of bonus bidding. In a report issued Nov. 26, the National Ocean Policy Study, a group within the Commerce Committee, recommended that the government undertake exploratory drilling to determine the probable worth of the offshore areas.

"The precipitous leasing of so many acres in a single year [could] cause coastal states and communities environmental harm and create administrative, financial, and technical problems beyond the capability of the federal government and the oil industry to handle," the report concluded.

Sen. Ernest Hollings (D, S.C.), chairman of the study group, charged Nov. 26 that the Interior Department's leasing program was a "fraud." "It will produce no energy in the next few years when we need it. It will only divert our attention from the sources that already have been found and that could be developed immediately," Hollings said.

Among the sources available for immediate production were "shut-ins," wells that had been drilled and then closed off. A shut-in oil producible zone was one capable of producing oil or gas in paying quantities, according to the government.

Hollings charged that oil companies were engaging in "speculative withholding" of oil from these shut-ins in order to maximize profits through a reduction in output. He cited a government study showing that the number of shut-in oil producible zones in offshore areas had increased from 853 in 1971 to 3,054 in 1973. During the same period, the number of active wells had declined from 5,704 to 3,814.

4 states to study offshore drilling. It was announced Dec. 4 that the governors of Maryland, New Jersey, Virginia and Delaware had formed the Atlantic Governors'

Coastal Resources Council to deal with common problems brought about by federal plans to lease offshore seabottom for oil and natural gas drilling.

U.S. wins Atlantic shelf title. The Supreme Court March 17, 1975 unanimously ruled that the federal government, not the individual states along the Atlantic Coast, had the exclusive right to exploit oil and natural gas reserves beneath the continental shelf seabed beyond the three-mile territorial limit. Ownership of the land rested with the U.S. government, the court said, as part of its jurisdiction over foreign commerce, foreign affairs and national defense.

The Atlantic coastal states had argued that their original royal charters had given them title to seabed resources out to 100 miles; that they had not surrendered ownership of the land when they entered the union; and Supreme Court decisions of 1947 and 1950 that conferred title on the federal government were in error.

The court's opinion, written by Justice Byron R. White, affirmed the recommendations of a court-appointed special master, who ruled that the high court was bound by its rulings of 1947 and 1950, which upheld federal claims to land off the California and Gulf coasts. Congress had modified the rulings in 1953, however, passing the Submerged Lands Act, which gave the states ownership of offshore resources out to the three-mile limit, or in the case of certain Gulf coast states out to a nine-mile limit. Beyond those limits, the act reserved title to the resources for the federal government. The Outer Continental Shelf Lands Act, also passed in 1953, authorized the government to lease tracts for exploitation.

The current Supreme Court ruling stemmed from a 1969 Justice Department suit against the State of Maine, which had begun preparations to lease 3.3 million offshore acres for private exploitation. Subsequently, 11 other states joined Maine in the suit—New Hampshire, Massachusetts, Rhode Island, New York, New Jersey, Delaware, Maryland, Virginia, North Carolina, South Carolina, and Georgia.

Energy Supply & Development

Energy reserves seen down. The U.S. Geological Survey, part of the department of the interior, published estimates of the country's energy reserves, cited in the New York Times May 8, 1975, which it called "considerably lower" than those it had released in March 1974.

The agency's study, directed by Vincent E. McKelvey, placed the "undiscovered recoverable resources" of crude oil at 50–127 billion barrels, natural gas liquids reserves (natural gasoline) at 11–22 billion barrels and natural gas at 322–655 trillion cubic feet. The March 1974 figures, criticized as being too optimistic, had combined estimates for crude oil and natural gas liquids at 200–400 billion barrels and had pegged natural gas reserves at 2000–4000 trillion cubic feet. Interior Secretary Rogers Morton, chairman of the Energy Resources Council, said the new Geological Survey figures were "sobering additional evidence of our need" for "a comprehensive energy program."

(In another study financed partly by the department of interior, the National Research Council of the National Academy of Sciences reported Feb. 11 that the U.S. would run out of oil and natural gas in 25 years. The council's study, which had taken nearly three years to complete, put crude oil reserves at 113 billion barrels and natural gas reserves at no more than 600 trillion cubic feet.)

In related developments, the American Petroleum Institute reported April 1 that despite a 35% increase in crude oil prices in 1974, proven reserves had declined in that year from 35.3 billion barrels to 34.25 billion barrels. It was reported by the American Gas Association the same day that proven reserves of natural gas had dropped in 1974 by about 5%, from 250 trillion to 237.1 trillion cubic feet.

A Feb. 28 FEA news release said domestic crude oil production in 1974 had declined 4.5% below the level for the previous year and that petroleum demand was off 3.8%. FEA Administrator Frank G. Zarb said the figures showed that "1974 domestic energy price increases have resulted in significant energy savings without causing painful shortages."

It was reported March 24 by the Oil and Gas Journal that U.S. crude oil refining capacity had reached 14.8 million barrels a day at the beginning of the year, an increase of 4.4% over the figure for 1974.

Oil shale, geothermal tracts leased. In two separate but related actions, the Interior Department moved toward the long-term development of federal energy resources in western states in early 1974.

The department announced Jan. 17 that it had accepted a top bid of $210.3 million for development of a tract of oil shale land in Colorado. The bid had been placed Jan. 8 as an equal-interest joint venture by Standard Oil Co. of Indiana and Gulf Oil Corp.

The per-acre price of the lease ($41,319 for 5,089 acres) was a record high for federal mineral lease sales.

In the first competitive leasing of federal geothermal resources, the department Jan. 22 received bids totaling $6.81 million for exploration of 20 tracts in California.

Shell Oil Co. was the highest bidder—$4.5 million for two tracts in Sonoma County, north of San Francisco. The Shell tracts were near the country's only existing commercial geothermal operation—Pacific Gas & Electric Co.'s installation on private land, which had been using natural underground steam for production of electricity since the mid-1950s.

The Interior Department announced March 4 that it had accepted the high bid of $117.8 million for a 5,094-acre oil shale tract in Colorado containing a potential of 723 million barrels.

The consortium submitting the winning bid included Atlantic Richfield Co., Oil Shale Corp., Ashland Oil Co. and Shell Oil Co.

A high bid of $75.6 million for a third tract was submitted March 12 by Phillips Petroleum Co. and Sun Oil Co. The leasehold, in eastern Utah, contained an estimated 244 million barrels.

The State of Utah had filed suit against Interior Secretary Rogers C. B. Morton March 4 for title to approximately 157,000 acres of shale land, including the two 5,000-acre tracts scheduled for federal leasing.

The suit contended that Morton had exceeded his authority in denying Utah's claim to the land in lieu of acreage deeded to the state when it gained statehood and later taken over by the federal government for national forests and other preserves.

In a related development reported April 15, the department accepted a bid of $75.6 million submitted by Phillips Petroleum Co. and Sun Oil Co. for a shale oil tract in Utah.

A high bid of $45.1 million for another Utah shale tract was submitted April 9 by a consortium including Standard Oil Co. (Ohio), Phillips and Sun.

U.S. OIL WELLS

Number drilled annually

	New production wells	Total wells drilled	Total wells in production
1950	24,000	NA	466,000
1955	32,000	NA	524,000
1960	22,200	40,700	591,000
1962	21,700	38,700	592,000
1964	21,000	38,600	588,000
1966	15,900	30,500	583,000
1968	13,800	26,300	554,000
1970	12,600	23,400	531,000
1971	11,400	21,400	517,000
1972	10,800	21,400	508,000
1973	9,555	19,600	497,000
1974 [1]	10,700	17,000	-------

[1] Preliminary.

NATURAL GAS

Number of U.S. wells drilled

	New production wells	Total number of producing wells
1950	NA	65,000
1955	3,600	71,500
1960	5,300	90,800
1962	5,800	97,000
1964	4,900	103,100
1966	4,100	112,500
1968	3,300	114,400
1970	3,800	117,500
1971	3,700	120,200
1972	5,100	121,200
1973	6,400	124,200
1974 [1]	7,000	------

[1] Preliminary.

The last of six shale oil tracts scheduled for leasing under the experimental federal program failed to attract bids during a lease sale June 11. The tract was in Wyoming, as was the fifth tract, which had also failed to get bids.

Interior Department and industry spokesmen attributed the lack of interest to the relatively low estimated potential production for the Wyoming tracts—about 170 million barrels each.

According to the Bureau of Mines, oil shale deposits in Colorado, Wyoming and Utah contained the energy equivalent of 600 billion barrels of oil.

Oil shale plant construction halted. Colony Development Operations announced Oct. 4, 1974 a suspension of plans to begin construction of an oil shale plant producing 50,000 barrels a day in Garfield County, Colo. Suspension of the project, scheduled to begin in the spring of 1975, dealt a major setback to hopes that the nation's plentiful oil shale supplies could be developed as a major source of domestic energy.

Colony spokesmen blamed the postponement on "current double digit inflation, tight money and the absence of a national energy policy, which establishes clearly the role of oil from shale in the national energy picture."

Colony was a joint venture of Atlantic Richfield Co., Shell Oil Co., Ashland Oil Inc. and Oil Shale Corp. Oil had been extracted from shale deposits in the Colorado area since 1971 under a Colony pilot project. The group had already invested $40 million in the program. Spokesmen said their initial cost estimates made in 1973 for building the facility had climbed from $450 million to a projected cost of more than $800 million by 1977.

Oil shale development costs debated—In its draft of the Project Independence blueprint, the Federal Energy Administration stated that the oil shale recovery process would use as much energy as was extracted from the shale deposits. The charge was disputed by the Interior Department, and rather than address the argument, the section on oil shale recovery costs was omitted from the final version of the blueprint.

In a new estimate of the costs involved in oil shale development, the U.S. Bureau of Mines said Dec. 2 that an initial capital investment of $522 million would be required to build an oil shale facility capable of producing 100,000 barrels of oil a day in Colorado. The analysis was based on 1973 wage and price figures and did not reflect current inflationary trends, officials said.

Alaska pipeline progress. The Interior Department Jan. 23, 1974 issued a permit to build the multi-billion-dollar trans-Alaska oil pipeline, expected to take at least three years to complete.

Interior Secretary Rogers C. B. Morton said the project marked the "first time we have ever imposed such strict environmental and technical" conditions on a right-of-way permit, and the first time permitees had been required to pay for environmental studies, supervision and monitoring.

Environmental challenges to the project had apparently ended Jan. 16 when attorneys for a coalition of ecology groups

NATURAL GAS

Production Cost & Selling Price

	Average cost per new well	Drilling cost per foot	Price/ mcf[1] (current cents)	Price/ mcf (1958 cents)
1964	$104,800	$18.57	15.4	15.2
1965	101,900	18.35	15.6	15.1
1966	133,800	21.75	15.7	14.7
1967	141,000	23.05	16.0	14.4
1968	148,500	24.05	16.4	14.1
1969	154,000	25.58	16.7	13.7
1970	160,700	26.74	17.1	13.3
1971	166,600	27.70	18.2	13.5
1972	157,800	27.78	18.6	14.0
1973	NA	NA	21.6	13.2

[1] Interstate gas—Average National Price Entering Pipeline.

Annual U.S. supply & demand
(In trillion cubic feet)

	Domestic consumption	Domestic production	Imports	Imports as a percent of demand
1950	5.9	6.3	---	----
1955	8.9	9.4	---	----
1960	13.0	12.8	.2	1.5
1962	14.3	13.9	.4	2.8
1964	15.7	15.3	.4	2.5
1966	18.0	17.5	.5	2.8
1968	20.0	19.3	.7	3.5
1970	22.6	21.8	.8	3.6
1971	23.5	22.6	.9	3.8
1972	23.5	22.5	1.0	4.2
1973	23.6	22.6	1.0	4.2

notified federal district court in Washington that they would not challenge the constitutionality of the law authorizing construction. Judge George L. Hart Jr. then dismissed the environmental suit which had been one of a series blocking the project since 1970.

A consortium of eight companies involved in the pipeline's construction revealed Nov. 1 that inflation, higher labor costs and sharp increases in the price of steel had caused the price for the pipeline to rise from $4.5 billion to $5.98 billion. The cost of tankers to haul the oil from the Alaskan port of Valdez was not included in the estimates, officials of the Alyeska Pipeline Service Co. said.

El Paso Co. said Sept. 24 that it had filed a formal application with the FPC for approval of a $6 billion 800-mile combination pipeline and ship transportation system to bring natural gas from the Prudhoe Bay field on Alaska's North Slope to the U.S. West Coast.

A 26-member group, whose spokesman was American Natural Gas Co., had filed an earlier rival petition to move Alaskan and Canadian gas through Canada and the U.S.

El Paso spokesmen said their proposal was preferable to the U.S.-Canadian venture (utilizing a Canadian pipeline route) because no "dependence on a foreign power" was involved. El Paso also claimed that its transportation system, under which natural gas would be transported to Valdez, Alaska, where it would be liquefied and shipped to a facility north of Los Angeles and reconverted to its gaseous state, would prove operational by the mid-1980s, "earlier than any other proposed project."

Coal gasification project. The Commonwealth Edison Co. of Chicago and the Electric Power Research Institute of Los Angeles announced plans to build a demonstration coal gasification plant near Pekin, Ill. (reported Feb. 26, 1974). Gas from the plant would be used for electric power production.

The plant, scheduled to be completed in late 1976, would use the type of high-sulfur coal whose use in power plants had been limited in recent years because of environmental rules. Controlled burning of the coal in special gasifiers would produce gas from which sulfur would be removed before the gas was burned for power production. The gas would have relatively low heat value compared with pipeline-quality natural gas, but would be sufficient, in the Illinois plant, to produce about 70,000 kilowatts.

Geothermal & solar energy programs. A geothermal energy research bill was approved by the Senate Aug. 20, 1974 and by the House Aug. 21. President Ford signed it Sept. 3. The measure called for a management team to be set up to coordinate a federal effort in locating and evaluating geothermal resources, determining technical problems and establishing demonstration projects exploring the commercial potential of producing electric power from underground heat.

The bill authorized agreements to be made with utilities and municipalities to build geothermal facilities for commercial energy. Appropriations of $50 million were authorized for guarantees on loans to acquire resources and build facilities.

A Solar Heating & Cooling Demonstration Act was passed by the Senate Aug. 12

COAL

Annual U.S. production & consumption

	Domestic demand	Domestic production	Exports [1]
1950	494.1	560.4	29.0
1955	447.0	490.8	54.1
1960	398.0	434.3	37.7
1962	408.8	438.0	40.2
1964	445.5	503.0	49.3
1966	497.7	546.8	50.7
1968	509.0	556.7	51.7
1970	523.9	612.6	72.4
1971	502.2	560.9	58.0
1972	525.7	596.5	56.0
1973	569.3	602.5	[2] 32.0

[1] Totals will not add because of stockpiling.
[2] Estimate.

Mine & Price Data

	Number of mines	Percent of coal surface mined	Average price per short ton
1950	9,429	23.9	NA
1960	7,865	31.4	$4.69
1965	7,228	35.1	4.44
1968	5,327	36.9	4.20
1970	5,601	43.8	6.26
1971	5,149	50.0	7.70
1972	4,879	48.9	8.25
1973	-----	-----	9.30
1974	-----	-----	[1] 12.85

[1] Average January-June.

and the House Aug. 21 and was signed by President Ford Sept. 3. The measure established a five-year, $60 million program to explore technology to use solar energy to heat and cool buildings. Research, development and demonstration would be handled by the National Aeronautics and Space Administration. Installation of units in homes and federal or private buildings would be supervised by the Housing and Urban Development Department.

Court blocks Md. refinery. Circuit Court Judge Ralph W. Powers ruled Dec. 20, 1974 that Steuart Petroleum Co. of Washington could not proceed with building its proposed $240 million oil refinery at Piney Point, Md.

Votèrs in St. Mary's County had voted 2-1 against the refinery in a July 23 referendum, despite strong support for the facility earlier expressed by Treasury Secretary William Simon, then administrator of the Federal Energy Office. Steuart, which had rejected the referendum outcome, resumed construction and the resulting series of suits and countersuits were settled by Powers' decision.

Producers & Consumers

Production & Trade

U.S.S.R. leads in production. According to figures compiled by the Venezuelan Mines Ministry for the first six months of 1974 and reported Oct. 7, the Soviet Union was the world's largest oil producer. Soviet production rose by 10% during the period to total 9,018,000 barrels a day, the ministry said.

The U.S. was the second largest producer with 8,995,000 barrels a day, a 1% drop from the first half of 1973.

The world's other major oil producers, as ranked by the Venezuelan ministry:

Saudi Arabia—8,336,000 barrels a day; up 11.7%.
Iran—6,131,000 barrels; up 5.4%.
Venezuela—3,113,000 barrels; down 7.1%.
Kuwait—2,846,000 barrels; down 6.1%.
Nigeria—2,259,000 barrels; up 31.4%.
Libya—1,887,000 barrels; down 17.7%
Canada—1,823,000 barrels; up 2.4%.
Iraq—1,743,000 barrels; down 8%.

Oil consumption falls. World consumption of oil fell 1.2% to 2.743 billion tons in 1974, the first decline since 1956, the British Petroleum Co.'s Statistical Review of the World Oil Industry reported June 2, 1975.

Oil production rose .2% to 2.851 billion tons. The largest increase, 40 million tons, came from the Soviet Union, China and Eastern Europe. The Middle East produced an additional 31 million tons, nearly a 3% boost. Output in the Western Hemisphere decreased 5.3%, or 40 million tons.

Oil-related trade. The International Monetary Fund reported March 9, 1975 that the oil cartel's trade surplus was $97 billion in 1974, only $22 billion in 1973. In contrast, industrial nations reported a $67 billion trade deficit, compared with a $21 billion deficit in the previous year. The poor and less-developed countries also were hard hit by rising oil prices; their combined trade deficit rose from $12 billion in 1973 to $26 billion in 1974.

Exports by a group of 17 oil-producing nations soared from $43.4 billion in 1973 to $133 billion in 1974 (despite a slight drop in the volume of oil exported). That was the greatest rate of growth recorded by any group of nations, according to the IMF.

The group's exports accounted for 17% of all world trade in 1974. Since 1967, its trade had expanded 10 times. The countries were Algeria, Bahrain, Brunei, Ecuador, Gabon, Indonesia, Iran, Iraq, Kuwait, Libya, Nigeria, Oman, Qatar, Saudi Arabia, Trinidad and Tobago, United Arab Emirates, and Venezuela.

Because of the new wealth derived from their oil exports, these nations were able to increase their imports 69% from $21.3 billion in 1973 to $36 billion in 1974.

World trade exports as a whole increased 48% from $518 billion in 1973 to $768 billion in 1974; imports rose 44% from $529 billion to $764 billion. (The growth in trade, however, reflected the inflated price of oil and other commodities rather than a real rise in volume.)

The U.S. remained the world's biggest exporter, recording a 38% rise in trade to $98.5 billion in 1974. West Germany, whose trade expanded 32% to $89 billion in 1974, ranked second and was the only major country to show a trade surplus.

Borlaug said Arab oil cutbacks had drastically curtailed fertilizer production, which required heavy energy use and which used petroleum by-products as a base. Japan had cut nitrate fertilizer output by half since the oil crisis began in October 1973. Most of the deficit was expected to hit India, Indonesia and Southeast Asia.

Massive famines predicted. Dr. Norman E. Borlaug, Nobel Peace Prize-winning developer of high-yield grains, warned that 20 million people in developing countries might die in the next year as a result of fertilizer shortages and climate shifts, it was reported Jan. 26, 1974.

Nigeria offers Africans cheaper oil. The Nigerian regime announced Feb. 27, 1975 that it would sell crude oil to African nations at lowered prices "in order to demonstrate African solidarity and ensure that economic development is not slowed in any African country because of the oil situation." The Nigerian offer stipulated, however, that the purchaser must have its own refinery and barred re-export of the oil.

The announcement was made by Victor A. Adegoroye, a Nigerian delegate to the United Nations Economic Commission for Africa, meeting in Nairobi, Kenya. Adegoroye reiterated his government's offer March 3; it apparently amplified a 1974 statement of intent.

Algerian Commerce Minister Layechi Yaker criticized the unilateral Nigerian action, which would, he said, be difficult to implement because of the attached conditions. In his remarks, reported March 3, Yaker maintained that the issue of providing the third world with cheaper oil "can only be dealt with at a meeting of OPEC [Organization of Petroleum Exporting Countries]. A common decision must be taken within OPEC."

Comecon sets new pricing system. Under a new raw materials pricing system agreed to in January 1975 by the Executive Council of the Council for Mutual Economic Assistance (Comecon), the price of Soviet oil sold to Comecon members would as much as double in 1975. The price would be about half the world market price of $10-$11 a barrel, an article in the Hungarian Communist Party newspaper Nepszabadsag confirmed Feb. 23.

At the January Comecon meeting in Moscow, a resolution was passed creating a new method for price negotiations on all basic raw materials during the next five-year plan period. Prices would be determined, the resolution declared, by contractual prices to be established on the basis of the prices of the major markets during the preceding five years. (For instance, prices in 1976 would be established on the basis of the 1971-1975 period; in 1977, on the basis of the 1972-1976 period; and so on. However, the 1972-1974 period would be used to establish the 1975 price.) Previously, a single five-year price had been determined by taking the average price of the five years preceding as the basis for the subsequent five years. The new system was established, Nepszabadsag commentator Istvan Foldes said, because "prices should not lose touch with the prices of the major markets, but should filter out any unjustified fluctuation due to profit-grubbing." To maintain the old method, "separating price movements from the development of world market conditions, would hamper the development of [Comecon] cooperation."

The new pricing system, taking effect in 1975, a year before the present five-year plan was to expire, would see the price of Soviet oil sold to the Comecon members rise from the present level of about $2.60 a barrel to more than $5 a barrel, still significantly below the world market price.

East German, Czechoslovak and Polish officials had earlier affirmed that the new oil prices would not be more than half the prevailing Western prices, according to a London Times report Feb. 23.

Soviet-Hungarian trade agreement—Nepszabadsag reported Feb. 23 that the Soviet Union and Hungary had concluded a trade agreement under the new pricing arrangements. The average price of aw materials and sources of energy would rise by 52%, the report said, but would remain "considerably below the prices in effect on the world market."

The price of one ton of oil delivered to Hungary was more than doubled from 16 ubles to 37 rubles ($52.48, at the official exchange rate, which was still only about 48% of the world market price). In barrel terms, the cost rose from less than $3 to more than $6. The Soviet Union granted Hungary 10-year credits to help pay the higher bill. Moscow had also offered to ship a further 760,000 tons of crude oil to Hungary in addition to the 6 million tons scheduled for delivery in 1975. (Hungary had in the past purchased about seven-eighths of its total crude oil imports from the U.S.S.R.)

Pipeline deal—Yugoslavia had signed a contract with Czechoslovakia and Hungary for construction of a 420-mile oil pipeline from the Adriatic Sea to serve industrial centers in the three countries, it was reported Feb. 13, 1974.

The pipeline, expected to cost about $350 million, would have an annual capacity of about 240 million barrels, to be imported from the Middle East.

The contract linked Yugoslavia with the Soviet-Hungary-Czechoslovak Friendship pipeline system. Soviet-Yugoslav talks on gas and oil were then held in Baku in March, and the U.S.S.R. agreed to start supplying Yugoslavia with natural gas during the five-year plan period scheduled to start in 1976.

Iraq agreed to provide Yugoslavia with long-term oil deliveries in exchange for extensive economic assistance, according to a pact signed in Belgrade June 14.

Latin American unity. The creation of the Latin American Economic System (SELA), a regional unit that would exclude the U.S., was formally proposed March 20, 1975 by Presidents Luis Echeverria of Mexico and Carlos Andres Perez of Venezuela, the heads of two of Latin America's three oil-exporting nations.

The two leaders issued a joint communique in Mexico City, during a state visit by Perez, inviting the heads of state of 24 nations in Latin America and the Caribbean to appoint representatives to organize constituent meetings for SELA.

The communique cited, among SELA's "broad aims," the goal of defending the prices of Latin America's raw materials.

Venezuelan President Perez visited Mexico March 17–22.

Perez said March 18 that Venezuela, Ecuador and Mexico, as Latin America's only petroleum exporters, had a commitment to aid other countries in the hemisphere that were affected by the high price of oil imports. During his visit Perez tried to persuade Mexico to join the Organization of Petroleum Exporting Countries (OPEC), but Mexico resisted in apparent fear of damaging its relations with the U.S., according to diplomatic sources cited by El Nacional of Caracas March 20.

President Guillermo Rodriguez Lara of Ecuador, visiting Venezuela March 13–17, had reaffirmed support for SELA and had arranged to intensify Ecuadorian-Venezuelan cooperation on oil policy.

Soviet-Latin American plan. Mexico, the Soviet Union, Cuba and Venezuela planned to cooperate in selling crude oil and petroleum to countries that lacked an oil-producing capacity, it was announced in Havana May 20, 1975.

The proposed agreement, revealed by Mexican Information Minister Jose Campillo Sainz, was in the form of a triangular trading company through which the oil commitments of one nation could be filled by another if the buyer were situated closer to the latter producer. Campillo said Mexico had granted Cuba a revolving credit line of $20 million to be used for Cuban purchase of Mexican machinery, equipment and industrial raw materials.

North Sea Oil & Gas

The discovery of natural gas fields in the province of Groningen, the Netherlands in 1959 and 1962 led to speculation that gas and oil might lie under the North Sea. Great Britain, Norway, Denmark, West Germany and the Netherlands, therefore, came to agreement assigning to each of the five countries exclusive North Sea areas for exploration and development. This division was made under the provisions of the 1958 U.N. Sea Conference convention that each nation had a right to the natural resources in the part of the continental shelf off its coasts.

Britain began to license North Sea areas in September 1964. The British Petroleum Co. discovered a giant gas field 40 miles off the Scottish coast in 1965, and North Sea gas started flowing to British consumers in 1967. By 1974, with North Sea gas coming ashore in Britain at an estimated rate of more than 400 million cubic feet a day, the North Sea was supplying about 90% of Britain's gas needs.

Oil was not found in the North Sea, however, until 1967, when a small strike was reported off Denmark. Britain's first oil find in the North Sea was made by British Petroleum, which discovered the distant but giant Forties Field in 1970. One of the world's largest oil fields, it is rivaled by the Brent Field, which Shell-Esso found 100 miles north of the Shetland Islands.

North Sea oil finally reached British shores June 18, 1975 with the delivery of petroleum from the small Argyll field off the northeastern coast of Scotland.

Among 1974–75 developments involving North Sea oil and gas operations:

GREAT BRITAIN

North Sea taxes. The government Feb. 26, 1974 announced plans to take 45% of gross revenues from North Sea oil and offered private companies incentives for development of the fields. Initial reaction from industry was moderately favorable, but left-wing members of the Labor Party charged that the government had tried to appease the multinational oil companies.

The announcement followed an intensive lobbying campaign by the major companies and warnings that too stringent proposals would discourage North Sea investment and development.

The proposed tax system called for royalties of 12.5% of the wellhead value of the oil produced, as well as 45% of gross revenues plus the standard corporation tax of 52% on remaining profits.

Among the incentives, aimed mainly at spurring development of the marginal fields, the government announced an increase in the tax write-off on capital expenditure from 150% to 175%, tax exemption on the first seven million barrels of oil a year, and cancellation of the revenue tax if the rate of profit in any field fell below 30% of the capital invested.

Paymaster General Edmund Dell estimated that the proposals would bring the government about £3 billion ($7.2 billion) a year from 1980.

New oil strikes. Oil discoveries in Britain's North Sea sector were made public in February 1974. The Continental Oil Co., as operator for a group that included the state-owned National Coal Board and the Gulf Oil Corp., announced Feb. 24 that a new find located 1.7 miles west of the Norwegian sector's Stratfjord field could have a flow rate of more than 10,000 barrels a day. Texaco had announced Feb. 18 a discovery off Scotland's coast where, in a test, oil had flowed at a rate of 7,605 barrels a day.

A group led by Burmah Oil April 2 reported what appeared to be the biggest oil discovery in the British sector of the North Sea. The find, located 100 miles east of the Shetland Islands, was in the Ninian field, where British Petroleum and Canada's Ranger Oil Co. had announced a discovery in January. Ninian's potential was estimated at 500,000 barrels of oil a day, a quarter of Britain's current daily consumption. About two-thirds of the field was in Burmah's concession area.

Phillips Petroleum Co. announced that a second exploratory well in its Maureen field in the North Sea had tested oil at over 10,000 barrels a day, it was reported May 29. Oil was first discovered in the field in 1973. The field was situated 170 miles northeast of Aberdeen, Scotland.

(The U.S. company Marathon May 2 announced discovery of significant natural gas reserves off the southern coast of Ireland in the Irish Sea. This was the first breakthrough for any company in that area.)

The British Petroleum Co. announced Oct. 24 it had found its fourth oil field in the British sector of the North Sea. The field was located northeast of Aberdeen, Scotland.

Texaco, Inc. announced Mar. 5, 1975 an oil discovery, located about 83 miles east of the Shetland Islands. The find, situated south of the main Brent field, had an oil flow rate of 5,785–7,485 barrels daily.

Amoco (U.K.) Exploration Co., a unit of Standard Oil Co. (Indiana), reported a find northwest of the Hutton oil field with a test flow rate of 6,540–8.160 barrels a day of oil and two million cubic feet of gas, it was reported April 10. The find was the third in its block. Amoco, with a 25.77% interest in the block, was operator for the group. Other participants included the Gas Council (Exploration), a branch of the state-owned British Gas Corp.; Mobil North Sea Ltd., a unit of Mobil Oil Corp.; and the U.S.-based Amerada Hess Corp.

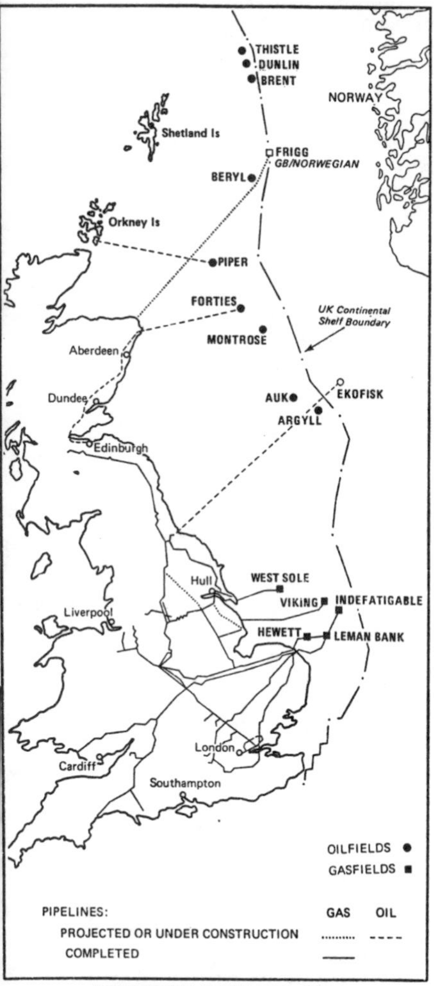

Pan Ocean Oil Corp., operator for a group of companies, had announced a significant oil discovery 150 miles east of the Orkney Islands, with some of the highest flow rates yet found in the North Sea, it was reported May 6. The discovery, named the Brae field, produced flow rates of more than 22,000 barrels of oil a day.

Pipeline deals signed—Seventeen oil companies, led by Exxon Corp. and the Royal Dutch-Shell Group, had contracted Sept. 2, 1974 to build a £200 million pipeline system to transport one million barrels of oil from five fields in the North Sea to the Shetland Islands, north of Scotland, where construction of an £80 million storage and shipping terminal was planned. The project would be completed by 1980.

BRITISH NORTH SEA
OIL & GAS FIELDS

Self-sufficiency by 1980 predicted. In a report to Parliament May 21, 1974, Energy Minister Eric Varley estimated that Britain would be self-sufficient in oil by 1980 because North Sea oil production would be in the range of 100–140 million tons a year (equal to about 2–3 million barrels a day). The production estimate represented a sharp increase over the 70–100 million ton forecast in the Energy Department's 1973 report, an indication of the accelerating pace of Britain's North Sea discoveries.

The report said total proven North Sea oil reserves now totaled 895 million tons in the 10 fields so far declared commercial. If "probable" and "possible" new discoveries were added, the North Sea reserves could reach 2.95 billion tons.

A government team led by Harold Lever, chancellor of the Duchy of Lancaster, began negotiations Nov. 28 with British Petroleum and Shell Oil companies on the government's claim for a 51% stake in 12 commercial oil fields in the North Sea.

North Sea oil control planned. The Labor government July 11 announced its plans to gain a majority interest in North Sea oil development, giving the nation a greater share of the profits and more control over production.

Energy Minister Eric Varley, announcing the program in the House of Commons, said the program was designed to give Britain "the full benefit from our newly developed wealth while leaving a substantial role for the oil companies.

The proposals provided for an additional tax on continental shelf profits and closing of various tax loopholes; creation of a Scottish-based national oil corporation which would obtain a majority interest in future licenses granted for North Sea fields, while companies currently operating in the North Sea would be "invited" to talk about majority state participation in existing leases. The program also called for extension of government powers to control offshore operations, including production; and immediate creation of a Scottish development agency to use offshore oil revenues to promote Scotland's economy, while a similar body would be established for Wales as oil development proceeded in the Irish Sea.

Conservative Party leader Edward Heath expressed opposition to direct government investment in the oil fields. He said Britain could obtain additional revenue through taxation and estimated that British taxpayers would have to pay almost $5 billion to buy majority shares in the fields.

To reassure the worried companies about commercial prospects in the North Sea, Energy Secretary Varley disclosed in the House of Commons Dec. 6 that the government would not curb the development of oil discoveries made up to the end of 1975; on finds made after 1975, he said he saw no production controls at least until 1982. He also said companies would be permitted to recover 150% of their capital investment in any field before production cuts were imposed.

The government Nov. 19 had introduced its bill on taxation of North Sea oil revenue, including stringent measures to prevent various loopholes. The bill omitted mention of the planned tax rate.

State oil company proposed. The government presented a bill to Parliament April 9, 1975 for the creation of a state-owned British National Oil Corp. that would buy a majority interest in all commercial petroleum and natural gas finds in Britain's North Sea fields.

The BNOC would also have the power to explore for, produce, refine and sell petroleum and its products. The bill would also control the rate of depletion of oil and gas reserves, require additional data from holders of oil production licenses, provide tighter control over the exploration and development of oil and gas deposits, regulate changes in the ownership of licenses and impose controls on the construction of pipelines and refineries. BNOC's financial resources and borrowing facilities would initially be set at £600 million ($1.5 billion), with the potential to rise to £900 million. The corporation would be exempt from the special petroleum revenue tax, although the government said it would not use this benefit to undercut privately owned oil companies.

Oil reserves estimate up. In an annual report to Parliament April 14, 1975, the government raised its estimate of proven oil reserves in Britain's North Sea sector to more than 1 billion tons, a 20% increase over the 1974 figure. It also said that the total recoverable reserves from the U.K.'s offshore fields could be 3–4.5 billion tons.

However, the report cut its estimate of production during the next few years. Britain's total North Sea oil output in 1975 was expected to be 1–3 million tons, down from the 1974 estimate of 5 million tons, which itself was reduced from the original forecast of 25 million tons. Production had been slowed by poor weather conditions and delays in construction of platforms, caused in part by labor disputes and material shortages. The estimate for 1976 was cut to 15 million tons, down from an earlier forecast of 23 million.

By 1980, production was expected to be 100–130 million tons, down from the 1974 estimate of a maximum 140 million tons.

Gas reserves discovered through 1974 were estimated at 27–45 trillion cubic feet. The report added that existing discoveries would support production of 5–6 billion cubic feet a day into the 1980s. Production in 1974 averaged 3.5 billion cubic feet.

Drilling activity in 1974 was the highest ever, the report said. Since May 1974, it said, 21 significant oil or gas discoveries had been made.

North Sea oil flow begins—The first oil from Britain's North Sea fields began being pumped June 11 from the Argyll field, off Scotland's coast, into a tanker. The crude was landed at a British Petroleum terminal in the Thames estuary June 18.

The Argyll field was owned by a consortium headed by the U.S.-based Hamilton Brothers concern. Texaco, Inc. of the U.S. was one of the partners.

The Argyll field's capacity was 40,000 barrels a day. It was one of Britain's smaller North Sea fields.

DENMARK

Gulf quits North Sea oil group. Gulf Oil Co., operator for the consortium developing the Dan field in the Danish offshore sector of the North Sea, had announced its withdrawal from the group (reported Dec. 10, 1974). Gulf's 30% share would be divided among the remaining partners: Shell, Chevron, Texaco and A. P. Moeller, the Danish shipping and industrial concern which held the concession to the Danish offshore sector. Gulf's decision was based on its evaluation of the zone's potential in relation to growing exploration costs.

The Dan field, the only commercial discovery so far in the Danish sector, had shown a flow of 3,500 barrels a day in its initial 1971 production test, but since then had proved disappointing, the report said. The consortium had also encountered problems with the chalk structure containing the oil.

NORWAY

Oil & gas policy. In a White Paper Feb. 15, 1974, the Norwegian regime outlined a restricted role for foreign oil companies in the oil and gas fields in Norway's sector of the North Sea. Foreign firms would be subject to close government control; they would be confined mainly to contracting for exploration and drilling and would be authorized as operators only in exceptional cases.

No new concessions would be granted in the next few years south of the 62nd parallel in the North Sea, the White Paper proposed. The Norwegian State Oil Company Statoil would become an independent operator and would be favored in exploitation of fields north of the 62nd parallel.

Production would be held at "a moderate tempo" to protect oil and gas resources and to permit gradual absorption by the Norwegian economy. Private foreign and domestic groups would be allowed "reasonable" profits.

The White Paper estimated that extra revenue from oil and gas in the early 1980s would reach between 10 billion–15 billion kroner ($1.8 billion–$2.7 billion) annually, equal to 25% of current total revenues. The increased revenues would enable the government to lower taxes, increase public expenditure in such areas as regional development and education, cut the standard workweek from 42½ hours to 40 hours, and raise Norway's aid to developing nations.

The report predicted that by 1975 oil and gas production would exceed domestic consumption and by 1978 would total about 35 million tons of crude and 30 billion cubic meters of natural gas annually, rising to about 50 million tons of oil and up to 50 billion meters of gas by 1982.

The government presented an economic program which would be partially financed by borrowings of $360 million abroad in 1974 against future income from North Sea oil, it was reported April 3. It was designed to fight inflation.

The government had presented the oil companies with stricter conditions for future concessions, the Oslo newspaper Aftenposten reported April 30. The government was reportedly demanding participation in concessions on a sliding scale of 50%–80% according to yield.

The government April 17 announced a big new North Sea oil discovery in the Brent field, northwest of Bergen. The first well had tested out at a maximum production of 10,500 barrels a day. The strike was made by Mobil Exploration, the operator for a group in which Statoil held a 50% interest.

The government announced Aug. 29 that a discovery by a group headed by Statoil and Mobil was the biggest oil and gas field found in Norway's sector of the North Sea so far.

The new field was located in Blocks 33/9 and 33/12, 95 miles west of the Sognefjord estuary, and bordered the British sector. The field, which displaced Ekofisk as Norway's biggest, was renamed Statfjord after previously being called "Norwegian Brent," after the adjoining British field.

The government May 10 decided to recommend to the Storting (Parliament) a plan under which gas from the Frigg field in the Norwegian sector of the North Sea would be piped directly by the French-Norwegian Petronord group to the Scottish coast for use by the British Gas Corp. To appease opponents of Norwegian gas being sold abroad, the government included an option clause requiring the Petronord group to lay a pipeline from Frigg to Norway "when technology permits."

Most oil companies had accepted exploration concessions granted in the Norwegian sector of the North Sea, the industries ministry said March 17, 1975. The only company to reject the offer was Chevron, a subsidiary of the Standard Oil Co. of California, the ministry said. Chevron said it had accepted the offer provided the proposed special tax laws were satisfactory, but the ministry had said such conditional acceptances would be treated as a refusal.

Among the U.S. companies involved in the concessions were units of Exxon Corp., Continental Oil Co., Standard Oil Co. (Indiana), Mobil Oil Corp., Amerada Hess Corp. and Texas Eastern Transmission Corp. Others firms included the Norwegian state-owned Statoil and French companies.

The government was reported March 24 to have authorized about 400 million kroner ($80 million) to help finance development of the vast Statfjord field, which had recoverable reserves of at least 3 billion barrels of oil and 3.5 trillion cubic feet of natural gas, according to revised estimates.

The industries ministry issued a revised survey reducing estimates for total production from Norway's Ekofisk field in the North Sea from 76.3 million barrels in 1975 to 56.5 million in 1975 and from 134 million barrels to 82.5 million barrels of gas and oil in 1976, it was reported June 3. The revised estimates meant that Norway would become a net exporter of oil and gas in 1976, not 1975 as planned. The survey said Norway's total production was now estimated to reach 507 million barrels in 1980, and, based on known finds, would gradually decline toward the year 2,000.

Investing the PetroProfits

Arabs invest in the West. The New York Times reported April 25, 1974 that the Arab oil producers were beginning to invest a small but significant amount of funds from their increased revenues in Western enterprises, including about $1 billion in the U.S. Some of the money was said to be flowing indirectly to the Soviet Union in payment for arms and might possibly be returning to the U.S. in payment for wheat.

Among the Arab investments reported by the Times: Several Kuwaitis paid $27 million for a property in Paris to be used for a large luxury office and bank building; The Kuwait Investment Co. in March bought Kiawah Island, 15 miles south of Charleston, S.C., for $17.4-million to develop a residential resort.

The New York investment banking firm of Lehman Brothers was reported seeking to interest the Arabs in Middle East development projects in such areas as food, education, housing and desalinization. The firm was also seeking U.S. government support for these projects. The Arabs in turn were establishing their own banks and joint ventures, especially with the French, and were using them as a vehicle to move part of their funds for the West, the article said. However, the expected massive flow of petrodollars to the West had not yet materialized, partly because the payments for oil from the large international oil companies had not yet reached the oil-producing states in sizeable amounts. Those firms normally paid their bills with lags of three-six months.

Later, large amounts of Arab oil money

were reported Aug. 16 to have been withdrawn from London accounts and reinvested in the U.S. This was the first such shift since the oil price raises of October 1973.

As a direct result of the withdrawals, the British pound lost .5¢ against the dollar in trading Aug. 16 on the London foreign exchange market to close at its lowest level in five months—$2.3405. Although the precise amount of petrodollars involved was not known, official government figures showed that about $4 billion had been deposited in Britain during 1974 by oil producing Arab nations.

An official of the Federal Reserve Bank of New York said Sept. 4 that oil producing nations in the Middle East were showing an "awakening interest" in the New York investment market. "We can see chunks of $100 million or so coming in," the official said, adding, however, that the inflow was "only a fraction, and a small fraction, of the total intake" from members of the Organization of Petroleum Exporting Countries.

Oil nations invested an estimated $7 billion in the U.S. and about $3 billion in Great Britain during the first eight months of 1974, according to the U.S. Treasury Department Sept. 19.

In a report submitted to the Permanent Subcommittee on Investigations of the U.S. Senate, the Treasury estimated that the surplus revenue of members of the Organization of Petroleum Exporting Countries (OPEC) was "somewhere between $25 billion and $28 billion" during the eight-month period, although financial reports were admittedly "fragmentary."

Of the $7 billion invested in the U.S., the report stated, "roughly" $4 billion was invested in various types of marketable U.S. Treasury and government agency securities. The remainder had been deposited in U.S. commercial banks and "a few hundred million dollars may have gone into corporate securities and real estate," according to the report.

At least $3 billion was invested in Britain in sterling, "some of which no doubt involved purchases of British government securities and some sterling deposits in British banks," the Treasury Department said.

According to the report: another $10 billion-$13 billion was being held in Eurodollar and other Eurocurrency deposits in banks outside the U.S., largely in London; at least $2 billion was invested "through direct placement loans to official or quasiofficial agencies, plus direct purchases of private securities and real estate"; $3 billion was used to aid developing nations through direct grants or loans and through multilateral lending institutions, such as the World Bank. (An estimated $500 million in World Bank bond purchases was included in the Treasury Department figure.)

The Treasury Department reported Oct. 30 that the oil nations had apparently increased their investments in the U.S. by another $1 billion in September, bringing their total for 1974's first nine months up to about $8 billion.

During the same nine-month period, the U.S. oil import bill increased by about $13.5 billion.

U.S. officials also offered an explanation for the Saudi Arabian government's sudden transfer of virtually all its gold held in reserve at the Federal Reserve Bank in New York. According to Treasury Undersecretary Jack F. Bennett, Saudi Arabia had activated a longstanding law requiring that all official gold be held within the country.

Figures released Nov. 7 by the Commerce Department indicated that the Saudi Arabian gold transfers continued through September, when 516,562 ounces valued at $21.8 million (at the official rate) were withdrawn from the New York Federal Reserve Bank. Since July, Saudi Arabia had withdrawn about 2.6 million ounces valued at $109.1 million.

The U.S. figures were confirmed by a Bank for International Settlements (BIS) statement Nov. 29 that oil-exporting nations had invested $27 billion in surplus oil revenues in U.S., Britain and Western Europe during the first nine months of 1974. Nearly half was believed invested in traditional reserve assets, such as government obligations.

According to the BIS breakdown, $8 billion was invested in the U.S., $4 billion in Great Britain, $12 billion in Eurodollar investments in London and $3 billion in Eurodollar deposits elsewhere in Europe.

Saudi Arabian invests in U.S. bank. The First National Bank of San Jose, Calif. disclosed Nov. 26, 1974 that Odnan M. Khashoggi, a Saudi Arabian businessman with extensive investments in the Middle East and the U.S., had purchased one-third interest in the bank for $14 million.

Khashoggi also had controlling interest in two banks in Walnut Creek, Calif. The San Jose bank's assets at the end of September were $306.1 million, with deposits of $276.4 million.

Kuwait buys into British firms. In what was believed to be the first purchase by an oil country of a major stake in a Western industrial enterprise, Kuwait bought a more than 10% interest in St. Martin's Property Corp., a British firm with large commercial real estate holdings in London. Kuwait acquired 2.1 million shares in the company Sept. 11, 1974 and an additional 100,000 shares Sept. 12.

Kuwait had entered a $248 million bid for St. Martin's Sept. 6.

Kuwait would buy eight million more shares valued at $14.2 million in Lonrho Ltd. of London under an agreement reported Nov. 19. The deal would increase to 14% the Persian Gulf state's holdings in the company, which already amounted to three million shares. Kuwait thus became the second-largest holder of the firm's 65 million shares.

Lonrho was a major British mining, agricultural and trading company with widespread holdings in Africa.

(The Wall Street Journal reported June 3, 1975 that a group of Dubai investors had acquired five million shares of Lonrho, valued at $17.7 million, in March and later had paid $6 million for 2.2 million more shares. Mohamed Al-Fayed headed the group.)

Kuwait buys stock in Daimler. The Bonn government disclosed Dec. 2, 1974 that Kuwait had purchased a 14.6% share in Daimler-Benz, West Germany's second largest automobile manufacturer after Volkswagen and maker of the Mercedes car.

The identity of the purchaser had been revealed only after pressure by the government and opposition parties on the Dresdner Bank, which arranged the sale of the 14.6% share, worth an estimated $320 million, held by the Quandt holding group.

The Dresdner Bank chairman, Juergen Ponto, said at a news conference Dec. 3 that "we assume that Kuwait has no intention of seeking a voice in running the business side of the company." But he acknowledged that Kuwait had given no guarantees.

Kuwait lends $40 million to Hungary. Hungary announced Dec. 4, 1974 that Kuwait had promised it a $40 million loan believed to be the first loan from an Arab oil-producing nation to a member of the Soviet bloc. The Hungarian National Bank said the eight-year loan at 10.5% interest would be used in projects to boost exports. The Kuwait International Investment Co. negotiated the bond issue which would be repaid in five equal installments beginning in 1978.

Iran raises investment in U.S. Iran's ambassador to the U.S., Ardeshir Zahedi, said his government planned to spend several billion dollars in the U.S. within the next five years on equipment for developing Iranian agriculture and technology and on training personnel, according to an interview published in the New York Times Dec. 21, 1974. Zahedi said U.S. atomic power expertise and equipment would account for part of Iran's outlay.

The groundwork for the U.S.-Iranian cooperation had been laid in talks Secretary of State Henry A. Kissinger had held in Iran Nov. 1-2. The joint communique had stressed "a broad program of cooperation in the political, economic, cultural, defense, scientific and technological fields." Zahedi said "if it weren't for American help in the past few years, we would not have existed as a country, and we are very grateful."

Shah Mohammed Riza Pahlevi urged U.S. businessmen Dec. 21 to become "more aggressive and dynamic" in competing for Iranian development contracts. Iran's current five-year economic plan, then in its second year, provided for expenditure of $80 billion–$90 billion on the purchase of capital goods and services to complete planned projects, the shah noted. Among the U.S. firms listed as seeking to do business in Iran were Westinghouse Electric Corp., General Electric Co., American Telephone and Telegraph Co. and Pan American World Airways.

The Iranian leader also invited thousands of Americans to come to his country to fill the critical need for manpower; openings, he said, were available for engineers, teachers, nurses, technicians and skilled workers.

Iranian loan to Denmark. Iran and Denmark signed an agreement Dec. 4, 1974 for a five-year $150 million Iranian loan to Denmark and agreed in principle that Denmark would build two hospitals and a housing project in Iran. The loan was the largest ever obtained by Denmark abroad. The protocol also welcomed an Iranian proposal of cooperation in oil exploration off Greenland.

OPEC reserves rose 8.95% in September. The International Monetary Fund said Nov. 1, 1974 that the reserves of nine members of the Organization of Petroleum Exporting Countries (OPEC) rose 8.95% in September to $34.7 billion. During August, the same group of nations showed a 6.56% increase in reserves. (Data was available only for Ecuador, Venezuela, Iran, Iraq, Kuwait, Saudi Arabia, Algeria, Nigeria and Indonesia.)

In a later release Nov. 3, the IMF reported that over a one-year period, Saudi Arabia advanced from 13th to fourth place in amount of total reserves. At the end of September, West Germany ranked first with $32.5 billion, followed by the U.S. ($15.7 billion), Japan ($13.2 billion) and Saudi Arabia ($11.5 billion).

Other nations experiencing large gains in reserves were Iran, up about 700% from the previous year to $6.3 billion; Nigeria, up nearly 1,000% to $4 billion; Venezuela, up 300% to $4.9 billion; Iraq, up nearly 400% to $3 billion; and Algeria, up more than 300% to $2 billion.

Industrial countries generally registered gains in reserves despite trade deficits because petrodollars were reinvested in these nations, the IMF stated. Inflated oil prices also caused world trade to continue to "grow rapidly," but the actual volume of trade increased at a much smaller rate, according to the IMF. During the second quarter, ending in June, exports were up 49% from the previous year, reaching an annual rate of $750 billion.

Control by petropowers feared. Various sources estimated that the surplus revenues of the oil-exporting countries would total $60 billion for all of 1974. The sheer volume and disruptive power of these petrodollars had caused concern among officials of the industrial nations that large-scale OPEC investments would entail OPEC control over vital national industries.

According to the New York Times Dec. 23, France and other countries had established investment controls requiring prior notification of, and formal application for, major share acquisitions. Belgium and the Netherlands also had machinery to prevent foreign take-overs, the Times said. West Germany and Italy reportedly were hampered in efforts to control unwanted investments because they lacked disclosure laws requiring that the identity of investors be revealed.

In the U.S., Secretary of State Henry Kissinger announced Dec. 7 that a study had been launched to determine whether there were sufficient safeguards against the foreign take-over of critical defense industries.

According to the Times Dec. 9, the study was triggered by Administration concern over Iran's abortive efforts to aid the financially troubled Grumman Aerospace Corp. Although, despite this fear, some Iranian participation in a credit program involving several other banks was approved, the Defense Department did not permit Iran to gain equity control over the firm, a maker of F-14 jet fighters for the Navy, as Iran had sought.

There had been other attempts by oil rich nations to buy into sensitive U.S. industries. During the spring a Lebanese consortium offered to buy 41% of the stock in Lockheed Aircraft Corp., the nation's largest defense contractor, but no formal proposal reportedly was tendered.

Lockheed was reported Nov. 30 to have rejected the bid by Arab financiers to invest $100 million in the company, but it was later announced that talks were held Dec. 1 in Washington on Iran's latest offer, involving a bid to buy 10 of Lockheed's $55 million C-5A military cargo planes and a plan to pay for reopening the C-5A's production line in Georgia.

It was reported Dec. 9 that Iran was holding preliminary talks with Pan American World Airways about extending aid to the airline, which was in financial trouble.

The Wall Street Journal reported Nov. 29 that Iran's state owned oil company, National Iranian Oil Co., was negotiating with three U.S. oil facilities to share their distribution facilities and retail marketing outlets in the U.S. The firms were Shell Oil Co., Crown Central Petroleum Corp. and Apco Oil Corp.

Iran ends rial-dollar link. Iran said Feb. 12, 1975 that it had severed the rial's direct tie to the U.S. dollar and was linking it instead to the International Monetary Fund's special drawing rights (SDR), which were based on 16 leading world currencies and were used for transactions between IMF countries. Banking sources in Teheran said the move was aimed at protecting Iran's oil income, which was paid in dollars, against any possible future dollar devaluation.

Under the new arrangement, the rial would be permitted to fluctuate within a margin of 2.25% on either side of the median exchange rate of the rial to the SDR, Iran's Central Bank reported. The rate was 82.2425 rials to the SDR.

Arab Monetary Fund formed. The governors of Arab central banks approved the formation of an Arab Monetary Fund at the end of a three-day meeting in Baghdad Feb. 24, 1975. The projected fund would have a capital of 750 million special drawing rights (SDR) units, equivalent to $1.75 billion; the unit would be called the Arab dinar. The money would provide easy-term loans to member states that had problems with their over-all balance of payments.

The conference also urged a world monetary systems to "discipline the financial policies of the major industrial countries" and guarantee the stability of the exchange rate of the Arab petrodollar. The conferees also urged world monetary reforms to deal with pending problems "in a way that guarantees the interests of the world community, including developing countries and Arab oil exporters," until a new system was adopted.

The Baghdad meeting was sponsored by the Arab League and was attended by representatives of the league, the Kuwait-based Arab Fund for Social and Economic Development, the Union of Arab and French Banks, the International Monetary Fund and the Organization of Arab Petroleum Exporting Countries.

Saudi Arabia plans London bank. Saudi Arabia May 8, 1975 announced plans to open a merchant bank in London in its program to develop broader economic and financial ties with other nations and gain experience in international money markets.

The bank, to be called the Saudi International Bank, would have authorized capital of $60 million. The governor of the Saudi Arabian Monetary Agency (SAMA), Saudi Arabia's central bank, said the new bank would not function as an arm of official policy, although it might perform certain services such as trade financing for the government.

SAMA, with a 50% interest, would be the principal stockholder. The Morgan Guaranty Trust Co. of N.Y. would have 20% and two Saudi commercial banks would have 2.5% each. Other stockholders, each with 5% interest, would be the Bank of Tokyo, Banque Nationale de Paris, Deutsche Bank, National Westminster Bank (Britain) and the Union Bank of Switzerland. Edgar C. Felton, a vice president of Morgan Guaranty, would become executive director, the announcement said.

An Arab consortium, headed by Prince Abdullah bin Musaid bin Abdul Rahman of Saudi Arabia, had purchased 25% of the shares of Edward Bates and Sons, London merchant bankers, from Atlantic

Assets, an investment trust, for more than $3 million, it was reported May 14. Prince Abdullah had joined the board of Bates. The consortium included the Pan-Arab Merchant Bank of Beirut, which also had an option to buy a further 15% of Bates' shares within three years. The Bates chief executive said the transaction would spur more Arab investment in advance technological projects in the U.S. and Europe.

Oil-money reserves rise seen ending. The total monetary reserves held by oil-producing nations apparently had stopped rising since the end of 1974 and were moving "unevenly," the International Monetary Fund reported May 4, 1975. At the end of February, the IMF said, the oil-exporting nations' reserves as a group totaled $48 billion, nearly unchanged from December 1974, but far above the $20 billion level recorded at the end of the 1st quarter of 1974.

Since the end of 1974, reserves had declined for four OPEC nations—Iran, Libya, Algeria and Indonesia. (The IMF noted, however, that reserves, which were defined as liquid funds such as gold, SDRs, and convertible foreign currency, did not include long-term investments made abroad.)

Reserves of petrodollars increased in the two-month period, but in some cases by small amounts, for Kuwait, Nigeria, Venezuela, and Iraq.

Reserves held by industrial nations increased over a 12-month period to total $124 billion at the end of March, nearly the same level prevailing in 1974 and above the $116.6 billion total for 1973 period. The increases reflected both surpluses in balance of payments accounts and borrowing on the Eurocurrency market, the IMF stated.

Saudi Arabia became the second largest holder of international monetary reserves, topping the U.S. which ranked third, the IMF reported July 3. West Germany, with $33.2 billion in liquid monetary assets was the leader. The U.S., which ranked number one until 1971, held $16.7 billion, followed by Japan with $14.6 billion.

Saudi Arabia's reserves were estimated at $20.5 billion, an increase of $9 billion since January, and more than five times its holdings of $3.9 billion at the end of 1973.

Total world monetary assets were $227 billion, compared with $186 billion at the end of the 1st quarter. Nearly all of the growth in reserves derived from OPEC nations. The reserves held by industrial nations did not decline because they were able to cope with balance of payments deficits through financing rather than by drawing on reserves, according to the IMF.

Country-by-Country Developments

ABU DHABI

Abu Dhabi gets oil firm control. Under an agreement signed in Abu Dhabi Sept. 3, 1974, the Persian Gulf emirate took a 60% share, up from 25%, of the Abu Dhabi Petroleum Co., retroactive to January. The remaining 40% was divided by the six Western oil companies that owned Abu Dhabi Petroleum. They were British Petroleum Co., Royal Dutch-Shell Group, Compagnie Francaise des Petroles, Exxon Corp., Mobil Oil Corp. and Partex.

The accord provided for Abu Dhabi to pay the firms $40 million for its additional share of the producing operations and to sell its share of crude oil to the firms for $11.90 a barrel.

ALGERIA

Oil developments. Finance Minister Ismail Mahroug said that Algeria's oil revenue had almost tripled between 1969 and 1973 (reported Jan. 7, 1974).

Algeria Jan. 19 signed an accord resolving a $5 million contract dispute with ELF-ERAP, the French state oil company, and renewing ELF-ERAP's exploration permits.

New gas liquefaction project. A consortium of six Belgian, French, West German, Austrian and American firms signed a contract Sept. 2, 1974 with Algeria's national petroleum company, Sonatrach, for construction of a natural gas liquefaction complex to process gas, produced at Hassi R'mel, in Arzew about 300 miles to the northwest. The gas output from the project—which would be the world's largest—would be mainly for the European market.

ARGENTINA

Libyan oil deal. An Argentine trade mission headed by Social Welfare Minister Jose Lopez Rega visited Libya Jan. 26–Feb. 4, 1974 and signed several agreements including one for Libya to supply Argentina with 3 million tons of oil in 1974 and as much or more in 1975.

The oil, which began arriving in Argentina in March, was bought at a very high price—$18.72 a barrel—but was of high quality, according to the Andean Times' Latin America Economic Report Feb. 22. The purchases were financed first by a group of Italian banks and subsequently by Chase Manhattan Bank of New York, the London newsletter Latin America reported May 17.

As its part of the oil deal, the Argentine mission to Libya offered to set up a vehicle assembly plant near Tripoli, help build an oil refinery and a liquid natural gas plant, and help construct half a million Libyan schoolrooms and 2,000 apartments.

A Libyan mission to Argentina signed agreements in Buenos Aires March 1 to import 30,000 metric tons of Argentine sugar, 96,000 tons of bread wheat, 15,000 tons of corn and 24,000 tons of barley. The Argentine National Bank later signed a $40 million agreement to export motor vehicles, irrigation pipes and beef to Libya, it was reported April 12.

Soviet power credit. Argentine sources in Moscow reported May 7, 1974 that the Soviet Union had pledged $600 million in credits to help Argentina double its power generating capacity by 1977.

The credit was granted during a visit to the Soviet Union by a 135-member Argentine mission headed by Economy Minister Jose Gelbard.

The Soviet credit would primarily finance purchase of Soviet equipment for hydroelectric projects at Salto Grande on the Uruguay River (undertaken jointly by Argentina and Uruguay) and at Alicura in Mendoza Province, and for expansion of the Chocon Dam and construction of two thermal projects in La Plata and Rosario.

Argentina and Czechoslovakia Feb. 24 had signed an agreement under which the Czechoslovak firm Skoda would supply turbines and generators for four Argentine thermal and hydroelectric power stations, and launch a joint venture with Argentina for a turbine manufacturing plant.

Nationalization continues. The Argentine government Aug. 23, 1974 ordered the nationalization of oil marketing, but postponed expropriating the foreign oil refineries. Gasoline service stations were ordered nationalized Aug. 28. Until these measures were taken, the state oil firm YPF had marketed about 50% of Argentina's oil. The nation was virtually self-sufficient in petroleum.

AUSTRALIA

U.S. oil interest purchased. The Petroleum and Minerals Authority bought part of the Australian property of Delhi International Oil Corp. under an agreement announced by the U.S. firm in Dallas, Texas Feb. 3, 1975.

The accord provided for the sale of 50% of Delhi's petroleum and natural gas reserves and 25% of other company holdings in Australia for a minimum payment of $18 million, and a possible additional sum of up to $14.5 million to be determined later on the basis of the value of the acquired reserves.

The agreement replaces a similar pact between Delhi and the Australian subsidiary of France's Societe Nationale des Petroles d'Aquitaine.

BELGIUM

Ban on Sunday driving ended. Belgium Feb. 15, 1974 ended the ban on Sunday pleasure driving introduced in late 1973 and eased in early January. It also reduced restrictions on public lighting and ended a 5% cut in electric supply.

Oil price freeze angers firms. The outgoing government of Premier Edmond Leburton froze oil prices until the end of April (reported Feb. 24, 1974). The move, intended to win votes in the forthcoming election, angered international oil companies, which cut off supplies for the nation.

The government justified the freeze on grounds that the companies supplied insufficient information on crude oil prices and operation costs to back their demand for a price increase of $50 a ton on refined products, including a rise of 16¢–20¢ a gallon in gasoline prices.

To protest the price curbs, some of the major multinational oil companies barred crude oil imports to Belgium, the Wall Street Journal reported March 12. Among the firms mentioned in the report were the British Petroleum Co. and the local subsidiary of Brussels-based Petrofina S.A., while a Belgian unit of Continental Oil Co. (U.S.) and an affiliate of the Cie. Francaise des Petroles of France were said to have suspended all distribution of petroleum products in Belgium.

Oil price truce reached. A $31 a ton across-the-board price increase for refined oil products went into effect April 1, while the caretaker government of Premier Edmond Leburton and the major international oil companies agreed to continue talks on the latter's demand for a $50 a ton increase to meet higher crude oil costs. The agreement was reached a week after the companies closed most refineries because of dwindling oil supplies resulting from their earlier halt to crude imports.

The companies were permitted to decide how to apportion the increase among the oil products. The price of gasoline and diesel fuel would rise 12.7¢ a gallon, which, with tax increases, would boost consumer prices to $1.20 a gallon for regular gasoline. The price of heavy fuel for industry would increase 14.6¢ a gallon, while heating fuel prices would go up only slightly.

Oil prices raised again. The government May 10 increased gasoline and heating oil prices 10%, effective May 15, ending the bitter dispute with the international oil companies over their price increase demands. The new increases, still lower than those demanded by the oil firms, would raise the price of high octane gasoline to the equivalent of $1.62 a gallon.

BOLIVIA

U.S. oil pacts signed. The state oil firm YPFB recently had signed oil and natural gas exploration and exploitation contracts with three U.S. companies, after signing four similar contracts with U.S. and French concerns during the previous 10 months, according to the Andean Times' Latin American Economic Report Jan. 4, 1974.

The latest contracts were with the Amerada Hess-Amoco partnership, for a one million hectare area in the Chaco, on the Paraguayan border, and with Sun Oil, for a comparable area on the Bolivian high plateau.

Bolivia's contract terms were borrowed from the pacts being signed by Peru and foreign concerns. Their central idea was that the foreign company financed and carried out all exploration in the contract area, and the oil was shared in kind between the company and YPFB on a prefixed percentage, at the well-head. This avoided complex tax and royalty calculations. The production split averaged out to roughly 52/48 in favor of YPFB, although payments such as social security reduced the company's net share to about 45%.

Each contract carried minimum work and expenditure requirements.

All oil and gas pipelines would belong to YPFB, which would charge pipeline fees to the contract company. If the pipeline were originally constructed to serve an area assigned to one company, the company would have priority over third parties in its use.

YPFB was also exploring on its own in various parts of the country—including the high plateau, the Tarija area in the far south and the Santa Cruz and Chaco regions—but its technical and financial capacity remained limited, the Times reported.

Most exports of Bolivian crude—an oil of very high quality—went to other South American countries. The main buyers were Argentina, Peru and Brazil.

U.S. oil loan. The Bank of America had organized a $35 million loan for the state oil company YPFB, the newsletter Latin America reported Jan. 31, 1975.

The government had announced Jan. 21 that it would reduce its relatively high oil export prices to "competitive levels." Bolivia exported only 35,000 barrels of light crude daily, but it charged a high price because of the quality of its oil. The price had been reduced from $16 a barrel to $14.65 per barrel in August 1974.

CANADA

Oil tax bill approved. The House of Commons approved a bill setting an oil export tax of $6.40 a barrel for February and March, but rejected a government request for authority to extend and adjust the tax every month to a maximum of $10, it was reported Jan. 9, 1974.

Opposition members said a permanent tax should only be imposed after long-term pricing policies had been determined, and after a scheduled federal-provincial energy conference was held. Members were also in dispute over federal equalization payments to non-oil-producing provinces, to match export tax funds going to oil producing provinces.

The new export tax rise would raise the cost of Canadian crude oil to U.S. purchasers to $10.50 a barrel, roughly in line with world prices.

The U.S. Senate, with only half a dozen members on the floor Jan. 29, approved a resolution warning Canada that the U.S. could retaliate for the $6.40 a barrel export tax. The resolution was supported by Majority Leader Sen. Mike Mansfield (D, Mont.) and Minority Leader Sen. Hugh Scott (R, Pa.).

Standby oil power voted. Commons approved a bill Jan. 11 giving the government standby oil emergency powers.

The bill provided for a five-member Energy Supplies Allocation Board. If the Cabinet declared a national emergency, the board would begin mandatory allocation of petroleum and other energy sources, subject to approval of Commons. New Democratic and Conservative members of Parliament had deleted a provision that would have given similar approval power to the Senate.

Oil output up in 1973. Crude oil production in the western provinces increased a record 15% in 1973 to 1,972,000 barrels a day (reported Jan. 9, 1974). Most of the increase occurred in Alberta, where production was running above rated capacity.

Exports to U.S. up—Exports of gasoline to the U.S. went up tenfold in the first 10 months of 1973 to 140.1 million gallons, it was reported Jan. 15. Fuel oil shipments rose 30% to 1.203 billion gallons in the same period, and crude oil exports increased from 280.5 million to 352.2 million barrels. Natural gas exports rose by 12.5% in value to $285 million. The figures did not reflect the effects of the Arab oil embargo, imposed in October 1973.

Natural gas price raised. The National Energy Board approved an 81% increase in the wholesale price of natural gas, imposed by Westcoast Transmission Co. under an agreement with the British Columbia government in November 1973, it was reported Jan. 14.

The price rise had been contested by El Paso Natural Gas Co., which had an ongoing contract with Westcoast for daily deliveries.

Quebec oil pipe OKd. The government decided Jan. 16 to allow an immediate start on a 520-mile oil pipeline from Sarnia, Ontario to Montreal, to be built without public financial assistance.

The $175 million project would be built by Interprovincial Pipe Line Ltd. as an extension of its present pipeline system, which brought oil from western Canada to

Ontario through the midwestern U.S. The 30-inch pipe would carry 250,000 barrels of crude oil a day to Montreal. Added pumping equipment could double the capacity in an emergency. Interprovincial would increase the capacity of the U.S. section of the line to supply the new extension.

Oil from the pipeline, which would be completed late in 1975, would replace about one-fourth the oil currently imported by Quebec and the Atlantic provinces, while Canadian crude oil exports to the U.S. would be cut by about 20%.

The government said it still hoped to build an all-Canada pipeline to bypass the U.S. by 1980, but rejected the route at the present time, because of an estimated additional cost of $200 million and additional two-year construction time. Also planned was an extension from Montreal to eastern Quebec or New Brunswick. Future plans depended in part on development of Alberta oil sands and Atlantic coast discoveries.

Partial energy compromise. Prime Minister Pierre Trudeau and the premiers of the 10 provinces reached agreement at an energy conference in Ottawa Jan. 22–23 for a continued freeze on Alberta oil prices until April and for subsidies to eastern provinces to cover oil import price increases. But the first ministers failed to agree on means to bring about a nationwide one-price system after April.

Under the compromise, Saskatchewan, a relatively poor province contributing 11% of Canadian oil output, would be allowed to raise its wellhead price from roughly $4 a barrel to $5–$6, with all the increased revenues going to the province. But Alberta, which produced 84% of the nation's oil, much of it exported to the U.S., would retain the current price during February and March.

To prevent a 12¢ a gallon increase on gasoline and heating oil prices in the eastern provinces, which would have resulted from price increases in February import contracts, the federal government would subsidize eastern oil refiners with $240 million, about $190 million from the current 50% federal share of the oil export tax (due to increase to $6.40 a barrel in February) and $50 million from general revenues. Eastern prices would remain 5¢–6¢ a gallon higher than they were before the recent massive increase in world oil prices. Prices in western Canada had been kept frozen, and were several cents lower.

The conference participants agreed that a one-price system should be imposed from April on.

Many of the conference proceedings were televised, but the final compromise was worked out at a private session. The federal government had wanted to extend the freeze for three months, the eastern provinces had called for a rollback of the recent price increases, and the oil-producers had hoped for a greater immediate share in the more than $2 billion that U.S. consumers were expected to pay in price increases and export taxes for Canadian oil in 1974.

Energy Minister Donald Macdonald told the conference Jan. 23 that Canadian oil self-sufficiency could not last unless an "intensive program" was begun to develop tar sands and heavy oil reserves, and to seek new oil fields in offshore and frontier areas. Private industry would need incentives to develop these new sources, he said. The federal government planned to allot the oil companies about 15%–20% of the new revenues.

Oil controls disputed. As part of its program to devise and implement a one-price system for oil throughout Canada, the federal government said March 2 that a national petroleum board to buy all oil traded interprovincially would be created if Alberta and Saskatchewan did not accept voluntary price constraints. The two provinces, which together produce more than 95% of Canadian crude, had opposed federal intervention in what they regarded as provincial affairs.

According to the Ottawa Citizen March 5, the provinces planned to control and collect all oil revenues through provincial marketing boards and royalties when the nationwide oil price freeze expired March 31.

Price accord. Trudeau and the provincial premiers met in the year's second energy conference March 27 and agreed on a new 12-15 month interim price for crude oil. The price per barrel rose April 1 from $4, the price at which oil had been frozen since late 1973, to $6.50.

Trudeau and the premiers agreed that the federal government would retain all the revenues from the oil export tax to finance the subsidy to consumers in the East; gasoline and heating oil prices would rise about 8¢ a gallon in central Ontario and the West. The western oil-producing provinces of Alberta and Saskatchewan would determine for themselves how much of the $2.50 increase would go to the oil companies and how much to their own treasuries. The export price of oil would remain unchanged.

Trudeau had met with Alberta Premier Peter Lougheed March 25 and with Saskatchewan Premier Allan Blakeney March 26 to discuss the price negotiations.

Gasoline prices jump. Gasoline prices across the country rose by as much as 14¢ a gallon May 15 as oil companies and retailers began adjusting prices at the end of the temporary freeze on petroleum prices negotiated at the March oil price parley.

Imperial Oil Ltd., Canada's major integrated petroleum company, increased its wholesale prices May 16 by 9.2¢ a gallon west of the Ottawa Valley energy line and by 3.7¢ a gallon east of the line. All the increases were within the guidelines which the federal government set to compensate for higher crude and other costs and, at the same time, equalize wholesale prices across the country.

Ontario bore the largest increase (in Toronto) and continued to see the highest gasoline pump prices. A number of Ontario legislators charged that Premier William Davis was responsible for the price rises. The province's energy minister acknowledged May 16 that the premier had not been briefed on the difference between "wellhead" and "citygate" prices, a distinction that substantially affected the meaning—and the cost—of the agreement ultimately agreed upon at the March meeting of Canada's provincial and national leaders.

The misunderstanding had been realized shortly after the conclusion of the talks during which a $3.80 a barrel wellhead price was set, but Canadian External Affairs Minister Mitchell Sharp and Energy Minister Donald Macdonald said, according to the Journal of Commerce April 9, that the differences were too minor to reopen the pricing deal for new talks. The difference represented an increase, to 9.2¢, of almost 2¢ a gallon over the extra costs predicted by Davis upon his return from the meeting.

Macdonald in U.S. Energy Minister Donald Macdonald visited Washington Jan. 30, Feb. 1, 1974 to discuss energy problems in Canadian-U.S. relations.

Accompanied by four deputy Cabinet ministers and other experts, Macdonald met with William Simon, director of the Federal Energy Office, and other officials. Both sides agreed to institute regular high-level meetings on energy issues, starting in Ottawa in about one month. Joint committees would be set up in specific areas, including a liaison group to deal with short-term supply problems, such as supplying Maine paper mills with Canadian oil and Vancouver with bunker oil from the U.S.

Macdonald told a press conference Jan. 31 that U.S. officials had advocated a comprehensive U.S.-Canada energy agreement, but that Canada's goal of self-sufficiency by 1980 precluded such a policy. Macdonald said Canada expected to reduce its oil exports to the U.S. in the next two years, followed by "a gradual withdrawal of Canadian oil from the American market" as Canadian consumption rose. He announced at a Feb. 1 news conference that the Athabasca oil sands would be developed for Canadian domestic use only, since the enormous capital investment needed to exploit the sands in quantities large enough to help fill U.S. needs would be too great for the Canadian economy to support. If cheaper offshore or Arctic discoveries were made, Macdonald said, exports to the U.S. could again increase.

Macdonald said Feb. 1 he had proposed in State Department talks the previous day that the two countries sign a bilateral pipeline treaty, to facilitate construction of a natural gas pipeline from Alaska through the Mackenzie Valley to the U.S.

Midwest. But he reported that U.S. officials had remained noncomittal on the proposal, and on an alternate trans-Alaska route proposed by El Paso Natural Gas Co. Both governments had expressed interest in a treaty guaranteeing unrestricted movement of oil through pipelines from Western Canada to Ontario through the U.S., and through Maine to Quebec, as a prerequisite to a Mackenzie Valley gas pipeline. Macdonald said Canada would allow increased natural gas exports in the near future, but not for the 16–20 year contracts previously obtained.

Macdonald said Feb. 1 he supported the conference of energy consuming nations to be held in Washington later in the month, and would support a reduction in the international price of oil. But he said Canada could not reduce its export tax on oil until then, since the tax was pegged to world prices paid for by eastern Canadian consumers, and to the domestic price received by U.S. producers for new oil.

Newfoundland-Brinco deal set. After two weeks of negotiations and a threat of nationalization, the Newfoundland government regained control of Labrador water power rights that it had formerly granted to Brinco Ltd. of Montreal, a resource development company it established in 1953.

Under terms of the March 28 agreement, the provincial government would pay $160 million cash for Brinco's 57% interest in Churchill Falls (Labrador) Corp. Ltd. and the company's water power rights. A Brinco subsidiary, Churchill Falls (Labrador) developed the $950 million hydroelectric project at Churchill Falls and sold all but a small amount of the electricity produced there to Hydro-Quebec; Brinco planned the same output distribution for another project.

The Newfoundland government took the position that the power should serve consumers within the province and offered March 15 to take over the company's shares. Premier Frank Moores introduced nationalization legislation March 18 after Brinco rejected the purchase bid, but negotiations continued until the compromise was reached.

Oil firms cut projects in tax dispute. The Amoco Canada Petroleum Co., a subsidiary of the Standard Oil Co. (Indiana) and one of Canada's most active oil and gas explorers, suspended all individual exploration indefinitely, it was reported May 23, 1974. Meanwhile, Texaco Canada, Ltd. announced drastic reductions in its exploration and development programs "pending clarification of budgetary treatment by the present government or a new government of . . . the oil industry generally."

A federal government proposal that would not allow provincial royalties on oil and gas production to be deductible for Canadian federal income tax purposes was the catalyst for the actions.

Although as yet not implemented (the budget and government were defeated May 8), the proposal was denounced by the Canadian oil and gas industry, which said many concerns would ultimately, in some cases, pay taxes on money they had never received.

Crude export levy raised. Ottawa increased June 1 the export surtax on western crude oil and refined products sold to the U.S. to $5.20 a barrel, pushing the cost of Canadian crude in the U.S. to $11.90 a barrel. The tax had been lowered from $6.40 to $4 a barrel in April when the Canadian price freeze on oil was lifted.

The June 1 increase was imposed, according to the National Energy Board, because world oil prices were "provisional" and "subject to retroactive adjustments"; Canada therefore had to "cover" itself, the board said. The tax revenues would be used to subsidize the cost of imported oil in the East.

Mackenzie Delta gas find. Canadian units of Gulf Oil Corp. and Mobil Oil Corp. announced a natural gas discovery in the Mackenzie Delta area of the Canadian Arctic, it was reported July 8. Northern Affairs Minister Jean Chretien said the discovery, the sixteenth oil or gas find in the region, added to reserves needed to support construction of the proposed Canadian Arctic gas pipeline.

Ontario approves power projects. The Ontario government approved electricity-generating projects valued at $2.6 billion, provincial Energy Minister Darcy McKeough announced July 11. The projects included a second nuclear generating station, an oil-fired generating station and the second and third of four planned heavy water plants at the Bruce generating station at Douglas Point on Lake Huron. Private financing was being sought for the heavy water plants.

Coal deals with Moscow, Japan, Europe. Kaiser Resources Ltd. of Vancouver signed an agreement with the Soviet Union's Ministry of Coal Industries July 19 for an exchange of technical expertise in hydraulic and open strip coal mining.

One of the world's largest strip operations, Kaiser Resources was 59% owned by Kaiser Steel Corp. of Oakland, Calif. The Vancouver firm held the North American and Australian licensing rights for an advanced hydraulic mining technology developed by the Mitsui Mining Co. Ltd.

It was reported March 13 that Kaiser Resources had signed an agreement with its Japanese customers to permit higher coking-coal prices to cover increased costs. (The firm exported 4.5 million tons of coking-coal annually to Japanese steelmaking concerns.)

Also reported March 13 were Kaiser agreements to sell thermal coal, used in energy generation, to Denmark and France and a "test shipment" to Ontario's government-owned Ontario-Hydro.

Natural gas price to U.S. raised. Energy Minister Donald Macdonald announced Sept. 20, 1974 Canada's decision unilaterally to raise the price of natural gas exported to the U.S. by 67%, to $1 per thousand cubic feet, effective Jan. 1, 1975. The government set separate prices for exported gas and gas sold domestically.

Macdonald said the increase would put Canadian gas in a "more equitable" relationship with other energy sources in the U.S. by pricing Canadian exports "in a competitive relationship" to U.S. alternatives.

The move would provide some $330 million in additional revenues from the 1 trillion cubic feet of natural gas which Canada exported annually to the U.S. The increased revenues would go to gas producers, the minister said, and not to pipeline companies. However, in the case of British Columbia, the additional funds would go entirely to the British Columbia Petroleum Corp., the provincial marketing agency.

Oil tax rebate set. A spokesman for the Department of Energy, Mines and Resources said Oct. 8 that Ottawa would rebate 50¢ of the $5.20 a barrel tax on heavy and medium crude oil exported to the U.S. The move was intended to help producers hurt by softened market conditions. Only about 100,000 of the 978,575 barrels of oil exported daily to the U.S. were of heavy or medium grade, however.

Phase-out of oil exports to U.S. Energy Minister Donald Macdonald announced Nov. 22 that the government would begin Jan. 1, 1975 an eight-year phase-out of oil exports to the U.S. The cut, set forth in a government statement based on a report issued by the National Energy Board (NEB), would reduce exports Jan. 1 to 800,000 barrels a day from the average level of 896,400 barrels of crude purchased by U.S. importers during the first eight months of 1974. (Current shipments were 977,950 barrels a day.)

Macdonald said a further reduction to 650,000 barrels a day would be imposed July 1, 1975 if the oil-producing provinces of Alberta and Saskatchewan concurred. Subsequent annual reductions would reduce exports to 5,000 barrels a day in 1983. The policy contemplated a complete phase-out thereafter.

The mid-1975 export cut would represent a "shut in," or production cut, of 250,000 barrels a day, if the provinces approved. This was the amount of oil the NEB estimated the Sarnia-Montreal pipeline, scheduled to be completed in 1976, would ship to the Eastern Canadian market which presently depended largely on imported oil.

The proposals issued by the NEB were designed to extend the period of national oil self-sufficiency by almost two years to the end of 1983 by accelerating export reductions. Macdonald said Ottawa had ruled out an immediate and total halt in exports to the U.S., the only country which imported Canadian oil, because "an immediate halt . . . would be disruptive of Canadian-U.S. relations" and would adversely affect northern U.S. refineries and Canadian producers.

The U.S. State Department said Nov. 23 it was "somewhat disappointed" at the Ottawa decision, but admitted that "we have known for more than a year of the likelihood that Canadian oil exports to the U.S. would be phased out around the end of the present decade." Hope was expressed, however, that Ottawa would not implement the mid-1975 production shut-in.

Alberta cuts royalties. Alberta Premier Peter Lougheed said Dec. 12, 1974 that the province would reduce provincial oil and gas royalties to help the petroleum industries overcome the effect of what he called "harsh and punitive" taxation imposed under the federal government's budget presented Nov. 18.

Federal Finance Minister John Turner Dec. 13 praised the premier's actions as "responsible and responsive." Lougheed estimated that the concessions would be worth at least twice the amount of the federal reductions. (The Nov. 18 budget had raised the industry's share of oil and gas production income to 29.5% from 24%; the Alberta move would raise it to 33%-34%. The federal budget had also provided that provincial royalties would no longer be considered deductible from federal income tax payments.)

Ottawa, provinces buy into Syncrude. Syncrude Canada Ltd., the consortium formed to extract oil from the Athabasca tar sands of northern Alberta, was rescued from possible death Feb. 4, 1975 when the federal government and the provincial governments of Ontario and Alberta agreed to invest $600 million for equity interests in the project. The future of the oil extraction development had been threatened when one of the consortium's four corporate members withdrew in December 1974.

Ottawa purchased a 15% interest in Syncrude for $300 million; Alberta acquired a 10% interest for $200 million; and Ontario bought a 5% interest for $100 million. The three governments thereby made Syncrude a public-private consortium, assuming the 30% interest of which Atlantic Richfield Canada Ltd. had divested itself.

Because $1 billion had been needed to salvage the project, the three remaining corporations agreed to increase their own investments by a total of $400 million, one half of which would be financed by the Alberta government. Imperial Oil Ltd. thus increased its interest to 31.25% from 30%; Canada-Cities Service Ltd. reduced its participation from 30% to 22%; and Gulf Oil Canada Ltd. raised its involvement from 10% to 16.75%. All three firms were subsidiaries of U.S. corporations.

The Alberta government also agreed to pay an estimated $600 million to build a pipeline to carry oil from the sands to market, an electrical plant to power the Syncrude development and "other infrastructure." These commitments brought total federal and provincial participation in the project to $1.4 billion.

According to J. A. Armstrong, chairman of Imperial Oil Ltd., the governments had, in joining the consortium, agreed to certain conditions: that Syncrude oil would sell at world prices, that taxation levels would be those which prevailed in 1973 when the project was begun, and that Syncrude would not be ordered to prorate its production.

New energy program. Energy Minister Macdonald announced an energy conservation program with a $1.3 million advertising budget, a speed limit of 55 miles per hour for federal vehicles and a reduction of parking facilities for federal employes to encourage greater reliance on public transportation, it was reported Feb. 7. The speed limit was to take effect immediately.

The program also involved consultations with provincial governments, industry and labor and would, on a long-term basis, be concerned with fostering a transition "from an economy obsessed with quantity to one that places a premium on quality and from a society focused on competition to one based more on sharing and compassion."

Export tax changes. It was announced Feb. 5, 1975 that, effective March 1, the export charge on Canadian crude oil, all of which was sold to the U.S., would be increased by 30¢ a barrel. The new levy would bring the price of Canadian crude to $12 a barrel.

Energy Minister Donald Macdonald said the rise "reflects changes which have occured in international crude oil and tanker freight markets over recent weeks and also the weakening of the Canadian dollar in terms of U.S. currency."

But two months later the National Energy Board (NEB) announced May 5 that the export tax on a barrel of crude oil would drop 80¢ effective June 1, lowering the tax to $4.20–$4.70 a barrel, depending on the quality of oil.

In announcing the tax cut, the NEB cited the need to make Canadian oil more competitive in U.S. markets.

Record price set for natural gas to U.S. The Canadian National Energy Board (NEB) and the U.S. Federal Power Commission March 18 approved the short-term export sale of as much as 55 million cubic feet a day of Canadian natural gas to the U.S. at prices averaging $1.61 to a maximum of $1.91 per 1,000 cubic feet. The gas, purchased from Pan-Alberta Gas, Ltd., would be supplied over 19 months, beginning Oct. 1, to the U.S. Pacific Northwest by the Northwest Pipeline Corp. of Salt Lake City.

NEB officials said the record high prices were granted as an exemption from the $1 per cubic foot border price. The main factor in the higher price was that the contract required Northwest Pipeline to pay for almost complete amortization over the 19-month delivery period of extra pipeline facilities required in Alberta and British Columbia to bring the gas into production.

Natural gas export price increased. Energy Minister Donald Macdonald announced May 5 that the export price of natural gas would rise by 60¢, in two installments, during 1975. The present price of $1 per 1,000 cubic feet would increase to $1.40 Aug. 1 and to $1.60 Nov. 1.

The announcement was criticized by the U.S. State Department May 6. The sole foreign importer of Canadian natural gas, the U.S. imported approximately one trillion cubic feet of gas annually from Canada, or about 45% of total Canadian production. The increased export price would cost U.S. purchasers an additional $583 million a year.

Ottawa's decision to implement a two-phase increase was a concession to the U.S., which had strongly opposed the scheduled single-phase increase.

In making the announcement, Macdonald also predicted future increases, saying the export price of Canadian natural gas would be equal to the world market price in two years.

Sarnia pipeline gets go-ahead. The Canadian National Energy Board authorized Interprovincial Pipe Line Ltd. to begin construction of the 520-mile oil pipeline between Sarnia, Ontario, and Montreal to provide eastern Ontario and western Quebec with oil from western Canada, it was reported May 21.

To be completed for the winter of 1976–1977, the pipeline would cost an estimated $185 million and would have a capacity, which could be doubled, exceeding 300,-000 barrels a day.

CHINA

Oil pipeline from north completed. The completion of a 715-mile oil pipeline from the Taching oil field to the Yellow River port of Chinwangtao was disclosed in a Peking radio broadcast Jan. 11, 1975. The broadcast, reported by a U.S. government monitoring service three days later, said a main line begun in 1970 had been completed in September, 1973 and a parallel line from Taching to Tiehling, north of Mukden, was finished in the autumn of 1974.

In 1974, China reported a 20% rise in oil production, which was estimated to total 60 million tons annually, a little less than half coming from Taching.

ECUADOR

CEPE takes Texaco-Gulf share. The state oil company CEPE June 6, 1974 took a 25% share of the operations of the Texaco-Gulf consortium in the northeastern jungle, despite the lack of a final agreement with the U.S. firms on details of the purchase.

The action gave CEPE a quarter of the consortium's daily production of 210,000 barrels of crude. Production had totaled 250,000 barrels daily until May 24, when the government ordered a 16% cut to conserve oil reserves.

Texaco Inc., which operated the consortium's properties, received a first compensation installment of $25 million from the Ecuadorean government July 2.

Besides the Texaco-Gulf share, CEPE took over 25% ownership of the Trans-Ecuadorean oil pipeline, also administered by Texaco. The pipeline, which reached 318 miles from the Texaco-Gulf oilfields to the Pacific port of Esmeraldas, was valued at $100 million.

The government had advised a group of three U.S. companies headed by Cayman Corp. that they could use the pipeline to transport crude extracted in the northeast. The group planned to begin construction of a $15 million, 50-mile interconnecting pipeline which would hook its fields up with the Texaco-Gulf pipeline, it was reported June 5.

Ecuador was forced to suspend oil exports July 8 because of damage done to the pipeline by heavy flooding in the northeast.

Oil taxes raised. Ecuador Oct. 8, 1974 increased the tax on the income of foreign oil companies by 8% and raised the royalty rate paid by the companies by .67%. The increases were retroactive to Oct. 1.

The moves conformed with a recent decision by the Organization of Petroleum Exporting Countries (OPEC) to increase by 3.5% the taxes and royalties paid by foreign oil companies in producing nations.

Ecuador was reported Jan. 17, 1975 to have again raised the foreign-oil-company income tax; the government share of petroleum output rose from $9.91 a barrel to $10.12. This raise also was in implementation of an OPEC decision.

Jarrin fired. Natural Resources Minister Gustavo Jarrin Ampudia was dismissed from the Cabinet Oct. 4, 1974 and replaced by a fellow navy captain, Luis Salazar.

According to reports, Jarrin had been under attack from conservatives for the aggressive nationalism he displayed in shaping the nation's oil policy and, since June, as general secretary of OPEC. Responding to President Ford's recent warnings to oil exporters to reduce their prices, Jarrin had declared: "OPEC will not accept a neo-colonialist policy of imperialism," the French newspaper Le Monde reported Oct. 8.

Other oil developments. The Ecuadorian government and a private Japanese consortium had signed an agreement to build a $91 million oil refinery as a first step toward a petrochemical plant, it was reported March 22, 1974.

Ecuador admitted Jan. 6, 1975 that it was coordinating its oil policy with Venezuela's.

The state oil firm CEPE had decided earlier to sell 50% of its 1975 production to Chile and Peru, solving its short-term marketing problems, it was reported Jan. 10. CEPE also would begin importing Peruvian natural gas.

CEPE's manager, Col. Raul Vargas, had denounced the major multinational oil companies for not making acceptable bids for an oil and gas exploration contract in the Gulf of Guayaquil, it was reported Jan. 17. Only one multinational—Northwest Pipeline Corp.—had made such a bid. The Texaco-Gulf consortium, the largest foreign oil producer in Ecuador, had refused to reinvest more than the legally required minimum in its operations in the eastern jungle, the report said.

EGYPT

Oil strike. Egypt was reported March 23, 1974 to have discovered an oil field in the Gulf of Suez near Ras Ghareb. The 57-square-mile find, the largest in Egypt, was expected to yield an initial 50,000 barrels a day, rising eventually to 300,000 million barrels daily.

These five Western oil firms were reported March 26 to have been granted concessions to explore for oil in Egypt: Mobil Oil Corp., Standard Oil Co. (Indiana), Continental Oil Co., Tripco Petroleum, and Ente Nazionale Idrocarburi of Italy. Mobil had previously signed an agreement with Egypt for an exploration program on 1.5 million acres offshore in the Nile Delta area.

FRANCE

Natural gas cut. The French government Jan. 2, 1974 ordered a 25% cut in industrial gas supplies, mainly in the southern and central regions of the country. Domestic consumers were also asked to reduce their use of gas.

The action was caused by a 15% reduction in Algerian gas deliveries because of a breakdown in a natural gas liquefaction plant in Algeria.

Gas consumption in France had risen by 25% in 1973, leaving little spare capacity in the nation.

(Great Britain agreed to give France one billion thermal units of its natural gas supplies from Algeria in exchange for other petroleum products, mainly light gasoline, the French newspaper Le Monde reported Jan. 5. The contract was negotiated before the temporary natural gas cutbacks were decided.)

Fuel prices raised. The government Jan. 10 announced sharp increases in the price of gasoline, heating and industrial fuel oil, effective the following day, to reflect recent price raises by oil-producing nations.

The price of a liter of high-test gasoline rose by nearly 30% to 1.75 francs (about $1.30 a U.S. gallon), the second highest rate in Western Europe after Italy. The price of heating fuel rose by 45%.

Energy plan announced. In an effort to reduce French oil consumption, the Cabinet March 6, 1974 adopted a plan that set limits on home heating, cut spending on highways and provided for construction starts of 13 nuclear power plants through 1975. The program was designed to save 10 million tons of oil a year.

The accelerated atomic power plant construction program would be maintained so that nuclear electricity would provide 70% of France's total electricity production by 1985. Immediate gas and coal price increases were also approved.

As part of the new emergency energy program, the Cabinet Sept. 25 ordered a "definitive" limit on spending for oil imports in 1975 and ordered reductions in industrial and home heating consumption.

Expenditure on oil imports during 1975 would be limited to a maximum F51 billion ($10.7 billion), representing a 10% reduction, at prevailing prices, from the volume of oil imports in 1973.

A national fuel consumption agency would be created to supervise rationing of heavy industrial fuel to factories and other users. Rationing for industry would also cover naphtha and various plastic products.

In homes and public buildings, the government ordered central heating to be limited to a maximum of 20° Centigrade (68° Fahrenheit). The Cabinet also ordered the state coal board to increase its production goal by 46 million metric tons over the next 10 years.

President Giscard d'Estaing, opening the annual Paris automobile show Oct. 4, virtually ruled out gas rationing and greater use of speed limits, which some Cabinet ministers favored. His statements revealed concern over France's slumping automobile business.

Parliament approves—The emergency energy conservation program was adopted by the National Assembly Oct. 5, giving the government unlimited powers to legislate by decree to meet the energy crisis. But when he introduced the bill Oct. 4, Premier Jacques Chirac had said the government would rely on voluntary restraint by industry and the public to achieve the domestic oil consumption cutback proposed in the bill.

Chirac noted that oil consumption since January had declined 4.7%, compared with a previous 9% annual increase. (However, France had an unusually mild 1973–74 winter and stricter speed limits.)

Prices & consumption. France raised energy prices Jan. 1, 1975. The range was from gasoline's 1.7% boost—to the equivalent of $1.56 for a U.S. gallon of high test—to 20% for natural gas for industrial users. The increase in electric power was 19.2% for industrial use and 11.7% for homes.

French consumption of petroleum products declined from 111 million tons in 1973 to 104.8 million tons in 1974, it was reported Jan. 14. Fuel for domestic use showed the sharpest decline—15.3%.

10-year energy plan adopted. The National Planning Council Jan. 28, 1975 approved a 10-year energy program to reduce the nation's dependence on foreign fuel from the 76% level in 1973 to 55% or 60% by 1985. Under the program, no one country would be permitted to supply more than 15% of France's total energy supplies.

The plan envisaged that by 1985 France's energy consumption would be limited to the equivalent of 240 million tons of petroleum, a 16% reduction from the 284 million-ton target previously set; domestic coal production would be stabilized at the 1973 level through the development of French mines and more coal imports.

The Planning Council adopted further energy decisions Feb. 1, including authorization to the state electricity board to begin construction of 12 new 1,000-megawatt nuclear power stations in 1976–77, down from the 14 originally planned.

Oil & gas pacts. A subsidiary of France's state-owned oil company ERAP and the French exploration company Aquitaine had agreed to acquire a 14.1% interest in a permit from the Italian government to drill for offshore oil in the Italian Adriatic, it was reported Jan. 29. The concession was held by Intercontinental Energy Corp., a U.S. exploration and production company.

Aquitaine Oil Corp., a U.S. branch of the French firm, announced it had purchased oil and gas reserves and mineral properties from Pruet & Hughes Co., a U.S. company based in Jackson, Miss., for $25 million, it was reported Jan. 15. The properties were located mainly in Mississippi and Alabama.

Oil imports down. France imported 34.5 millions tons of oil in January–May, a 23.9% decline compared with the same period of 1974, it was reported June 20. Consumption of domestic heating oil and heavy industrial fuels had declined in May by 14.6% and 22% respectively compared with a year earlier.

GREAT BRITAIN

New energy department. Prime Minister Edward Heath Jan. 8, 1974 created a Department of Energy, with power over the nation's programs for offshore oil, electricity, coal, gas and nuclear energy.

The new department would be headed by Lord Carrington, outgoing secretary of defense. His defense portfolio was given to Ian Gilmour, formerly his assistant.

Major coal find. National Coal Board Chairman Derek Ezra Jan. 24 confirmed Britain's biggest coal find of the century, which he said would be comparable in size and importance "with the oil discoveries so far made in the British sector of the North Sea." The find was made between Selby and York in eastern England.

Coal-mine strike. Following a coal-miner work slowdown that had begun in November 1973, had cut coal deliveries by about 40% and had played a major role in bringing on an economic crisis, the National Union of Mineworkers began a nationwide strike Feb. 10, 1974. After the resignation of Heath's Conservative government and the installation of a new Labor government, the dispute was settled by a pay agreement March 6, and coal production was resumed March 11.

The contract, providing for an overall 30% increase in wages that would cost the equivalent of about $230 million, nearly doubled the 16½% increase worth $100 million offered by the Heath government.

The NUM accepted the pay offer, by a 25–2 vote, after 12 hours of negotiations with the government-run National Coal Board.

The settlement would reportedly provide increases ranging from $15.43 a week for the lowest paid surface workers to $37.50 for night-time pit workers, with weekly wages to rise to $74 for surface workers, $83 for lower-grade underground workers, $104 for daytime pit miners and $122 for night-time pit workers. The contract included an increase from the equivalent of over $400 to $1,000 in the lump sum payment given to miners who retire voluntarily.

Ironically, the sum total of the accord was virtually the same as that of the Coal Board's original offer plus the recommendations of the Pay Board's report on wage relativities issued March 6.

Because of the coal settlement, the new energy secretary, Eric Varley, March 7 ended Britain's three-day workweek. But curbs on heating and lighting in shops and offices would continue until coal stocks returned to normal.

Varley estimated that the three-day workweek, imposed Jan. 1 because of the miners' initial slowdown and continued after the strike began Feb. 10, had cost the nation around $4.6 billion in lost production and wages.

Budget triples gasoline tax. The Labor government of Prime Minister Harold Wilson presented a new budget Nov. 12, 1974. It provided for a tripling of the tax on gasoline.

Delivering the budget message to Parliament, Chancellor of the Exchequer Denis Healey said that the main tasks were to promote investment so that productivity would increase, and to protect the balance of payments against its growing oil-related deficit.

Announcing a national campaign against waste, Healey said, "Our present pattern of prices, subsidies and taxes simply does not fit a world in which the price of imported oil has increased fivefold in less than a year." He said the government planned to phase out subsidies for most of the deficit-ridden nationalized industries, a move that probably would result in increases in rail and air fares, electricity rates, and natural gas and coal prices. The first big increase for consumers would be the tripling, effective Nov. 18, of the value-added tax on gasoline, from 8% to 25%. This would raise the price of medium grade gasoline about 16¢, to $1.15 a U.S. gallon. (The value-added tax was a form of sales levy set on each stage of the production and distribution of goods.)

Fuel conservation measures. Energy Secretary Eric Varley announced an energy conservation program Dec. 9 designed to reduce the nation's energy use by 10% and save £700 million a year in imported oil costs.

The program called for a reduction in highway speed limits, effective Dec. 14 by statutory order, a ceiling of 68° Fahrenheit on the heating of office and factory buildings and, effective Jan. 1, 1975, a daylight-only use of electricity for external displays and advertising.

Britain's Price Commission Nov. 14 had rejected 30 applications for price increases from oil companies. According to an unidentified source, the commission had concluded that the increases, based on higher crude oil costs, had "jumped the gun" on the cost increases.

Government aids Burmah Oil. The government announced Dec. 31 that the Bank of England would guarantee a $650 million one-year loan to Britain's second largest oil company, Burmah Oil, which said it expected to show a big 1974 loss in its oil tanker operations. In exchange, the Bank of England would acquire Burmah's "unpledged" shares in the British Petroleum Co. (BP) and Shell Transport & Trading Co., the British unit of the Royal Dutch/Shell group. Burmah held a 21.5% interest in BP and a 2% share in Shell Transport. The government already owned 49% of BP.

Burmah also agreed in principle to turn over to the government 51% of its stake in North Sea oil and gas fields.

Shell & BP cut prices. Shell and British Petroleum May 20, 1975 announced a 2⅓¢-a-gallon cut in the scheduled price of gas oil and diesel fuel because of "competitive pressures." It was the first cut in scheduled oil prices for more than five years and came, according to the Financial Times of London May 21, "at a time of increasing pressure in the 'middle distillate' range of oil products in Europe following a mild winter and a continuing surplus of supply."

GREECE

Energy measures. The government announced a series of strict energy conservation measures Jan. 11, 1974.

The price of premium gasoline was raised 65% to $2.50 a gallon, while regular gasoline was increased 75% in price to $2.05 a gallon. A maximum speed limit of 50 m.p.h. was imposed, as well as an alternate weekend driving ban based on odd and even license plate numbers.

In addition, street lighting was cut by one-half, store lighting banned after 9 p.m. and television broadcasting cut off after 11 p.m.

Oil and gas found. Commercial quantities of oil and natural gas were found off the island of Thassos in the northern Aegean Sea by the Oceanic Exploration Co. of Denver, Colo., it was announced Feb. 14, 1974. It was estimated that the deposits could provide about 500,000 tons of crude, which would satisfy 5% of Greece's domestic requirements.

(Greece presently imported all its oil.)

The well was also capable of producing some 40 million tons of natural gas which, according to Industry Minister Constantine Kypraios, would be used either for industrial purposes or to fuel a thermoelectric plant in Kavala, the mainland port north of the deposits.

Greece's sovereignty over submerged areas in the Aegean Sea was disputed by Turkey, it was reported Feb. 25, after Turkey had awarded to two U.S. companies exploration rights in the eastern Aegean, in areas it considered international waters. "The Aegean Sea is not a Greek lake," Ankara officials contended.

In their claims, the two governments presented conflicting interpretations of the doctrine of the continental shelf, the criterion adopted by the 1958 Geneva convention for determining territorial waters. Turkey, which signed but did not ratify the document, held that continental shelf jurisdiction did not extend to the Greek islands; Greece insisted that its more than 350 Aegean islands had a legitimate continental shelf.

GREENLAND

Offshore leases. The Danish government April 8, 1974 granted the first concessions for oil and natural gas exploration off the west coast of Greenland. Six groups comprising 20 companies from nine nations won the leases for the 15,000 square mile area.

Leasing conditions set an initial exploration period of up to 16 years, with production rights for up to 30 years. The Danish government could claim up to 80% of the proceeds in royalties, taxes and production sharing.

U.S. companies involved in the grants were Chevron, Mobil, Gulf, Amoco, Cities Service, Murphy, Atlantic Richfield and Ultramar. Other firms included British Petroleum Co., the National Iranian Oil Co. and the French firms of Total and Aquitaine.

GUATEMALA

Oil discovered. Outgoing President Carlos Arana said June 25, 1974 that oil had been struck at a well in the Tubelsalto area near the Mexican border. Production was expected to reach 3,000 barrels daily, or 15% of Guatemala's domestic oil consumption.

The strike was made by Recursos del Norte, a mostly German-financed company, which planned to drill at least five more wells in the area. A score of multinational firms had searched for oil in Guatemala since 1956.

INDIA

Offshore oil discovered. The discovery of an oil well in the Gulf of Cambay, 115 miles northwest of Bombay, was reported by Indian Petroleum Ministry officials Jan. 17, 1975.

The ministry estimated the field had a reserve of one billion barrels, enough to yield 10 million tons a year for 15–20 years. Two U.S. firms already had contracts for oil exploration elsewhere in the Arabian Sea and the Bay of Bengal.

IRAN

New oil, gas reserves found. The discovery of new oil and natural gas deposits in Iran was announced July 31, 1974.

The government-owned National Iranian Oil Co. said "gigantic" oil reserves had been found near Ahwaz in southwest Iran. Discovery of a major natural gas deposit in the Persian Gulf containing an estimated 70–100 trillion cubic feet was announced by Nissho-Iwaii & Co., a Japanese trading firm that had conducted drillings with three other foreign companies.

National Iranian Oil July 30 had signed two separate contracts with West Germany's Deminex group of companies for exploration and exploitation of two oil-rich areas near Shiraz and Abadan.

Oil revenues quadruple in '74. Iran's oil revenues quadrupled to almost $20 billion in 1974, according to a report released Feb. 10, 1975 by National Iranian Oil Co. Production in 1974 was about 2.7% higher than in 1973.

The announcement followed an earlier report by a Teheran newspaper Jan. 25 that Iran's oil exports had dropped by more than 10% in January compared with December 1974 and that the downward trend would probably continue in the months ahead. The English-language newspaper Kayhan International, which generally reflected government views, attributed the export drop to a decline in demand for Iranian crude by several major Western oil companies. Oil exports in 1974 averaged more than five million barrels a day.

IRAQ

Major oil strike reported. A major oil discovery on the western fringes of Baghdad was announced by Iraqi Higher Education Minister Ghanem Abdel Jalil, the Beirut newspaper Al Anwar reported Feb. 8, 1975.

Jalil, who also was chairman of the board of directors of the Iraq Petroleum Co., was quoted as saying the field contained "proven deposits bigger than those of Iran and Kuwait," which would make Iraq's reserves "the world's second biggest, after Saudi Arabia." Iraq, which produced two million barrels a day, already had proven reserves of 31.5 billion barrels.

Iraq cuts oil output. The Middle East Economic Survey reported June 21 that Iraq had decided to lower oil production and petroleum exports and would concentrate instead on oil exploration for the next five years.

The publication quoted Iraq's secretary general of the committee of oil affairs, Adnan Hamdani, as saying that the country's output target had been reduced to about 200 million tons a year by 1982 from 325 million by 1980. A total of $1.5 billion would be earmarked for oil exploration in a five-year period, Hamdani said.

IRELAND

Terms for offshore oil. Irish Industry & Commerce Minister Justin Keating April 29, 1975 announced terms for new offshore oil and gas exploration licenses that would net the state about 80% of the revenue produced from wells. The revenue would come from the current corporation profit tax of 50%, royalties ranging from 8%–16% depending on production volume, licensing fees and the government's right to acquire up to 50% equity participation in successful wells.

With a 50% equity share, the government would undertake to contribute an equal portion of exploration and development costs.

Properties already being developed by companies would continue under previously agreed, but still undisclosed, terms.

Oil spills. Three oil spills in five months had occurred at the Gulf oil terminal at Whiddy Island in Bantry Bay, according to the World Environment Report April 14. The latest spill from an oil tanker involved only an estimated 240 gallons, compared with spills of 651,000 gallons in October 1974 and 150,000 gallons in January.

ITALY

Record oil, gas find. The state-owned Ente Nazionale Idrocarburi (ENI) announced Italy's biggest oil and gas find, at Malossa in the Po basin, it was reported Oct. 15, 1974. Tests reportedly revealed minimum reserves of at least 50 billion cubic meters of natural gas and 40 million tons of crude oil.

The find could supply up to 6% of Italy's energy needs, currently 85% imported.

ENI announced it had completed laying the world's deepest underwater gas pipeline, covering a more than nine-mile stretch across the Strait of Messina between Sicily and mainland Italy, it was reported Sept. 24. At its deepest point, the pipe was 1,800 feet below sea level. It was the first stage of a project which would pipe natural gas from Algeria to La Spezia, in northern Italy.

JAPAN

Industry fuel curtailed. The government issued an order Jan. 11, 1974 requiring industry to cut its use of oil and electricity by 5%–15% until the end of February. The electricity curbs went into effect Jan. 16; restrictions on oil started Feb. 1.

The directives superseded the government's "administrative guidelines" of November 1973 and put into effect instead the tighter restrictions announced in December. Under the new order, industry was placed in three categories—5%, 10% and 15%. The 5% cuts applied to transportation, food, pharmaceutical and data-processing industries. The 10% categories included newspaper publishing and broadcasting, and most other industries faced 15% reductions.

But the government Feb. 9 ordered an easing of the restrictions because more oil had become available than had been expected. The current 15% cutback was reduced to 10%.

South Korean accord. Under an agreement signed in Seoul Jan. 30, 1974, Japan and South Korea were to jointly develop seabed petroleum resources in the East China Sea.

The area, in which the claims of both nations overlapped, was a 366,000 square mile tract south of the Korea island of Cheu Cheju-do and west of the Japanese island of Kyushu.

Three oil firms—Gulf, Texaco and Royal Dutch Shell—were already conducting exploration in the region. Under terms of the agreement, they were to be joined by three Japanese companies.

China charged Feb. 5 that the accord infringed on Chinese sovereignty in the East China Sea. The Foreign Ministry warned Japan and South Korea that they would "bear full responsibility for all consequences" resulting from their agreement.

Soviet coal & gas deals. Japan and the Soviet Union had initialed a memorandum which would initiate a joint project to explore and extract the vast coal deposits in the Siberian Yakutsk region, it was reported March 10, 1974.

The U.S.S.R. and Japan April 26 signed an agreement confirming their intention to prospect for natural gas in the Yakutsk region of eastern Siberia. The project was contingent on U.S. participation.

Under the accord signed by Hiroshi Anzai, board chairman of the Tokyo Gas Co., and Vladimir N. Sushkov of the Soviet Ministry of Foreign Trade, Japan was to provide up to $100 million in credit (to be matched by the U.S.).

Tokyo and Moscow June 3 signed a pact under which Japan would grant the U.S.S.R. the equivalent of $450 million in credits in exchange for 104.4 million tons of Soviet coal from 1979 to 1980.

The U.S.S.R. had announced ending the Tyuman oil pipeline project, a planned Soviet-Japan deal which Tokyo had considered vital to the maintenance of its fuel supplies.

The Soviet news agency Tass said that Tokyo would provide $100 million to finance a joint five-year search for oil and natural gas on Sakhalin, an island that figured in a territorial dispute between Japan and the U.S.S.R., it was reported Feb. 12, 1975.

KUWAIT

Kuwait gets Gulf Oil, BP control. The Kuwaiti Parliament and Cabinet ratified May 14, 1974 an agreement giving Kuwait a 60% share in the Kuwait Oil Co., jointly owned by Gulf Oil Corp. and British Petroleum Co. Ltd. An increase in the 60% share was to be negotiated after Dec. 31, 1979. The agreement was retroactive to Jan. 1.

According to details of the accord disclosed by the Kuwaiti government Jan. 31, Kuwait would pay $112 million for 60% of the company's operations, and there was no provision for Gulf and BP to buy back the majority control. The pact was for five years and gave the government the right to raise its participation seven percentage points a year to a total of 95% at the end of 1979. The 60% acquisition would cover all operations, including exploration, production and refining.

A new firm would be formed that would be headed by a Kuwaiti. It would operate under a joint-management committee of Kuwaitis and Gulf and BP personnel.

Gulf & BP pay higher prices—Gulf and BP disclosed July 18 that they had agreed to pay $10.95 a barrel for Kuwait government participation oil during the third quarter of 1974, under threat of having their access to that oil cut off permanently.

Payment of the higher price—which amounted to 95% of Kuwait's posted price of $11.55 a barrel—was criticized by the U.S. and British governments, which feared it would cause new increases in world oil prices, it was reported July 20.

Gulf said it had agreed to the higher price July 16, after "it was made clear to Gulf and BP officials that unless each company was immediately prepared to purchase substantial quantities of the government's share of the oil at not less than $10.95 a barrel, the government wouldn't make any of its share available to that company in the future."

Gulf and BP each agreed to purchase 350,000 barrels of Kuwait crude a day at the $10.95 price during the current third quarter.

The agreement followed Kuwait's rejection July 9 of bids for the oil from some 40 U.S., Japanese and European companies. The influential Middle East Economic Survey said the companies had offered an average $10.737 a barrel, but other oil industry sources quoted by the London Times July 10 put the average bid at $10.25 a barrel.

Kuwait controls Arabian Oil Co. Kuwait and the Japanese-owned Arabian Oil Co. signed a "participation" agreement Aug. 25, 1975 giving Kuwait 60% of the firm's shares and assets. Under the terms of another accord, Kuwait was to sell Arabian Oil all of the government's share of crude until the end of the third quarter of 1974 at $10.95 a barrel.

Gulf, Shell end oil pact. The Gulf Oil Corp. and the Royal Dutch/Shell group announced Jan. 4, 1975 an end to their long-term agreement under which Gulf supplied Shell with crude oil from Kuwait. The accord, however, was replaced by a new agreement under which Shell would continue to buy a smaller amount of Gulf's total available Kuwait crude in 1975. A Shell official estimated the 1975 purchases at 60 million barrels, compared with 190 million in 1974.

The old contract, entered into in 1947, was being terminated because of "the changed relationships" between oil producers and international companies, the two firms said.

New oil minister. A new Cabinet formed Feb. 9 included Abdul Muttaleb al-Kazimi as oil minister. He replaced Abdel Rahman al-Atiki, who retained his other post as finance minister.

LEBANON

Trans-Arabian Pipeline closing. Four U.S. oil firms said April 10, 1975 that they were closing the Trans-Arabian Pipeline (Tapline) because of supertanker competition and losses of more than $100 million a year, partly because of a dispute with Lebanon over oil prices and transit fees. Tapline had a capacity of 465,000 barrels a day and routed oil for Europeans via the Lebanese coast. It also was a major source of oil for Lebanon, Syria and Jordan.

The four companies that owned Tapline were Exxon Corp., the Standard Oil Co. of California, Texaco and the Mobil Oil Corp.

LIBYA

Occidental-Libyan pact. A new 35-year agreement between Occidental Petroleum Corp. and Libya for oil exploration was announced Feb. 7, 1974 by company chairman Armand Hammer. The agreement covered an area of 19 "blocks" of about 11 million acres in an area which contained all of the country's producing oil fields. Under the agreement, 81% of the oil produced would go to the Libyan government and 19% to Occidental.

Libya gets Exxon, Mobil control. Under an agreement signed in Tripoli April 16, Libya gained 51% control of the operations of Exxon Corp. and Mobil Oil Corp. The accord was in conformity with a Libyan nationalization law decreed Sept. 1, 1973.

Libya halts Exxon production. Libya ordered Exxon Corp. to stop all its oil production in the country, affecting a daily output of 225,000 barrels, or 17% of Libya's total petroleum production, it was reported Oct. 10. Two other smaller operations also were directed to close—W. R. Grace & Co. and Atlantic Richfield Co.

The action against Exxon was related to the firm's dispute with Italy's state-owned Ente Nacionale Idrocarburi (ENI). Exxon had stopped shipping liquefied natural gas to Italy because of a disagreement with ENI over prices. As a result, Exxon was forced to close its liquefied gas plant. After the plant's closure, Libya refused to permit Exxon to "flare" gas produced in association with oil in the oil fields. (Flaring was a process in which gas was burned or wasted at the wellhead.) Libya then ordered the closedown of all oil and gas production in which Exxon affiliates had interests.

Exxon had signed a $90 million oil exploration agreement with Libya the previous week. Exxon said the accord provided for both offshore and onshore drilling.

Libya compensates British Petroleum. British Petroleum Co. (BP) accepted about $40.4 million from Libya as compensation for the nationalization of the firm's property in 1971, the company announced Nov. 25, 1974.

Libya had agreed that it owed BP $144.8 million, but deducted counterclaims of $104.4 million, which reflected "taxes, royalties and other claims" by Libya, the company said. BP had filed an original compensation claim of $580 million.

Libya seizes 3 U.S. oil firms. Libya announced Feb. 11, 1975 the nationalization of three U.S. oil firms—Texaco Inc., the California Asiatic Oil Co., a subsidiary of Standard Oil of California, and Libyan American, a subsidiary of Atlantic Richfield Co.

Tripoli radio called the action a "severe blow to American interests in the Arab world." The broadcast did not give a reason for the take-over. It said employes of the companies could stay on their jobs if they wished.

Libya had taken 51% control of the three firms along with several others Sept. 1, 1973. The government had warned at the time the companies would come under complete Libyan control unless they agreed to terms of nationalization.

MEXICO

Oil investment set. The government announced plans to invest $5.5 billion in the oil and electricity industries in 1974–76, it was reported Jan. 10, 1974.

Mexico currently imported about 8% of its oil—45,000 barrels a day—from the U.S. and Venezuela. The state oil firm

Pemex planned to open 2,000 wells in 22 states by 1976, boosting production from 550,000 barrels a day to more than 700,000 barrels daily.

National Patrimony Minister Horacio Flores de la Pena said Mexico would continue to import natural gas from the U.S., after it became self-sufficient in oil, to meet 35% of its needs.

Major oil deposits found. The discovery of important new oil deposits in southeastern Mexico was reported by U.S. officials Oct. 11, 1974 and confirmed by the Mexican government Oct. 12.

U.S. officials said reserves of up to 20 billion barrels of crude oil had been found in the Mexican states of Chiapas and Tabasco, near the Gulf of Mexico. One official said the area might be as rich in petroleum as the Persian Gulf.

Mexico confirmed the discovery Oct. 12 but claimed the 20 billion barrel estimate was wildly exaggerated. Horacio Flores de la Pena, Mexico's national patrimony minister, said: "We ourselves . . . do not know how much there is. We are now working 40 wells, each of which produces an average of more than 5,000 barrels a day. Only when we have 100 wells will we know."

Reports of the discovery raised hopes in the U.S. that Mexico, which recently became a net oil exporter, might sell its new oil at prices below those set by the Organization of Petroleum Exporting Countries (OPEC). However, Flores de la Pena said Oct. 15 that Mexico sought observer status in OPEC and that it would sell its oil "at OPEC prices."

"Mexico rejects any suggestion that she might play a role of weakening the common front of the oil exporting nations," Flores de la Pena declared. "As the pioneering nation in total control over oil resources [Mexico nationalized its oil industry in 1938], Mexico will never be the Trojan horse of the multinational corporations."

Asked how much of Mexico's oil exports might go to the U.S., Flores replied: "It will depend on the price offered by the United States compared to that offered by other countries."

Antonio Dovali Jaime, director of the state oil monopoly Pemex, said Oct. 15 that Mexico would seek "the maximum diversification of purchasers" for its oil, giving "special priority to the needs of developing countries, above all those of Latin America."

(Flores de la Pena said Oct. 15 that Cuba was the first country to have been offered oil from the new deposits. But the offer, made Sept. 12, was rejected by Cuban Premier Fidel Castro because Mexico asked payment at world prices in cash, the Miami Herald reported Oct. 19.)

President Luis Echeverria Alvarez said Oct. 12 that Mexico must be "profoundly nationalist, anti-imperialist," in its oil policy. "We must cultivate that patriotism lost by those with a colonial mentality," he declared. Echeverria said Oct. 19, on the eve of a meeting with U.S. President Ford, that Mexico would not make any concessions to the U.S. regarding oil.

Mexican officials were suspicious of the timing of the U.S. reports of the discovery, coinciding with U.S. efforts to reduce world oil prices and coming shortly before President Ford's meeting with Echeverria, the London newsletter Latin America reported Oct. 18. In an apparent attempt to apply pressure to Echeverria, Interior Secretary Rogers C. B. Morton, chairman of the U.S. Energy Resources Council, said Oct. 16 the meeting would be an excellent opportunity for the U.S. to persuade Mexico to sell its oil below OPEC prices.

Having operated its own oil industry for 36 years, Mexico had the technical knowhow to exploit its new resources. However, this would take at least two or three years, according to press reports, because of a world shortage of oil production equipment. Dovali Jaime said Oct. 15 that Mexico was working with Venezuela to produce some of the necessary equipment.

Flores de la Pena said Oct. 15 that Mexico planned to increase its oil refinery capacity so that by 1976 it would no longer be exporting crude petroleum.

MOROCCO

Gas & shale deposits. In his national address March 3, 1974, King Hassan confirmed the presence of sizeable shale oil deposits in Morocco's middle Atlas Mountain region east of Rabat. The deposits were large enough, the King said, to satisfy the country's needs "for centuries."

Natural gas was found when a wildcat well was sunk south of the shale sites, the Journal of Commerce reported March 12, but the find was considered insignificant.

THE NETHERLANDS

Special powers act adopted. The First Chamber (upper house of Parliament), by 52-17 vote Jan. 8, 1974, passed a special powers act giving the government authority to control wages, prices, rents, dividends and layoffs caused by the energy crisis. The act, designed to alleviate the effects of fuel shortages, would require the government to give Parliament advance notice of any planned economic measures. The lower house of Parliament had approved the bill in December 1973.

Gasoline rationing. Gasoline rationing went into effect in the Netherlands Jan. 12, 1974 but was abolished Feb. 4 and replaced by a ban on Sunday pleasure driving every other week. Announcing the decision Jan. 23, Economic Affairs Minister Ruud Lubbers said the oil companies had informed him that the gap between supply and demand was only 15% currently. Such a situation, he said, negated the need for rationing.

The rationing system, whose purpose had been to cut gasoline consumption by 20%, allocated each driver 15 gallons for the first three weeks, a more liberal ration than originally set. Extra fuel was allotted for professional use other than commuting. An initial ban on Sunday driving had been dropped earlier.

The government had imposed the rationing to conserve oil supplies reduced by the Arab oil squeeze. However, Premier Joop den Uyl conceded Jan. 11 that the oil situation had "improved slightly" in January, with refinery production rising from 68% to 76% of capacity. A key union official had said that 12.5% more oil had arrived in the port of Rotterdam during the first week of January than the first week of December 1973.

A union official had said Jan. 1 that oil supplies to Rotterdam had declined by 25% in December, compared with September, before the start of the energy squeeze. The mayor of Rotterdam said the same day that non-Arab nations, including Iran, Nigeria and Venezuela, were sending more oil to the Netherlands—"none of this under the counter."

In a related development, prices of gasoline and other petroleum products were raised Jan. 1, including a 1.8% increase in heavy fuel prices.

Holland again raised the prices of gasoline and other oil products, effective March 1, reflecting increases in the cost of crude oil. The price of super gasoline was increased nearly 4¢ a liter to 35¢, while a similar raise boosted the price of regular gasoline to 34¢ a liter. (One U.S. gallon equals 3.79 liters.) Fuel oil for domestic use was also increased.

NICARAGUA

Oil & gas. President-elect Anastasio Somoza Debayle announced the award of 30 oil and gas exploration concessions to foreign companies, including Texaco Inc. and El Paso Natural Gas Co. of the U.S., it was reported Oct. 11, 1974.

The firms, spurred by the discovery of substantial oil deposits in Mexico, were exploring Nicaragua's Pacific and Caribbean coasts. Explorers already had found seven apparent gas structures in the Pacific, one of which was "about seven times the size of the North Sea structure," Somoza said.

A high Nicaraguan government source said Somoza, whose family's wealth was estimated as high as $500 million, had invested heavily in the explorations, according to the Miami Herald.

NIGERIA

Oil take-over declared. The Ministry of Mines and Power issued a decree, backdated to April 1, 1974, announcing the assumption of a 55% interest in five oil exploration companies operating there, it was reported May 19. Nigeria's total daily output was 2.3 million barrels.

Nigeria had previously held no interest in either of two U.S. companies included in the take-over—Gulf Oil Corp. and Mobil Oil Corp. Spokesmen for the firms said the 55% figure would be accepted only if an entire participation package could be negotiated. Such an agreement had been reached with the Royal Dutch-Shell Group/British Petroleum Co. joint venture (the largest producer and one of the concessions included in the decree).

In a related development, Gulf Oil Corp. announced that it would invest over $975 million in a proposed Nigerian liquefied gas project, it was reported March 8. The U.S. was expected to provide the major market for the 500 million cubic feet of gas to be produced daily. Lagos estimated that the gas project would cost the Nigerian government $1.169 billion.

Texaco production halted—Texaco Inc. confirmed that its oil production in Nigeria had been halted since May 8 under orders of the Nigerian government, the Wall Street Journal reported June 7. According to the Lagos Daily Express, Texaco had reportedly engaged in the illegal export of crude oil. A company spokesman, however, said the halt had been ordered because Texaco was producing at less than a fifth the scheduled rate of 50,000 barrels a day. Standard Oil of California's production was also halted for that reason, Texaco said. Neither firm was a major Nigerian producer.

Eximbank loan for oil project—The U.S. Export-Import Bank extended a $1.8 million loan to a Nigerian subsidiary of Ashland Oil Inc. to finance the purchase of $4 million worth of U.S. oil exploration equipment and services, it was reported May 30.

Ashland had announced the discovery of a second oil well that tested a production rate of 4,900 barrels a day, it was reported March 26.

Cheaper oil for Africans. Nigeria announced July 31, 1974 that it would sell oil to African countries at a cheaper rate than it offered to other nations. The decision contrasted with the position taken by the Arab oil-producing nations which had refused to provide African countries with oil at reduced prices.

Oil prominent factor in growth. Nigeria's economy grew at a rate of 10.2% annually during the 1970–74 development plan (reported Aug. 10, 1974). Planners had predicted a rate of 8.3%.

In a nationwide broadcast March 31, head of state Gen. Yakubu Gowon had announced a $4.7 billion budget calling for massive investments in internal development, including a liquefied natural gas industry.

According to the Ministry of Mines and Power, about 85% of the budget would be funded by oil revenues, it was reported July 30. With 1973 oil revenues totaling $3.3 billion and 1974 earnings expected to reach $8 billion, a budget surplus in excess of $2 billion was anticipated. Nigeria could afford, the ministry said, to halve oil production and still maintain its economy for another 20 years with petroleum earnings. Daily output was scheduled to reach three million barrels a day within the next five years.

Oil production cutback seen. Nigerian oil production had been cut back by almost 25% since November 1974, according to a New York Times report May 11, 1975. Late monthly reports showed that Nigerian production was about 1.7 million barrels a day, compared with the 1974 peak of 2.3 million barrels a day. The Nigerian National Oil Corp. refused to discuss the cutbacks, but sources said they had been ordered by the government.

NORWAY

No curbs. Norway Jan. 30, 1974 canceled plans for gasoline rationing and lifted the weekend driving ban in effect since early December 1973.

Texas oil concern bought. The Norwegian Oil Corp., Ltd., a publicly-owned exploration concern, June 26 purchased Oil and Gas Futures, Inc., of Texas for $16 million. The Texas firm, an exploration company with concessions in Louisiana, Texas and Oklahoma, produced about 7,000 barrels a day.

Norwegian Oil was currently a partner in an oil field in the British sector of the North Sea. The president of the firm, Erik Olimb, said June 26 that Norwegian Oil wanted "to develop the petroleum expertise which today largely lies in American hands."

Barents Sea talks with U.S.S.R. In a joint communique March 25, 1974 after a week-long visit by Norwegian Premier Trygve Bratteli to the Soviet Union, the Norwegian and Soviet governments said they would open negotiations in the fall, on their overlapping claims to the continental shelf of the Barents Sea between the Norwegian island of Spitsbergen in the Arctic Ocean and the Soviet mainland. Extensive oil and gas deposits were believed to exist in the area.

The talks were held in Moscow Nov. 25–29, but no agreements were announced.

The Soviet Union and Norway had signed an agreement in Oslo March 7 guaranteeing the Soviet Union certain rights at an all-year airport to be constructed on Spitsbergen. Service would begin by 1975. Under the accord, the Soviet Union would have landing rights for civil aircraft and could place up to six technicians at the airport. A 1920 international treaty had established the demilitarized status of Spitsbergen under Norwegian sovereignty and had given the signers, including the Soviet Union since 1935, the right to exploit the island's resources. The Soviet Union was believed to have reversed its long opposition to construction of an all-year airport because recent oil exploration on the island and the surrounding continental shelf had revealed the necessity for improved facilities.

Oil deposits found in Norwegian Sea. Indications of extensive petroleum deposits had been found separately by Soviet and U.S. scientists at two widely separated sites in the Norwegian Sea, off the coasts of Norway and Iceland.

The U.S. research ship Glomar Challenger, which had been drilling into the ocean floors for six years, had found indications of oil and natural gas in deep holes drilled in August–September 1974 in the Voring Plateau, which extended westward some 400 miles off the northwestern coast of Norway, the French newspaper Le Monde reported Oct. 10. The Norwegian government protested the ship's activity on an "extension of the continental shelf," asserting it had denied the ship permission to drill. The U.S. National Science Foundation, which subsidized the project, said the area was an "international zone."

According to the London Times Oct. 7, the area was part of the Norwegian shelf, whose outer limit had never been fixed. Norway laid claim to a shelf that extended as far out as drilling could be conducted.

The other oil find was made by Soviet scientists aboard a research ship northeast of Iceland, according to the Oct. 28 New York Times report. They discovered oil-bearing sediment at the southwestern end of the Jan Mayen Ridge, which extended toward Jan Mayen Island in the Norwegian Sea.

New oil tax law. The Storting (parliament) May 29, 1975 approved a new oil tax law providing for a system of standard prices and a surtax on oil companies' North Sea production.

Unidentified sources called "informed" said the measure would bring the government about 100 billion kroner ($20 billion) from oil exploitation during the next 20 years.

Finance Minister Per Kleppe said during the parliamentary debate that the big increase in the price of crude oil stemmed mainly from tax increases in other countries.

The government had submitted in February new tax proposals for oil companies operating in the Norwegian sector of the North Sea, providing for a special 25% tax rate on oil income in addition to the nation's regular 50.8% corporate income tax. Industry had raised strong objections to the government's original proposal for a 40% special rate.

PERU

Oil & coal developments (1974). The Japan Petroleum Development Corp. announced it had agreed to lend $330 million to Petroperu, the Peruvian state oil firm, in return for a stable supply of crude oil and petroleum products over 10 years, it was reported May 31.

Mines Minister Gen. Jorge Fernandez Maldonado had said May 17, after returning from a visit to Tripoli, that Peru would buy 15,000 barrels of oil daily from Libya to assure adequate supplies of crude oil and refined products until 1976.

A Polish technical mission reported that Peru had coal reserves of at least 50 million tons in the Alto Chicama district 400 miles north of Lima, it was reported April 26.

IPC compensation deplored. The government criticized the U.S. for paying $22 million to International Petroleum Corp. (IPC), a subsidiary of Exxon Corp., from funds provided by Peru in 1974 for the compensation of expropriated U.S. firms.

In a message to the U.S. embassy reported Jan. 2, 1975, Peru noted that the agreement under which it had turned over $76 million in compensation funds had specifically barred payments to IPC. The U.S. argued that the agreement allowed it to disburse the funds as it saw fit.

In a related development reported Jan. 14, H. J. Heinz Co. said it had received another $1.6 million in compensation for a Peruvian subsidiary that was nationalized in 1973. The firm already had received $6 million in compensation.

QATAR

Qatar gains control of 2 firms. Under an accord signed in Doha Feb. 20, 1974, Qatar gained a 60% share in the operations of Shell Oil Co. of Qatar and the Qatar Petroleum Co. The new participation accord, effective from Jan. 1, replaced one negotiated in 1973 which had given the government an immediate 25% interest in the two foreign firms, increasing by stages to 51% in 1984.

SAUDI ARABIA

Saudis to get Aramco control. Saudi Arabia was to increase its 25% share of the concessions and assets of the Arabian American Oil (Aramco) to 60% under an interim pact announced June 10, 1974. The accord was reached in discussions that had started in Geneva June 4 between Aramco officials and Saudi Petroleum Minister Sheik Ahmed Zaki al-Yamani. The pact was to be retroactive to Jan. 1.

Saudi government sources said June 12 that the agreement was part of a plan to lower the price of crude oil. The accord would have the immediate effect of increasing Saudi Arabian oil production available to the government for direct sales to customers.

SOUTH-WEST AFRICA

U.S. oil companies leaving Namibia. The five oil companies operating in South-West Africa (Namibia) had decided to terminate their offshore prospecting ventures there, according to the New York Times Jan. 29 and Feb. 19, 1974. The firms had been under pressure from various church groups, including the Church Project on U.S. Investments in Southern Africa, to withdraw their operations from the disputed territory.

The decisions reached by Continental Oil Co., Getty Oil Co., Phillips Petroleum Co. and Texaco, Inc., reported in the Jan. 29 article, were followed by that of Standard Oil Co. of California, reported Feb. 19.

In making its announcement, Phillips said its decision to withdraw was "attributable to the issues of sovereignty which have been recently accentuated. It now appears likely that substantial expenditures would have been required before those issues of sovereignty could be clarified sufficiently for Phillips to proceed." Standard Oil said: "There seems to be an adequate supply of oil in the world. Priorites have changed. We are tight for money and have to apply it in places that carry the highest priority."

SOVIET UNION

Energy output up. A Central Statistics Board report, published in Izvestia Jan. 24, 1975, asserted that in 1974, Soviet petroleum output had risen to 459 million tons. The 1974 goal for petroleum, the U.S.S.R.'s major source of foreign currency earnings, had been 452 million tons.

Other energy figures for 1974: Coal 3% more than in 1973, and natural gas production was given as 261 billion cubic neters, a 10% increase over 1973, but still 9 billion cubic meters short of the origi-nal goal of the 1971–75 five-year plan. According to a Jan. 2 report, Soviet Oil Minister Valentin Shashin announced that Moscow was planning an increased oil output of 490 million tons for 1975.)

Bonn asks energy projects. West Ger-nan Economy Minister Hans Friderichs said Bonn had proposed two major de-velopment projects during the third meeting of the German-Soviet economic commission, which ended Jan. 18, 1974.

Under one plan, West Germany would aid construction of an atomic power center in the Soviet Union in exchange for ong-term deliveries of electricity. The other plan would bring Iranian natural gas o West Germany via a pipeline across the U.S.S.R.

Exports lag—Soviet oil exports to West Germany, which had lagged behind contracted amounts throughout 1973, had been halted, it was reported Jan. 17, 1974. Only 20 million barrels had been de-livered in 1973, compared with an ex-pected 31 million. Soviet officials said echnical difficulties had caused the short-fall. Natural gas deliveries to West Germany totaled 800 million cubic meters n 1973, 80 million less than planned.

The West German government-controlled company Veba A.G. announced March 12 it would cease to purchase Soviet oil April because of shortfalls in deliveries and excessive price demands.

Although Veba had contracted for 3.4 million tons of crude in 1973, only 2.86 million were delivered. One-fourth of Veba's total supply came from the Soviet Union.

Energy arrangements with Italy. The pipeline bringing gas from the Soviet Union to Italy became operative, it was reported April 7, 1974; annual flow was scheduled to reach 6 billion cubic meters. The pipeline construction and gas supply contract had been signed in 1969. Under the agreement, the Soviet Union became Italy's major supplier of natural gas.

Italy's state-owned oil complex Ente Nazionale Idrocarburi (ENI) signed a barter agreement, estimated to be worth $960 million, to supply the Soviet Union with six petrochemical factories and about 24 heavy-duty gas-pumping sta-tions, it was reported May 9. ENI would be paid mostly in crude oil, natural gas and products of the petrochemical plants.

U.S. Eximbank loan. The U.S. Export-Import Bank announced May 21, 1974 that it had approved a $180 million bank credit, at 6% interest, to help finance a $2 billion Soviet natural gas and fertil-izer complex. The largest single such loan to date, it brought Eximbank credits to the Soviet Union to nearly $470 million. A consortium of private banks, headed by the Bank of America, would provide a matching loan at a "blended" interest rate of 7.8%.

The project would comprise the con-struction of eight ammonia and urea fertilizer plants at Togliatti and Kuibyshev and a 1,200-mile gas pipeline from the plant sites to Grigoryevka, a new Black Sea port near Odessa. The plants were scheduled to be completed by December 1978.

Worked out by Armand Hammer, chairman of the Occidental Petroleum Corp., the project called for the import by the U.S. of Soviet fertilizers in ex-change for superphosphoric acid from the U.S. In addition to supplying the needed fertilizer, Eximbank President William J. Casey said the loan would also finance the purchase of $400 million in U.S. goods and help to conserve the domestic natural gas supply.

Siberian gas exploration—The Occidental Petroleum Corp., the El Paso Natural Gas Co. and several Japanese busi-nessmen signed a $400 million two-year agreement with the Soviet Union to jointly explore for natural gas in the Si-berian Yakutsk region, it was announced in Paris Nov. 22. Implementation of it depended on the U.S. Export-Im-port Bank providing $100 million in credits to match a Japanese Eximbank credit.

Siberian gas & oil pipelines ready. A 2,000-mile pipeline connecting the huge Medvezhnye natural gas field in the Tyumen region of northwest Siberia with

Moscow was put into operation ahead of schedule, the Soviet Communist Party newspaper Pravda reported Oct. 25. The Moscow linkup would double the gas supply available for the Soviet capital.

The transmission line, with an annual capacity of 14.5 billion cubic meters, was the westernmost extension of the gas pipeline network branching from the west Siberian fields. The new line traversed developed areas of the Urals and central Russia so that the gas could also be tapped by consumers along the route.

The government paper Izvestia reported Nov. 5 that a pipeline had been completed from Kuibyshev in the Volga-Urals oil fields to the Black Sea tanker loading terminal of Novorossisk, providing a direct outlet for Siberian crude oil to world export markets.

The Soviet Union began testing its first oil supertanker Nov. 6 in the Black Sea.

Rail line linked to new oil policy. Soviet Minister of Oil Production Valentin Shashin said May 27, 1974 that the Soviet Union had ruled out construction of an oil pipeline from Irkutsk in central Siberia to Kakhodka on the Sea of Japan. Instead, he said, the U.S.S.R. would build a railway to link up with the existing Trans-Siberian network at Komsomolsk on the Amur River. Explaining the shift, Shashin stressed that "anything you want" could be shipped by rail, while a pipeline could transport only oil.

Shashin enunciated what was essentially a more nationalistic oil policy for the U.S.S.R., also precluding foreign participation in oil resource development. He admitted only two possible forms of foreign participation in the Soviet oil business—in the exchange of technical information and in the sale of oil equipment—but he emphasized that the U.S.S.R. was fully capable of carrying out any oil projects on its own.

Shashin said the Soviet Union planned no increase in oil exports before the early 1980s.

Shashin Nov. 5 told a visiting U.S. senator, Walter Mondale (D, Minn.), that he would find the Soviet Union "ready to repay technical cooperation" in increasing Soviet oil production with "a certain part of the oil extracted."

Shashin noted that two U.S. firms, Union Oil Co. of California and Standard Oil of Indiana, "had shown an interest in this Soviet proposal," but he added that cooperation was also being encouraged with other countries, particularly France, Great Britain and Japan.

In conceding the need for U.S. technology, Shashin said new American methods could raise the extraction rate of some Soviet oil fields from the 10%-15% possible with traditional methods to "50% or even higher."

Energy saving encouraged. In a series of calls published Oct. 14 to commemorate the anniversary of the 1917 October Revolution, the Communist Party Central Committee appealed to workers to economize on the use of "primary materials, gasoline, electrical energy, metals and other materials."

Despite the economy program, a recently increased oil production target and release of statistics which placed the U.S.S.R. above all other nations in oil production, Moscow foresaw no early increase in petroleum exports. Pyotr S. Neporozhny, minister of energy and electrification, said Sept. 26 that the Soviet Union "cannot increase exports in the near future" because of the time it takes to develop new fields. He also noted that the Soviet Union had to shift its electrical industry to emphasize more coal and nuclear power.

Iran natural gas pact signed. The U.S.S.R. and Iran signed an agreement which raised the price of natural gas exported to the Soviet Union by 85%, from 30.7¢ to 57¢ per 1,000 cubic feet, retroactive to Jan. 1, it was reported Aug. 19, 1974. Talks had opened in Teheran Aug. 6 after earlier talks, held June 24-26 in Moscow, had broken off, reportedly with the Soviets' rejection of Iran's proposed doubling of the gas price. Iran supplied the Soviet Union with 10 billion cubic meters of natural gas annually and Moscow sold a comparable amount to Europe.

Following Tehran's unilateral decision to double the gas price July 2, Iranian government sources stressed that the price increase was actually modest and represented a gesture of goodwill since the real market value of natural gas would have allowed quadrupling the price to $1.20 per 1,000 cubic feet.

The Tehran sources also noted that the Soviet Union sold gas to West Germany, France and Austria at 57¢ per 1,000 cubic feet.

(The Iranian government said July 10 that 200 feet of the pipeline that carried natural gas to the Soviet Union had been blown up near Resht on the Caspian Sea. The flow of gas was interrupted for a full day.)

Afghan gas price raised—The Soviet Union agreed to pay more for Afghanistan's natural gas under a June agreement signed in Moscow during a visit by President Mohammad Daud Khan, the Journal of Commerce reported July 12. Under the agreement, the price rose from 19.5¢ to 34¢ per 1,000 cubic feet. Afghanistan was to supply the U.S.S.R. with about 2.8 billion cubic meters of gas during 1974.

Siberian gas & oil finds. The Soviet news agency Tass Aug. 9, 1974 reported a natural gas strike consisting of several trillion cubic meters on the Yamal peninsula in the Arctic Ocean. Four deposits had previously been found in the region.

A major new oil deposit was discovered in the Tyumen region of Western Siberia, Tass reported April 9, 1975. The 45th found in the past five years, it had an expected yield of 7,700 barrels of oil a day.

In related developments, Boris Shcherbina, Soviet minister for the construction of oil and gas industry enterprises, said the U.S.S.R. planned to increase its oil production in Western Siberia, the nation's major oil producing area, by 30 million tons in 1975, to about 180 million tons, it was reported March 27.

Texas utility buys Soviet coal method. Texas Utilities Services, Inc. signed an agreement with V/o Litsenzintorg, the Soviet foreign licensing agency, for the purchase of Soviet technology for developing underground gasification of lignite coal deposits in east Texas, it was announced March 17, 1975. Texas Utilities, a three-company system that sold electricity to four million customers in Texas, had paid $2 million for the technology.

Magneto-hydrodynamics progress. Soviet scientists said in Moscow March 27, 1975 that they had succeeded in using a rocket-type engine and gigantic magnets to produce large amounts of cleaner, cheaper electricity by a process called magneto-hydrodynamics. The pilot plant produced 12.4 megawatts of power, the scientists said, providing electricity for 100,000 Muscovites for thirty minutes during the preceding week.

William Jackson, an official of the U.S. Energy Research and Development Administration, said the Soviet Union was five years ahead of the U.S. in the field of magneto-hydrodynamics. He urged more intensive bilateral cooperation in that area.

SWEDEN

Gasoline rationing. Sweden Jan. 8, 1974 became the first West European nation to implement gasoline rationing. It allocated 26 gallons for each motorist until Feb. 28. Rationing of heating oil also took effect Jan. 8, with a 25% cut in deliveries. Electricity rationing was planned for February.

But the government ended the gasoline rationing Jan. 30, a month ahead of schedule, because of adequate supplies. Rationing of heating oil for home and industry continued temporarily. Gasoline rationing had failed because the local authorities had granted so many extra coupons to "needy" cases that if all had been redeemed consumption would have increased.

Energy plan. The government March 10, 1975 presented to the Riksdag (parliament) a broad energy program calling for expansion of nuclear and hydroelectric power facilities as part of a plan to reduce the nation's dependence on imported petroleum.

The program proposed construction of two nuclear plants by 1985 to complement the 11 currently in operation or under construction. In 1978, the government would review its energy needs after 1985. The proposal represented a setback for Sweden's power authorities, who had recommended construction of 13 additional nuclear power stations by 1985. That plan, however, ran into sharp opposition from the public.

According to a recent poll cited in a March 10 report, 43% of Sweden's population opposed nuclear power plants, while only 20% supported them. The opposition Liberal Party had recently declared its opposition to nuclear power expansion.

The government's new program also called for a 10% increase in hydroelectric power-generating capacity, increased coal imports, energy conservation measures, creation of a national petroleum purchasing agency and allocation of funds for a pilot plant to process Swedish uranium ore.

The plan aimed at holding the rise in energy consumption to 2% annually until 1985. The expansion of nuclear and hydroelectric capacities was designed to reduce the use of petroleum to 60% of energy consumed in Sweden in 1985, from 73% in 1973. Nuclear power would supply 12%, an 11% increase, while hydroelectric power would decline from 14% to 12% of the total. Sweden would still consume 11% more petroleum in 1985 than in 1973, according to the plan. All the nation's oil was imported.

SWITZERLAND

Fuel curbs end. Switzerland Feb. 20, 1974 discontinued the 20%-to-25% cut in deliveries of gasoline and heating oil to retailers ordered in November 1973 at the height of the energy crisis. President Ernst Brugger said the nation was "swimming in oil."

An 80 m.p.h. highway speed limit went into effect March 14, replacing a 60 m.p.h. limit introduced in November 1973.

URUGUAY

Energy conservation & supply. A series of government-ordered measures to conserve energy took effect Jan. 13, 1974. They included a 30% drop in public consumption of electricity, mandatory vacations for industrial workers in January–March, and turning the clocks ahead by 90 minutes.

Uruguay imported all its oil and depended for 50% of its electricity on thermal, oil-burning power stations.

The government later decreed severe new domestic energy regulations, including adjustment of working hours for maximum use of daylight, prohibition of commercial lighting and lights in the corridors and vestibules of apartment houses, and a 30% reduction in power consumption in private homes (against consumption in October 1973), it was reported May 29.

The Soviet Union was awarded a $50 million contract April 28 to install six turbines at the Salto Grande hydroelectric project on the Uruguay River, undertaken jointly by Uruguay and Argentina. The Soviet bid reportedly was opposed by the Uruguayan government, but was accepted because it was 30% lower than competing bids from firms in the U.S., Japan, Yugoslavia and other countries.

The state oil firm Ancap terminated crude oil import contracts with Exxon Corp. and the Royal Dutch/Shell Group, becoming Uruguay's sole importer of oil, it was reported June 4. The government also opened bidding for offshore oil exploration contracts, under terms calling for foreign companies to take all the risks and, if commercial quantities were found, for the government to repay them with a percentage of the production.

VENEZUELA

Prompt oil take-over urged. Outgoing President Rafael Caldera recommended Jan. 1, 1974 that Venezuela's oil and other key industries with foreign holdings be nationalized as soon as possible.

He did not indicate what steps his government might take to advance the expiration date of oil concessions held by foreign, principally U.S., firms. He had pledged previously not to make any important moves without consulting President-elect Carlos Andres Perez.

Caldera said in his nationwide New Year's message that the Mines and Hydrocarbons Ministry, "through its investigations and studies of the future of the Venezuelan petroleum industry, has seen more and more clearly that this activity, fundamental to our economy, must pass into the hands of the national public sector."

Juan Pablo Perez Alfonso, a former mines minister and founder of OPEC, Jan. 12 proposed immediate nationalization of the industry without compensation for virtually abandoned oilfields and with appropriate compensation for productive installations. He renewed his call for a cut in oil production from the present 3.3 million barrels a day to 800,000 barrels daily, to protect dwindling oil reserves. Perez Alfonso asserted current and future oil prices would guarantee the state treasury the income it needed.

Outging Mines Minister Hugo Perez La Salvia announced Jan. 15 that the departing Caldera government had already prepared detailed plans for nationalizing the $2 billion oil industry without compensation, possibly earlier than provided in the oil companies' 40-year leases. A spokesman for the ruling COPEI party later said the party was preparing legislation for early nationalization of the industry to be presented to the new Congress soon after it met in March, it was reported Feb. 12. COPEI, which lost the December 1973 general elections, would be in the opposition in the new government.

Royalty oil price doubled. The government Feb. 6 doubled to about $14 a barrel the price it charged petroleum companies for royalty oil. The increase, retroactive to Feb. 1, affected the one-sixth of each firm's output which accrued to the government either in cash or in kind.

Posted prices for Venezuelan oil remained stable in February, but not because of pressure from the Organization of Petroleum Exporting Countries (OPEC), according to the government. Mines Minister Hugo Perez La Salvia had denied Jan. 15 that there was an OPEC agreement to freeze oil prices until April 1, asserting there was instead an agreement by six OPEC members from the Persian Gulf to hold prices in January–March.

Increases in posted prices for Venezuelan oil through December 1973 had increased the nation's international monetary reserves to $2.418 billion, according to International Monetary Fund statistics reported Feb. 2.

U.S. resolution denounced—Jesus Soto Amesty, chairman of the Senate Foreign Relations Commission, charged Jan. 30 that a U.S. Senate warning to Venezuela on its oil prices was an "act of arrogance." The U.S. body had passed a resolution the day before saying further oil price increases by Venezuela, Canada and Middle East states might result in "reciprocal economic action" by the U.S.

"For many years," Soto asserted, "[developed countries] have sacked underdeveloped people throughout the world, carrying away their raw materials at laughable prices."

"Moreover," he continued, "still at present the importing countries have to bear the tremendous expenses that the United States incurs in countries such as Vietnam, Korea or Cambodia, paying at high prices for ... indispensable articles. Thus, when the exporting countries unite to charge a just price for their oil, they are doing nothing less than defending themselves from the voracity of giants."

Central America pledge—The government Feb. 1 guaranteed preferential oil supplies to Central American nations to counter shortages they were suffering as a result of the Arab oil squeeze. The decision followed three days of talks between oil officials and representatives of Guatemala, El Salvador, Honduras, Costa Rica and Nicaragua.

Perez vows oil nationalization. Carlos Andres Perez was inaugurated president of Venezuela March 12, 1974. He vowed in his inaugural address to seek a national consensus to expropriate the largely U.S.-owned oil industry during his five-year administration.

Speaking before a joint session of Congress and representatives of 72 countries, Perez declared: "We are going to fulfill the old aspiration of our people that our oil be Venezuelan. More than a new law, it is necessary to obtain the agreement of all Venezuelans about what we can and should do to obey this unique and singular command of history." He did not set a time limit on the nationalization debate. "The softer the tone of our voice, the easier it will be to be heard and understood," he said.

(The opposition Popular Electoral Movement [MEP] had introduced what was described as a radical nationalization bill March 2, the day the new Congress convened. Perez' Democratic Action Party [AD], which held a majority in both houses of Congress, had called the MEP bill "demagogic.")

In an apparent bid for a Venezuelan leadership role in Latin America, Perez said, "We must take up the defense of Latin American rights, trampled by the economic totalitarianism of the developed countries. Venezuela now has the opportunity to offer Latin America, with the backing of oil, efficient cooperation to carry out the common struggle for independent development, decent prices for other raw materials and a just and balanced participation in world trade."

U.S. reaction—U.S. oil firms operating in Venezuela March 12 reacted to Perez' statements on oil with what the Wall Street Journal called "stoical acceptance."

The companies, which had preferred Perez among the 12 candidates in the December 1973 election, reportedly felt they would receive "fair" compensation for their Venezuelan properties.

Perez had told foreign newsmen March 11 that it was "logical and suitable" for the U.S. firms to have a continuing role in Venezuelan oil marketing and technology. However, he had added that Venezuela would maintain its current high oil prices even if other countries reduced theirs.

Production cut. Mines Minister Valentin Hernandez announced April 6, 1974 that the government had ordered a 5% cut in oil production, effective April 15, to conserve large volumes of natural gas which were produced jointly with oil but were burned off at the oilfields because of insufficient gas processing facilities.

Government sources quoted by the Washington Post April 9 said the reduction would be closer to 7%, or 200,000 barrels a day. Venezuela produced some 3.3 million barrels of crude daily, more than half delivered to the U.S. However, the sources said the cutback would not affect the U.S. or other traditional markets.

Creole Petroleum Corp., the Exxon Corp. subsidiary and Venezuela's largest oil producer, said April 8 that it would reduce foreign oil sales by 100,000 barrels a day, or nearly 6% of its production, in compliance with the government order. Creole had announced previously that its investment in its 1974 production program would be 30% higher than in its 1973 program. Creole President Robert Dolph said the firm would maintain its 1973 production levels, which averaged 1.5 million barrels a day, it was reported Feb. 13.

The government Aug. 21 ordered petroleum output cut by an additional 100,000 barrels a day to reduce what the government called an oversupply which was driving oil prices downward.

Finance Minister Hector Hurtado said Sept. 6 that production would be reduced by a further 10% in 1975, to 2.6 million barrels daily, and would fall to around two million barrels daily by the end of the decade. Oil industry sources asserted Aug. 27, when the planned cutback was first reported, that none of Venezuela's traditional customers—principally the U.S. East Coast and Canada—would suffer shortages because of it.

A furor had been touched off in Venezuela July 10 when it was learned that U.S. Treasury Undersecretary Jack Bennett had told a U.S. House of Representatives subcommittee the previous day that any new oil production cutbacks by exporting countries would be "clearly regarded by the U.S. and by all other consuming countries . . . as a counterproductive measure" and would have "political and economic implications" for the exporters.

Mines Minister Hernandez sent Bennett a cable July 10 reminding him that

Venezuelan actions regarding oil were "guided solely by the national interest and based firmly on the exercise of sovereignty . . . [which], regarding specifically the defense of non-renewable natural resources, has been reaffirmed on different occasions by the highest organs of the United Nations." Hernandez added that Venezuela had always kept and would continue to keep its "international commitments."

The 1975 production cut was announced Feb. 5, 1975 when the government said it would cut daily oil production by 200,000 barrels within a week, to about 2.6 million barrels daily. However, government oil revenues would not be decreased, officials said. Based on an annual production of 2.6 million barrels a day, Venezuela's proven reserves of conventional oil would theoretically last for 20 years, the Journal of Commerce reported Feb. 6. Hernandez had said Jan. 20 that the nation's reserves had risen by 4.8 billion barrels in 1974, to a total of 18.6 billion barrels.

Mines Ministry spokesmen reported March 9 that Venezuelan oil production had fallen to 2.64 million barrels per day, its lowest level since 1958. Production had decreased by 18.7% in the past year, they said.

Creole had announced Jan. 8 that it would invest $114 million in its operations in 1975, a $14 million increase over 1974 expenditures. Creole President Robert Dolph said the company had undertaken the spending program, despite the impending nationalization of Venezuela's oil industry, because "we have an obligation to maintain production at an adequate level." Creole was confident that in the long run it could "buy more oil from Venezuela if the production level is high," Dolph asserted.

Cia. Shell de Venezeuela, the local subsidiary of Shell International Petroleum Co., had announced earlier that it would invest $50 million in the country in 1975, it was reported Jan. 9.

The government had begun talks with the foreign oil companies on their future role in supplying technical and administrative aid to the nationalized oil industry, it was reported Jan. 8.

Oil nationalization bill submitted. The government March 11, 1975 submitted to Congress a bill to nationalize the oil industry, which had earned $8 billion for Venezuela in 1974.

The bill, a revised version of a draft prepared by a government-appointed commission in 1974, was immediately criticized by opposition politicians, notably by leaders of the Social Christian Party (Copei). The critics complained about an article added to the draft which allowed the government, with the approval of Congress, to negotiate "agreements of association" with private companies to export Venezuelan oil.

The left-wing People's Electoral Movement (MEP) charged the article allowed the formation of "mixed enterprises" in the nationalized industry. Copei asserted the entire industry "should be left in the hands of the state."

President Carlos Andres Perez defended the bill in an address to Congress March 12, on the first anniversary of his accession to office. He said the controversial article had been added to "open all possible options for the success of the government's efforts in this most difficult task which has irreversible consequences for the nation's destiny."

Perez stressed that nationalization of the oil industry was not a hostile act against any foreign power. "We extend our hand to the United States and Europe to express our support if they wish to build a world of cooperation," he said. "Our international policy is and will be one of irreproachable dignity."

Turning to the enormous wealth Venezuela was accumulating because of high petroleum prices, Perez pledged austerity in the government's management of the oil funds and he urged Venezuelans to avoid waste.

The bill said the 22 nationalized companies would be compensated by a sum to be anounced within 45 days after Perez signed the legislation. The government would make an offer to the firms, and if they refused it the Venezuelan Supreme Court would set the compensation sum.

The bill said the sum would originally be determined by taking the net value of the companies' properties, plant and equipment, and subtracting the "value of the petroleum extracted . . . outside the limit of their concessions," the "amount of social benefits and other rights" denied to the companies' workers, and whatever the companies "owed the state."

Government officials said March 13 that four state firms would be created to replace the foreign companies currently operating the oil industry. The subsidiaries of Exxon Corp., Gulf Oil Corp. and Shell International Petroleum Co. would each be replaced by a state firm, and the remaining companies would be replaced by the fourth state concern.

Original draft—The first draft of the oil nationalization bill had been published Aug. 20, 1974. It set no nationalization dates, but said the oil industry would "return to the hands of Venezuelans within a year."

Oil companies owned by U.S. and other foreign interests would receive compensation not to exceed the net book value of their fixed assets, according to the draft. This value was arrived at after deducting depreciation and amortization. The draft added that further deductions should be made to cover social benefits to the industry's 22,000 Venezuelan workers and amounts owed by the firms to the government and its agencies.

At least one member of the drafting commission, Alfredo Paul Delfino, president of the businessmen's federation Fedecamaras, objected to the law's provisions. Paul Delfino charged Aug. 18 that the draft was "premature, unnecessary, inconvenient, punitive and restrictive." Fedecamaras denounced the bill Aug. 21, but it later gave the draft guarded support Sept. 3. Virtually all other sectors supported nationalization of the oil industry, although there was some disagreement on how and when to accomplish it.

The nationalization draft was published amid a number of mild diplomatic incidents between the U.S. and Venezuela. An uproar was caused in Caracas over U.S. President Gerald Ford's comments Sept. 18 that oil-producing nations were using petroleum as a political and economic weapon. President Perez sent Ford a cable Sept. 20 rejecting the charge and accusing the U.S. and other industrial nations of keeping "more than half of humanity in a precarious economic state and growing poverty" by traditionally paying low prices for raw materials.

Perez' protest followed newspaper reports in August that U.S. Ambassador Robert McClintock had, at the request of local subsidiaries of Exxon Corp. and Mobil Oil Corp., dissuaded a representative of the U.S. Federal Energy Administration from interviewing Venezuelan oil officials in July as part of a study on the conduct of multinational corporations during the energy crisis.

This alleged interference was denied by McClintock and denounced by Venezuelan government officials, politicians and newspapers. Mines Minister Valentin Hernandez promised an "exhaustive" official investigation of the incident; he met Aug. 21 with Robert Dolph, president of Exxon's Creole Petroleum Corp., and Aug. 22 with McClintock, who accused the Venezuelan press of fabricating the issue.

Draft revised—The presidential commission Oct. 17, 1974 approved a revised version of its oil nationalization bill.

The new version added provisions which gave the government total control of foreign marketing of Venezuelan crude oil and petroleum derivatives; barred mixed companies in the nationalized industry, but allowed state firms to make contracts to carry out their activities; urged immediate occupation of the petroleum industry if no agreement were reached on compensation for foreign corporations; and established job stability for oil workers.

Regarding compensation, the draft bill said the government would offer the foreign firms a sum, and if the firms rejected it, the matter would be referred to three experts—one chosen by the attorney general, one by the oil companies and one by the Supreme Court. After hearing from the experts, the court would set the compensation figure. In computing its original offer, the government would not consider any revaluation of oil company assets made during the previous 15 years, the draft stated.

The commission released a set of recommendations for a new government energy policy Nov. 1. It called for a further reduction of oil and gas production to prolong exploitation of these resources; a shift to other forms of fuel and energy wherever possible; increased exports of heavy crude oil and light refined derivatives; minimum possible earnings from oil exports to avoid upsetting the budget and the nation's finances; and increased exports of petrochemicals and other oil derivatives.

Oil tax raised. The Venezuelan government Oct. 1, 1974 raised the income tax paid by foreign petroleum companies by 3.5%, to 63.5% of the posted price per barrel. The measure, retroactive to Jan. 1, would increase the companies' 1974 tax bill by $440 million and boost the government's 1974 oil earnings to over $10 billion, according to press reports.

Mines Minister Valentin Hernandez said the increase conformed with a recent decision by the Organization of Petroleum Exporting Countries to raise oil taxes. He added that Venezuela had made the increase retroactive to cut "excess profits" made by the oil companies earlier in 1974.

Hernandez said the government received $9.06 of the average $10.14 for which each barrel of Venezuelan oil currently sold. The remaining $1.08 went to the oil companies.

Hernandez announced Jan. 21, 1975 that the oil-company income-tax rate was again being raised—from the current 63.5% to an effective 70%, retroactive to Jan. 1. At the same time, posted prices for crude and refined oil were cut by an average 84¢ a barrel, to $13.61 a barrel, and freight premiums attached to all oil exports were reduced by 77¢ a barrel.

Sources in Caracas said the tax increase would raise government oil revenues by 38¢ a barrel, the Wall Street Journal reported Jan. 22.

Mines Minister Valentin Hernandez said Feb. 4 that Venezuela would hold its oil export price at $13.61 per barrel for the remainder of 1975.

Foreign deals. Costa Rican Foreign Minister Gonzalo Facio announced Feb. 18, 1974 that the Venezuelan government had agreed to finance a $200 million oil refinery to be built in Costa Rica and would provide it with crude oil at preferential prices. The national oil firms of El Salvador, Guatemala, Honduras and Nicaragua, as well as those of Venezuela and Costa Rica, would participate in the project.

Mines Minister Hernandez said Jan. 10, 1975, after meeting in Caracas with Canadian Energy Minister Donald Macdonald, that Venezuela would guarantee oil supplies to Canada. Canada bought 70% of its fuel from Venezuela, representing 30% of Venezuela's oil exports, according to the French news agency Agence France-Presse.

Venezuela would sell Peru 40,000 barrels of crude daily in 1975–76 under a long-term financing agreement, according to reports in the newsletter Latin America March 14 and the newspaper El Nacional of Caracas March 15. Venezuela would purchase half of the oil from Ecuador and resell it to Peru, which could not afford to pay cash for the oil as Ecuador required.

Central Bank President Alfredo Laffee had said May 1, 1974 that Venezuela had offered to help other Latin American countries buy oil. Finance Minister Hector Hurtado had said earlier that Venezuela would funnel aid to its neighbors through the Andean Development Corp., the Central American Integration Bank and the Caribbean Development Bank, it was reported April 17.

Among domestic investments, the government set aside $185 million–$230 million to purchase oil tankers for a national fleet Oct. 14.

Other developments. An important oil strike in Southern Lake Maracaibo was announced Feb. 15, 1974 by Mobil Maracaibo C.A., an affiliate of the U.S.-owned Mobil Oil Corp. and the Venezuelan state petroleum firm CVP. A Mobil well, which reportedly set a Latin American depth record at 18,583 feet, struck light gravity crude which flowed at a rate of 5,400 barrels a day, more than 10 times the average in Venezuela. It was the first major strike in the 617,750 acre zone in the southern part of the lake granted to Mobil and other foreign firms in 1971 under the government's experimental "service contracts."

The state petroleum firm CVP announced Sept. 9 that a new oil deposit with a capacity of 150 million–500 million barrels had been discovered near Lake Maracaibo. Creole Petroleum Corp. had announced discovery of a new light oil deposit in the Maracaibo Gulf July 21, and it was reported Nov. 22 to have discovered new deposits in the lake.

The government rejected offers from the U.S. and France to help it develop the Orinoco heavy oil belt, it was reported April 5, 1974. Mines Minister Hernandez said: "We are not interested in negotiating with anyone for the Orinoco strip. All the work of research and subsequent exploitation will be carried out by our country." Venezuela had the capital to buy any technical help it wanted on the open market, the newsletter Latin America reported.

To provide new managerial talent for the nationalized oil and steel industries, the government budgeted $19 million a year to send Venezuelan students abroad for training, it was reported Sept. 29.

British Petroleum would build a $100 million plant in eastern Venezuela to extract protein from petroleum in a joint venture with the government and several privately owned animal feed concerns, it was reported Jan. 14, 1975 truction

VIETNAM

Oil found off South Vietnam. The Saigon government said Aug. 28, 1974 that oil deposits had been found Aug. 25 in the China Sea, 190 miles south of the South Vietnamese town of Vung Tau. The discovery was made by Pecten Vietnam, a U.S. oil company. Company officials said several months of further drilling and testing were necessary before it could be determined whether the oil reserves had commercial value.

Pecten was one of 11 foreign oil firms that had signed contracts with Saigon June 27 for oil concessions off the coast.

The Viet Cong delegation in Paris declared Aug. 29 that any foreign oil company drilling in South Vietnamese waters was illegal. The Saigon government had no right to grant concessions to such firms since the country's natural resources "belong to the people," the statement said.

Cambodian government sources said Sept. 4 that South Vietnam had filed a note with the Pnompenh regime protesting oil exploration by a French company 60 miles southwest of the Cambodian port of Kompong Som. Saigon claimed the drilling was being held in contested waters.

WEST GERMANY

Sunday driving ban canceled. The government said Jan. 9, 1974 that it would not implement the alternate Sunday driving ban scheduled to begin Jan. 19, but would retain speed limits on freeways and secondary roads.

Official figures indicated that gasoline supplies would be higher in February than a year ago, but because of increased demand they would fall 5.6% short of normal requirements in January and 3.2% in February. A government spokesman said the shortage of gasoline in November was due "not to lower deliveries" [caused by the Arab oil squeeze] but to suddenly increased demand."

An Economics Ministry spokesman said Jan. 9 "we really don't know where the oil companies got this oil . . . But it is clear that the Arabs haven't cut back as much as they said they would." He said oil deliveries to West Germany were 6%–10% less than in January 1973. However, home-heating oil deliveries were running 16.8% less than in January 1973 and industrial oil deliveries were 13.5% below the previous year's figures. West Germany received 75% of its oil from Arab producers.

(The Cabinet bowed to public pressure March 13 and removed the motorway speed limit of 62 m.p.h. imposed when the oil crisis began. Instead, it set a non-binding "advisory" speed limit of 81 m.p.h.)

Energy research program approved. A four-year DM1.5 billion energy research program was approved by the Cabinet Jan. 9. The largest single sum, DM616 million, was earmarked for research into ways to extract gas and oil from coal and extend the use of coal. Funds would also be allotted for seeking ways to increase coal production, improving prospecting methods for oil and natural gas, and converting, transporting and storing energy.

Oil shale found—Oil shale deposits estimated at about two billion tons, one of the largest exploitable finds in central Europe, had been discovered on the north German plain, between Braunschweig and Wolfsburg, according to a Feb. 8 report.

Oil merger planned. The Cartel Office said Jan. 8 it had rejected the government's plans to create a single government-run oil company from Gelsenberg AG and Veba AG. The government had sought to buy a 48.3% share, held by the Rheinisch-Westfalische Elektrizitatenwerken AG, in Gelsenberg, which would then merge with the 40% government-owned Veba. The Cartel Office said the combined concern would hold dominant market positions in electricity production, mineral oil and very light fuel oil, petrochemical raw materials and inland shipping.

But Economy Minister Hans Friderichs Feb. 4 issued a special decree overruling the Cartel Office's objections to the merger. Friderichs' decree retroactively approved acquisition by Veba of 51% control of Gelsenberg, increased from the initial 48.3% share holdings it sought.

Oil price dispute. The West German subsidiaries of Exxon and Shell oil companies yielded to a government demand that they rescind their latest price increases, it was reported April 21. The subsidiaries of Texaco and British Petroleum, also served with the price cut order, failed to respond.

The Federal Cartel Office, accusing the oil companies of misusing their dominant position in the oil market, had threatened to issue temporary injunctions against the regionally differentiated price increases of early April, which had raised the retail price of premium gasoline by 1.5¢–3¢ a gallon to about $1.40.

None of the companies agreed to lower the price of diesel fuel by at least 8¢ a gallon, which the Cartel Office had requested April 11. Diesel oil prices had more than doubled since the fall of 1973, to $1.50–$1.60 a gallon.

The diesel oil order followed public hearings March 22-April 2 by the Cartel Office into various price increases by the oil companies since October 1973. However, the office failed to prove their charges against the four foreign firms and two West German oil companies, Veba AG, and Gelsenberg AG. The six were accused of jointly controlling the West German market in violation of the antitrust law and of raising prices excessively. The foreign subsidiaries were also accused of paying excessive prices to their parent concerns for crude oil and thus funneling profits out of West Germany. The government had not set any fuel price controls during the oil crisis.

The oil companies attributed price increases to a sharp cutback in their fuel oil refining because of the mild winter and fuel conservation measures. This led to a reduction in domestic gasoline production, they said, and the additional cost of importing refined gasoline.

The Cartel Office had sent a report on evidence of oil price manipulation to the cartel office of the European Economic Community with the aim of initiating a European inquiry, it was disclosed Feb. 8.

'74 oil import cost up, volume down. West Germany's crude oil imports declined 7.2%, to 102.5 million metric tons, in 1974, but their value increased 153% to $9.85 billion, the Economy Ministry said Feb. 25, 1975. Oil imports from Arab nations fell 10.9%, to 71.1. million metric tons, while their value rose 146% to $6.79 billion.

ZAIRE

State takes oil firms. The government created a state-owned company to take over all petroleum product distribution, it was reported Jan. 16, 1974.

The companies affected were the Royal Dutch-Shell Group, Texaco, Mobil Oil Corp., British Petroleum Co. and Petrofina of Belgium. The Gulf Oil Corp. offshore wells, and the exploration activities of Texaco and Royal Dutch-Shell were apparently unaffected.

Atomic Power

Indian Blast Ignites Controversy

Controversy flared after India, in May 1974, detonated what it described as a "peaceful" atomic device. India, in accepting nuclear assistance from the U.S. and Canada for peaceful purposes, had pledged that the aid would not be used for weapons, and Indian officials said India had lived up to this commitment. But the Indian A-blast awakened fears that well-intentioned atomic assistance to energy-needy nations could add to international danger by putting the materials for nuclear weapons into untrustworthy hands.

India explodes A-device. India successfully exploded its first nuclear device May 18, making it the sixth nation to conduct such a test.* The government immediately declared it had "no intention of producing nuclear weapons," while most other nations expressed misgivings.

A government statement announced that India's Atomic Energy Commission (AEC) had conducted "a peaceful nuclear explosion experiment" about 330 feet underground. The blast took place in the Great Indian Desert in Rajasthan State. The statement said India's nuclear program was designed for "peaceful uses" such as mining and earth moving and reiterated the nation's "strong opposition to military uses of nuclear devices."

Later May 18, AEC Chairman H. N. Sethna said the device had been in the range of 10–15 kilotons (equal to 10,000–15,000 tons of dynamite), about the size of the bomb exploded at Hiroshima. He said, "It was a 100% Indian effort and the plutonium required for the explosion was produced in India." The AEC said it had used an implosion device in its experiment.

India had not signed the 1968 nuclear nonproliferation treaty, but had been forbidden by the U.S. and Canada, which had long aided India in peaceful nuclear energy projects, from using the plutonium produced by its reactors for any purpose other than fueling the reactors themselves. The Indian regime had signed the 1963 test ban treaty barring explosions on land, in the air or underwater.

Prime Minister Indira Gandhi told newsmen the explosion was "nothing to get excited about. We are firmly committed only to the peaceful uses of atomic energy."

Initial domestic reaction seemed to be favorable, with even government critics expressing pride at the scientific achievement. The explosion came amid grave economic unrest, with the nation still hit by a nationwide railway strike.

The nation's atomic energy program had been begun by Gandhi's father and

*The other nuclear nations were the U.S., the Soviet Union, France, Great Britain and China.

India's first prime minister, Jawaharlal Nehru, with the stated intention of generating electricity to spur industrialization. Nehru had strongly opposed the use of atomic energy for military purposes. However, according to a leading Indian commentator, "A qualitative change came about in India's stand after China exploded its first bomb in 1964." India "did not consider its policy of unilateral renunciation of nuclear weapons as something that was binding for all time to come irrespective of other international developments," he said.

India was reported capable of constructing a nuclear device in the 1960s but opposed doing so. In 1970, however, the government had agreed to a request of the parliamentary atomic energy commission to study the cost of manufacturing an atomic bomb.

Foreign reaction—Pakistan, which had fought three wars with India since 1947, led a number of nations in expressing concern over the explosion.

The Pakistani Foreign Ministry issued a statement May 18 saying "It is an incontrovertible fact . . . that there is no difference between tests for so-called peaceful purposes and military purposes. The technology is the same." Prime Minister Zulfikar Ali Bhutto May 19 accused India of "brandishing the sword of nuclear blackmail." He said nuclear weapons could be used as "a means of pressure or coercion against nonnuclear countries."

Pakistani Foreign Minister Aziz Ahmed said in Washington May 20 that the explosion was "a new threat to our security" and announced that Pakistan would seek protective guarantees from the U.S. and other major powers against nuclear attack by India. Ahmed said his government would not use its own limited reactor-capacity to develop nuclear weapons.

Pakistani Prime Minister Zulfikar Ali Bhutto told his nation's National Assembly June 7 that Pakistan would develop its nuclear program in response to India's explosion. He added that Pakistan's program would be restricted to peaceful purposes.

Pakistan's Atomic Energy Commission reported June 16 it had detected radioactivity in various parts of Pakistan from India's underground nuclear explosion.

A U.S. State Department spokesman May 18 reiterated U.S. opposition to "nuclear proliferation because of the adverse impact it will have on world stability." However, a spokesman for the U.S. Atomic Energy Commission expressed doubt May 20 that the fissionable material used in the explosion had originated in the U.S. He said enriched uranium supplied by the U.S. to India since 1969 had been subject to the International Atomic Energy Agency safeguard system, which required accountability of the material. The U.S. had supplied about 141,000 kilograms of enriched uranium for use in a U.S.-built electric power reactor.

The official Soviet news agency Tass reported the blast without criticism May 18, asserting that the test had no military significance.

Canadian External Affairs Secretary Mitchell Sharp announced May 18 that Canada would re-examine its nuclear arrangements with India. Ivan Head, chief foreign policy adviser to Prime Minister Pierre Trudeau, said May 20 that the government regarded the test as a violation of a 1971 understanding between the two nations. He said that during a visit by Trudeau to India in 1971, Prime Minister Gandhi had "guaranteed peaceful use of nuclear energy according to our definition, which did not extend to explosions." He added that Trudeau had "made it clear to Mrs. Gandhi that Canada would regard any nuclear explosion" by the Indians as a "nonpeaceful act." Head denied, however, that the two nuclear reactors or the technological assistance Canada supplied to India had enabled the Indians to transform the waste from the reactors into plutonium, since Canada could not itself produce plutonium.

India came under heavy criticism from the 25-nation disarmament conference in Geneva May 21. The Indian delegate, Brajesh C. Mishra, dismissed the complaints, reiterating that "India has no intention of becoming a nuclear-weapon power."

Gandhi unit scores test—India's prestigious Gandhi Peace Foundation, in a letter published in most Indian English-language newspapers, denounced the nation's May 18 nuclear explosion as "a cruel joke on the people of this country," it was reported May 30. In contrast to the generally unanimous domestic praise for the test, the foundation wrote that the nation's precarious economic situation, including gross underutilization of industrial capacity and available resources, should have ruled out "the search for a new source of energy of doubtful immediate use."

Founded in the 1950s, the group's charter members included the later Prime Minister Jawaharlal Nehru, father of Prime Minister Indira Gandhi.

Canada suspends A-aid to India. Canada suspended nuclear aid to India May 22, 1974 because of India's A-test.

Canadian External Affairs Minister Mitchell W. Sharp said his government would review its other aid programs "to be sure that our priorities are the same as the Indians." He added that Canada was also asking other governments for joint consideration of "the broad international implications" of the explosion."

Sharp said, "What concerns us about this matter is that the Indians, notwithstanding their great economic difficulties, should have devoted tens or hundreds of millions of dollars to the creation of a nuclear device for a nuclear explosion."

The aid suspension would affect all shipments of nuclear equipment and material to India as well as technological information. Food and agricultural aid would continue.

Canada had asked India to send high officials immediately for "a complete exchange of views" on the situation, Sharp said. He indicated that Canada particularly wanted to know the source of the plutonium used in the test.

News reports suggested that Canadian style nuclear reactors in India, built with Canadian aid, were more suited to the production of plutonium than were U.S.-designed reactors there.

A spokesman for the Canadian Ministry of External Affairs said May 28 that Canada was pressing "very actively" for international action to prevent any further spread of the capabilities to produce nuclear weapons. The statement was made following a news conference in Ottawa by visiting Pakistani Foreign Minister Aziz Ahmed, who called for international talks on "some way to prevent India from going ahead" with development of nuclear explosions.

(In January 1975, the Canadian Department of Industry, Trade and Commerce officially cancelled an export permit covering an estimated $1.5 million in nuclear-related material and equipment sales scheduled for use in a heavy-water atomic reactor in India.)

Indian defense. Prime Minister Indira Gandhi said May 25, 1974 that "allegations and apprehensions" that India was developing nuclear weapons were unfounded. In response to criticism that India was too poor a nation to spend resources on nuclear development, she said the same objection had been made "when we established our steel mills and machine-building plants." Such things were necessary, she said, because it was "only through the acquisition of higher technology" that poverty could be overcome. Indian newspapers denounced Western critics of the explosion.

Gandhi was reported to have sent a letter to Pakistani Prime Minister Zulfikar Ali Bhutto May 22, reiterating that India's explosion did not pose any threat to Pakistan's security.

Continuing her defense of the test against foreign criticism, Gandhi told foreign correspondents June 15 that India might subscribe to a ban on "all atomic tests if everybody else agrees to it." She said India had opposed the 1968 nuclear nonproliferation treaty because "we thought it was discriminatory and unequal."

U.S. halts uranium aid to India. The U.S. had suspended delivery of enriched uranium fuel to India until New Delhi pledged not to use the atomic fuel in any nuclear explosion, officials of the U.S. Atomic Energy Commission (AEC) disclosed Sept. 7, 1974.

Shipment was halted on an Indian order of enriched uranium for its atomic power plant near Bombay, built with U.S.

assistance. Under a 1963 accord, the U.S. had promised to provide fuel for the reactor over 30 years.

The first of about four planned shipments under the order was made shortly after the Indian nuclear explosion in May, but further deliveries were stopped until New Delhi met the U.S. demands.

A spokesman for the AEC said Sept. 16 that India had agreed to give the specific assurances sought by the U.S. The issue had been discussed in Vienna, Austria in recent days by AEC Chairman Dixy Lee Ray and Homi N. Sethna, chairman of the Indian Atomic Energy Commission. The two were attending the annual conference of the International Atomic Energy Agency (IAEA), which met Sept. 16–20. (The IAEA general conference Sept. 16 admitted North Korea and Mauritius as new members of the agency, raising total membership to 107.)

Curbs set for nuclear sales. In an attempt to prevent nuclear weapons proliferation, major exporters of nuclear materials, except for France and China, had compiled a list of equipment they would supply to other nations only under assurances it would not be diverted for explosive use, the New York Times reported Sept. 24. The decision was set forth in governmental letters to the International Atomic Energy Agency in Vienna.

The list detailed what the supplier countries considered to be special fissionable material and equipment for using it.

Among the supplier countries agreeing to the new curbs were the U.S., Great Britain and the Soviet Union.

Canada asks new atom curbs. Allan J. MacEachen, Canada's secretary of state for external affairs, addressed the U.N. General Assembly Sept. 25, 1974 and called for tighter international controls on nuclear materials.

MacEachen proposed that all nations put their inventories of fissionable materials held for peaceful purposes under the supervision of the International Atomic Energy Agency. Canada had already discussed the proposal with other major suppliers, including the U.S., Great Britain and the Soviet Union.

Concern about the spread of nuclear arms production was also expressed in the Assembly Sept. 25 by the foreign ministers of Ireland, Sweden and Denmark.

Other International Developments

U.S. A-aid to Egypt & Israel. During a visit to the Middle East in June 1974, President Richard M. Nixon promised Egypt and then Israel that the U.S. would provide them with peaceful nuclear aid.

The accord with Egypt was signed by Nixon and Egyptian President Anwar Sadat in Cairo June 14. The agreement was contained in a four-section joint statement. It said both governments would start negotiations soon on "an agreement for cooperation in the field of nuclear energy under agreed safeguards. Upon conclusion of such an agreement, the United States is prepared to sell nuclear reactors and fuel to Egypt." Meanwhile, a provisional pact was to be worked out at the end of June to enable Egypt to purchase nuclear fuel to produce electricity for industrial purposes.

White House Press Secretary Ronald Ziegler said the agreement would not enable Egypt to develop military nuclear capability.

Officials of the U.S. State Department and the Atomic Energy Commission expressed confidence that the nuclear assistance to Egypt would not lead to that country's possession of nuclear weapons. On the basis of an AEC briefing, Rep. Melvin Price (D, Ill.) said Egypt had pledged that it would not use the fissionable materials for peaceful nuclear explosions.

The U.S. had similar cooperative atomic agreements with about 35 countries, including Israel.

The arrangement with Israel was made during a two-day visit by Nixon.

A comprehensive communique issued June 17 before Nixon's departure said Israel and the U.S. would soon negotiate an agreement in the field of "nuclear energy, technology, and the supply of [nuclear] fuel from the United States under agreed safeguards," similar to the pact concluded with Egypt June 14. A provisional agreement was to be negotiated later in June that would provide for "the further sale of nuclear fuel to Israel," the statement said.

Nixon briefed Congressional leaders on his Middle East trip June 20. He assured them he had not reached any secret agreements or understandings with the Arabs or Israel and sought to ease any Congressional apprehension about the U.S. nuclear aid agreements with Egypt and Israel. Sen. Robert C. Byrd (D, W. Va.) later told newsmen that he was "concerned about possible misuse" of nuclear reactors the U.S. would provide under the accords. The President justified this assistance on the ground that it would encourage Egypt to work with the U.S. toward peaceful relations with Israel, Byrd said.

Egypt & Israel sign pacts—Provisional contracts providing Israel and Egypt with $78 million worth of uranium fuel for U.S. nuclear reactors were signed in Washington June 26. Down payments of $660,000 were received from Egypt and $726,000 from Israel. The remainder was to be paid over a 10-year period.

Both nations were to receive initial shipments of 115,000 pounds of uranium each. Each nation was to get a total of 184,000 pounds of uranium. The contracts were provisional, depending on the type of safeguards the U.S. was able to negotiate with both countries.

Atomic Energy Commission (AEC) Chairman Dixy Lee Ray signed for the U.S., Electric Power Minister Ahmad Sultan for Egypt and Ariel Amiad, managing director of the Israel Electric Corp. Ltd, for Israel.

U.S. sets conditions on aid—U.S. negotiations with Egypt and Israel on providing atomic assistance had slowed as the result of a new American proposal requiring both countries to agree to place all future nuclear facilities under international inspection as a condition for receiving U.S. power plants, the State Department disclosed Oct. 1. Egypt favored the proposed controls, while Israel expressed reservations.

The U.S. had previously required inspection by the International Atomic Energy Agency (IAEA) only over the atomic reactors and fuels the U.S. supplied to a foreign country. In the case of Egypt and Israel, the U.S. was proposing a broader agreement requiring international inspection of all atomic power plants and fissionable material the two countries might receive in the future from any nation to prevent the plutonium produced as a by-product from being diverted into the manufacture of atomic weapons.

U.S. officials regarded the international control issue as the principal reason for Israel's delay in responding to the proposal. Israel was wary of such controls since it would open up to inspection its Dimona reactor, which was capable of producing enough plutonium for a few atomic bombs a year. It had only reluctantly agreed to token inspection of the Dimona facilities by U.S. officials.

Israel informed the U.S. that it was not then interested in Washington's offer of an atomic power plant, but dropped its objections to Egypt's receipt of one from the U.S., it was reported by the U.S. State Department Dec. 16.

The Israeli position was outlined by Foreign Minister Yigal Allon in talks with Secretary of State Henry A. Kissinger in Washington Dec. 9. Officially, Israel made it known that it was not prepared to make the relatively large commitment required in building the atomic power plant. Administration officials, however, were said to believe the reason for Israel's lack of interest was its reluctance to place all its atomic facilities under international inspection.

The State Department said Dec. 17 that the U.S. was prepared to provide Egypt with a nuclear reactor if a parallel sale to Israel did not materialize. At the same time, the U.S. was ready to supply reactors to the two countries "in tandem or separately," a spokesman said.

A Soviet publication had reported Dec. 2 that Moscow had agreed to provide Egypt with a 460,000-kilowatt nuclear reactor. The report by the Soviet Weekly, an English-language publication, did not mention whether the accord called for safeguards against the production of nuclear weapons, as did the U.S. proposal to Egypt and Israel.

Egyptian, Israeli & other U.S. A-aid. Rep. Melvin Price (D, Ill.), chairman of the U.S. Joint Congressional Committee on Atomic Energy, gave the U.S. Congress June 18, 1974 the following background data on U.S. involvement in Egyptian and Israeli nuclear development and on other U.S. nuclear agreements:

First. Egypt has a small research reactor obtained from the U.S.S.R. It is a 2MW thermal reactor which began operation in 1961. Such a reactor produces plutonium, but in such small amounts as to have no practical value as a source of material for weapons. The reactor is not subject to International Atomic Energy Agency safeguards.

Pursuant to a cooperative agreement with Israel, for nuclear research—which excludes power reactors—the United States furnished Israel with a small research reactor of 5MW thermal capacity using highly enriched uranium. Such a reactor does not produce a significant quantity of plutonium. This reactor began operation in 1960. It is subject to IAEA safeguards.

Israel also has a 26MW thermal natural uranium, heavy water moderated reactor, known as the Dimona reactor. This reactor was obtained from France, and began operating in 1963. Such a reactor is capable of producing up to about 8 kilograms of plutonium per year. It is not subject to IAEA safeguards.

Second. Egypt has signed, but not yet ratified, the Non-Proliferation Treaty. Israel is not a signatory to that treaty.

Third. Following is a listing of U.S. nuclear international agreements:

A. BILATERAL AGREEMENTS FOR COOPERATION IN THE CIVIL USES OF ATOMIC ENERGY

COUNTRY, SCOPE, EFFECTIVE DATE, AND TERMINATION DATE

Argentina: Research and Power; July 25, 1969; July 24, 1999.

Australia: Research and Power; May 28, 1957; May 27, 1997.

Austria: Research and Power; Jan. 24, 1970; Jan. 25, 2000.

Brazil: Research and Power; Sept. 20, 1972; Sept. 19, 2002.

Canada: Research and Power; July 21, 1955; July 13, 1980.

China, Rep. of: Research and Power; June 22, 1972; June 21, 2002.

Colombia: Research; March 29, 1963; March 28, 1977.

Finland: Research and Power; July 7, 1970; July 6, 2000.

Greece: Research; Aug. 4, 1955; Aug. 3, 1974.

India: Power (Tarapur); Oct 25, 1963; Oct. 24, 1993.

Indonesia: Research; Sept. 21, 1960; Sept. 20, 1980.

Iran: Research; April 27, 1959; April 26, 1979.

Ireland: Research; July 9, 1958; July 8, 1978.

Israel: Research; July 12, 1955; April 11, 1975.

Italy: Research and Power; April 15, 1958; April 14, 1978.

Japan: Research and Power; July 10, 1968; July 9, 2003.

Korea: Research and Power; March 19, 1973; March 18, 2003.

Norway: Research and Power; June 8, 1967; June 7, 1997.

Philippines: Research and Power; July 19, 1968; July 18, 1998.

Portugal: Research; July 19, 1969; July 18, 1979.

South Africa: Research and Power; Aug. 22, 1957; Aug. 21, 1977.

Spain: Research and Power; Feb. 12, 1958; Feb. 11, 1988.

Sweden: Research and Power; Sept. 15, 1966; Sept. 14, 1996.

Switzerland: Research and Power; Aug. 8, 1966; Aug. 7, 1996.

Thailand: Research; March 13, 1956; March 12, 1975.

Turkey: Research; June 10, 1955; June 9, 1981.

United Kingdom: Research; July 21, 1955; July 20, 1976.

United Kingdom: Power; July 15, 1966; July 14, 1976.

Venezuela: Research and Power; Feb. 9, 1960; Feb. 8, 1980.

Viet-Nam: Research; July 1, 1959; June 30, 1974.

30 agreements with 29 countries.

B. AGREEMENTS FOR COOPERATION WITH INTERNATIONAL ORGANIZATIONS

ORGANIZATION, SCOPE, EFFECTIVE DATE, AND TERMINATION DATE

European Atomic Energy Community (EURATOM): Joint Nuclear Power Program; Feb. 18, 1959; Dec. 31, 1985.

Euratom: *Additional Agreement to Joint Nuclear Power Program; July 25, 1960; Dec. 31, 1995.

International Atomic Energy Agency (IAEA): Supply of Materials, etc.; Aug. 7, 1959; Aug. 6, 1979.

*The 1955 research agreement with Denmark was allowed to expire on July 24, 1973; cooperation was folded in under the Additional Agreement.

Poland's first A-plant. Warsaw and Moscow signed an agreement on cooperation to construct Poland's first atomic power station, the Soviet newspaper Sotzialisticheskaya Industria said March 2, 1974. The 440,000-kilowatt plant would be built on the Baltic Sea, near Lake Zarnowiecki.

Eurodif A-fuel progress reported. Directors of Eurodif, the European consortium planning to build a uranium enrichment plant using the gaseous diffusion method, decided Feb. 8, 1974 to build the plant at Triscatin, near the French military nuclear plant at Pierrelatte, in the Rhone Valley.

The five nations participating in the consortium were France, Italy, Spain, Belgium and Sweden. However, Sweden announced March 20 that it would withdraw from Eurodif to join the rival European consortium, Urenco, which planned to build a plant that would use the centrifugal enrichment technique. The other participants in Urenco were Great Britain, West Germany and the Netherlands.

The Eurodif consortium signed a $500 million contract June 27 to supply Japan with a sizable amount of enriched uranium, beginning in 1980, for electric power companies. Under the agreement, Eurodif would become Japan's second largest supplier of enriched uranium, after the U.S.

French Industry Minister Michel d'Ornano announced Jan. 2, 1975 the signing in November 1974 of an accord under which the Iranian government would lend the French Atomic Energy Authority $1 billion for 15 years in exchange for a 10% stake in the Eurodif enrichment plant.

Iran and France would establish a joint company that would own 25% of the Eurodif plant, nearly half of France's 52.8% stake in the project. France would hold 60% of the new company, Iran 40%, and Iran would receive 10% of the enriched uranium from the plant, originally estimated to cost the equivalent of about $1.6 billion and now projected at more than $2.5 billion.

France had also agreed to form with Iran a second company to study the financing and building of a second uranium enrichment plant. The $1 billion loan was in addition to the equivalent amount Iran had begun to deposit with France as advance payment for two of the nuclear plants it was buying from Creusot-Loire, the French heavy industry group, and for other French equipment.

French nuclear fuel arrangements. The French State Atomic Commissariat and the counterpart Italian agency had signed a nuclear cooperation agreement to enable the two nations, with West Germany as a minority partner, to build a fast-breeder commercial experimental reactor, it was reported June 13, 1974.

A French breeder reactor plant attained its maximum output of 250,000 kilowatts March 13, three months after linking up with the state-owned Electricite de France. It was the world's first breeder plant to reach maximum output. A Soviet breeder reactor was already in commercial operation, but at reduced power.

France and the U.S.S.R. signed a five-year contract under which the Soviet Union would supply France with enriched uranium, the Journal of Commerce reported Nov. 25. France would provide the natural uranium to be enriched under the pact. First deliveries were to begin in 1979.

France sets ties with IEA. France had agreed to institute informal contacts with the 16-nation International Energy Agency (IEA), which it had refused to officially join, it was disclosed Dec. 22.

A U.S. official who had attended the IEA's meeting in Paris the previous week said the agency's chairman, Viscount Etienne Davignon of Belgium, had been authorized to brief Jean-Pierre Brunet, the French Foreign Ministry's chief economic officer, after each agency meeting. Brunet in turn would submit proposals for the agency's consideration that France regarded as essential in preparing for the projected conference between oil-consumer and oil-producing nations in 1975.

In another action aimed at closer French-IEA cooperation, President Valery Giscard d'Estaing permitted the agency to be established within the 24-nation Organization for Economic Cooperation and Development (OECD), of which France was a member. He also allowed the Executive Commission of the European Economic Community (EEC) to have observer status with the IEA. France, along with other OECD and EEC members, had the veto power to block cooperative action with other international agencies.

U.S. legislation. A bill amending the U.S. Atomic Energy Act was passed by the Senate July 11, 1974 and House Aug. 1 and was signed by President Gerald R. Ford Aug. 17.

The major alteration permitted the Atomic Energy Commission (AEC) to increase the amount of nuclear material it distributed to groups of nations for peaceful purposes. The amount could be increased above statutory ceilings if both houses of Congress did not disapprove the increase within 60 days.

The provision retained Congressional control over such distribution. The AEC had sought elimination of the requirement for statutory approval of the distribution levels.

The bill also permitted export of small amounts of nuclear material for peaceful purposes to countries not having a nuclear-cooperation pact with the U.S.

A bill allowing Congress to take a position on international agreements negotiated by the U.S. for cooperation in peaceful nuclear technology was approved by both houses of Congress Oct. 10 and by President Ford Oct. 26. The measure arose from Congressional concern over the Nixon Administration's decision in June to sell nuclear reactors and fuel to Egypt and Israel for peaceful purposes.

The bill applied to future proposed agreements involving reactors of more than five megawatts of thermal power and the fuel for the reactors. Such agreements would have to be submitted to Congress and would become effective unless both houses passed a concurrent resolution disapproving them. During House debate Oct. 10, Rep. Bob Eckhardt (D, Tex.) had objected to the procedure because a concurrent resolution was subject to presidential veto.

Nuclear reactors sold by Canada. Ottawa and Buenos Aires reached an agreement under which Argentina pledged not to use a new Canadian nuclear reactor to make explosive devices, the Canadian government said Sept. 24, 1974. Argentina had agreed to purchase a 600-megawatt Candu reactor in 1973, and financing on the project, providing for a $129.5 million Canadian loan, was determined in April.

South Korea and Canada signed an agreement for the South Korean Power Co.'s purchase of a Canadian Candu nuclear reactor, it was reported Jan. 30, 1975. The pact included stipulations that South Korea sign a treaty not to use Candu technology to manufacture explosive devices, as India had done in 1974.

Finnish-Soviet A-pact. Finland and the Soviet Union Oct. 16, 1974 signed a 10-year energy cooperation agreement providing for the delivery to Finland of two 440 megawatt nuclear power stations which would go into operation in 1981–82. The agreement was signed in Helsinki by Jermu Laine, Finland's foreign trade minister, and S.A. Skachkov, chairman of the Soviet State Committee for Foreign Economic Relations, during a visit to Finland Oct. 14–17 by Soviet President Nikolai Podgorny.

(The accord would raise to six the number of atomic power stations under construction in Finland, with the first to go into operation in 1976, the French newspaper Le Monde reported Oct. 17.)

The energy accord also included an agreement in principle, pending further negotiations, for the two nations to cooperate in building a 1,000 megawatt nuclear power station in Finland and a Soviet undertaking to provide uranium supplies for the stations it delivered. The U.S.S.R. also expressed willingness to double the quotas of its electricity supplies to Finland from four billion kwh a year beginning in 1980 to 8 billion kwh toward the end of the 1980s.

(Under the sixth five-year Finnish-Soviet trade agreement, signed Sept. 12 in Helsinki, the U.S.S.R. agreed to export to Finland 6.5 million tons of crude oil a year and 2 million tons of oil products, with electricity exports to increase to the 4 billion kwh envisaged in 1980.

(According to the annual report on the Finnish economy issued by the Organization for Economic Cooperation and Development, Finland ranked seventh worldwide in per capita energy consumption, with most of its fuel imported for energy-intensive, export-oriented industries—such as paper and metals manufacturing, it was reported Aug. 22.)

U.S. blocks U.S.S.R. A-plant purchase. Approval for West Germany's sale of a nuclear power station to the Soviet Union, agreed by Chancellor Helmut Schmidt in 1974, had been withheld by the U.S., supported by Great Britain, at meetings of a Paris-based coordinating committee of the Western allies, the New York Times reported Jan. 15, 1975.

Iranian atomic & related deals. The shah of Iran, visiting France June 24–26, 1974, placed orders for an estimated $4 billion to $5 billion worth of items over 10 years.

Under an agreement signed June 27 by foreign ministers of both countries, Jean-Pierre Fourcade and Hushang Ansari, Iran agreed to make a $1 billion advance deposit with the Bank of France to cover the orders, a payment that would substantially ease France's oil-related balance of payments deficit.

Under the contracts, France would supply Iran with five 1,000-megawatt nuclear reactors worth about $1.1 billion. Delivery would be completed by 1985. France would also supply the atomic fuel, train Iranian scientists and technicians and establish a nuclear research center. The French Foreign Ministry said the agreement had implied safeguards, apparently a reference to the clause pledging both parties to respect their "international obligations." Iran had signed and ratified the 1968 nuclear nonproliferation treaty which provided for a safeguard system maintained by the International Atomic Energy Agency.

The other agreements cited in a joint communique issued June 27 provided for France to build a subway in Teheran, supply a steel plant to be built by Creusot-Loire, build a liquified natural gas plant with other nations, participate in construction of a natural gas pipeline and build 12 large tankers through a French-led consortium. Other potential projects were envisaged, including the creation of a petrochemical industry in Iran and electrification of Iranian railroads. The communique also said Iran would increase its oil shipments to France, but did not specify the amount. (Iran currently supplied 12% of France's oil.)

At a press conference in Paris June 27, the shah confirmed that the French accord also involved military sales but declined to give details other than the purchase of fast motor boats. The shah also called for nationalization of the entire oil industry, which would permit transactions on a state-to-state basis. Defending the sharp increases in oil prices, he said, "We're trying to defend ourselves against your rampant inflation. You are going to blow up, and you're going to blow us up with you."

The shah also reaffirmed Iran's readiness "to turn our area into a non-nuclear zone, that is, an area where no nuclear weapons should be used or stored." (The Iranian government had denied June 24 that the shah had said Iran intended to have nuclear weapons soon. The shah had been quoted in the French business magazine Les Informations June 23 as saying Iran would have nuclear weapons "undoubtedly and sooner than is believed." Earlier in 1974, Iran had created an Atomic Energy Agency and a former president of Argentina's Atomic Energy Commission, Rear Adm. Oscar Armando Quihillalt, had assumed a role as an adviser on atomic energy.)

At a luncheon with French businessmen June 26, the shah had proposed construction of a natural gas pipeline from Iran to Europe, saying that Iran could supply 50% of West Europe's gas needs.

The U.S. Atomic Energy Commission signed an agreement to supply Iran with at least $130 million worth of nuclear fuel processing over the next 20 years, it was reported July 2. The agreement was subject to the signing of safeguard and inspection agreements to prevent the spread of nuclear weapons.

The U.S. and Iran March 4, 1975 signed an agreement providing for $15 billion in non-oil trade between the two countries and the expenditure by Iran of $7 billion for the purchase of six to eight large nuclear power plants from the U.S. in the next decade. The accord, described as the largest of its kind, was signed by Secretary of State Henry A. Kissinger and Finance Minister Hushang Ansary after two days of talks.

Of the $15 billion, an increase of about 30% over current Iranian spending in the

U.S., $5 billion would be spent on usual imports, $5 billion on arms and $5 billion on development projects in Iran.

Kissinger said the nuclear reactors for Iran would be subject to "the safeguards that are appropriate" under the Nuclear Non-Proliferation Treaty, which Iran had signed.

U.S. exports to Iran in 1974 had totaled $1.7 billion, including about $300 million in weapons and grain, while U.S. imports from Iran came to $114 million.

Australia agreed to the sale of uranium to Iran under favorable conditions, according to a joint communique issued in Teheran March 20 following a four-day visit to Iran by an Australian delegation headed by Deputy Prime Minister James Cairns. Other aspects of the accord included Iranian agreement in principle to finance joint ventures in agriculture, industry and mining in Australia.

U.S. approves uranium export. The U.S. Nuclear Regulatory Commission (NRC) April 18, 1975 approved the export of 1.4 million pounds of uranium for processing in Great Britain and the Soviet Union. According to the Wall Street Journal April 25, the U.S. State Department had determined that such a transaction would "not be inimical to the common defense and security" of the U.S. and its allies, and authorized the NRC to approve the uranium-export license sought by Edlow International Co. of Washington, D.C. The U.S. had previously banned uranium shipments and transshipments to the U.S.S.R.

The uranium oxide, or yellow cake, milled from uranium ore extracted from mines in Wyoming and New Mexico would be transformed into uranium hexaflouride in Britain and the hexaflouride gas would then be processed in the Soviet Union into pellets rich in uranium 235, the isotope which provided the power for nuclear electric plants. Enriched uranium also yielded plutonium, the prime ingredient in atomic bombs. The ultimate customer for the enriched uranium was West Germany's Kraftwerk Union AG, the nuclear unit of Siemens AG.

Uranium enrichment contract with UK—Great Britain's Central Electricity Generating Board said Feb. 25 that the Soviet Union would process 1,100 tons of British-supplied uranium into nuclear fuel for delivery between 1980-89.

The first Anglo-Soviet agreement of its kind, the contract was valued at about $42.4 million and had emerged from discussions held during Prime Minister Harold Wilson's trip to Moscow earlier in February.

Soviet A-plant for Libya. The Soviet Union May 30, 1975 signed an accord to provide Libya with a nuclear research capacity suitable for the development of irrigation systems, the Arab Revolutionary News Agency reported. The accord, specifying a nuclear reactor in the 2-10 megawatt range, had been signed for Libya in Moscow by Maj. Omar Abdullah Moheshi, the minister for planning and research.

Brazil-West German nuclear pact. West Germany June 27, 1975 signed an agreement to supply Brazil with a complete nuclear industry by 1990. The accord came in the face of strong opposition by the U.S., which argued that the contract would supply Brazil with the fuel technology to produce nuclear explosives.

Under the pact, worth at least $4 billion, West Germany would sell Brazil eight nuclear reactors which would produce a total of 10,000 kw of electricity, and would deliver a uranium enrichment plant, a fuel fabrication plant and facilities for reprocessing fuel waste products into plutonium. West Germany would also help prospect for and exploit Brazil's uranium deposits in return for access to the reserves.

The agreement was signed in Bonn by West German Foreign Minister Hans-Dietrich Genscher and his Brazilian counterpart, Antonio Azeredo da Silveira.

U.S. State Department spokesman Robert Anderson said June 27 that the U.S. "had concerns about aspects of the Brazilian-West German pact that could have a potential for contributing to the spread of nuclear weapons."

Brazil had not signed the Nuclear Non-Proliferation Treaty; West Germany had signed and ratified it.

As part of the contract, Brazil had agreed to submit to controls by the International Atomic Energy Agency to insure that nuclear fuel material obtained from the German facilities was not diverted for nonpeaceful purposes. U.S. officials had disclosed June 3 that Washington had been instrumental in winning these controls, saying a special treaty incorporating the IAEA safeguards would be signed. The officials, who also admitted the U.S. had sought to dissuade West Germany from making the sale, said Washington had refused to sell nuclear facilities to Brazil because the Latins insisted on inclusion of plutonium and uranium enrichment plants. These would give the Brazilians the capacity to make nuclear weapons. The controls accepted by Brazil did not extend to future Brazilian-built processing facilities.

Brazilian officials vowed that the nuclear facilities would be used solely for peaceful purposes. The nation, which lacked oil and coal resources, had depended mainly on hydroelectric power for industrial development.

Brazilian Energy Minister Shigeaki Ueki said West Germany would lend Brazil $1.5 billion to develop a nuclear industry, it was reported July 2.

Brazilian-French agreement—France and Brazil July 5 signed a $2.5 million contract under which Brazil would purchase a French experimental nuclear reactor to be used in research for the construction of a new type of fast neutron, or breeder, reactor. Breeder reactors produced more fuel than they consumed.

U.S. Developments

Reactor hazards found slight. According to a study released by the Atomic Energy Commission Aug. 20, 1974, the risks from nuclear power plant accidents were smaller than from other man-made or natural disasters. The study concluded that, given the 100 conventional water-cooled plants expected to be in operation by 1980 (51 were currently operating), the chance of an accident involving 10 or more fatalities was one in 2,500 a year; an accident involving 1,000 or more deaths carried a risk of one in one million a year.

AEC officials emphasized that the study dealt only with the safety of the commercial reactors themselves, and not with risks involved in mining, manufacturing or transporting nuclear materials.

The report stated that of the approximately 15 million persons expected to be living in the vicinity of the first 100 reactors, one might be killed and two injured in every 25 years. The study noted that power plant reactors could not explode like nuclear weapons because of the fuel used.

AEC Chairman Dixy Lee Ray said Aug. 20 that there was "no such thing as zero risk," but in terms of the study the nuclear industry "comes off very well." She said the risks cited in the study were acceptable and urged that plant construction be continued. Ray said the study had been commissioned by the AEC, but the agency did not "influence" its findings.

Reactions by critics of the nuclear industry were less favorable. Daniel Ford of the Union of Concerned Scientists contended that the study was, "for all practical purposes," an "in-house" effort by the AEC, which had consistently advocated increased nuclear production. Consumer advocate Ralph Nader labeled the report "fiction," expressing doubt that such statistical projections were valid or possible.

Both Nader and Ford asked why—if the industry was so safe—it did not accept full liability for payment of damages in the event of accidents.

The report was criticized in a document released Nov. 23 by the Union of Concerned Scientists (UCS) and the Sierra Club, a conservationist group, as speculative and unreliable.

Speaking for the two groups, Dr. Henry Kendall, a physicist, said in Washington that the AEC's safety claims "are a conceit based far more on their enthusiasm for the nuclear power program than on solid and convincing scientific proof." The UCS and Sierra statement said the AEC study prepared by Dr. Norman C. Rasmussen contained a number of flaws. The safety analysis used by Rasmussen to estimate the probability of an accident, they said, had been developed and then abandoned by the aerospace industry and the federal government because it had been found to drastically underestimate existing hazards.

The AEC report was also criticized for the low number of projected casualties based in part on the successful evacuation of persons living near an atomic plant. A major accident at a nuclear plant could kill or seriously injure 126,800 people, 16 times the casualties estimated by the AEC report, the scientists said. They emphasized that there were no "adequate plans or means to evacuate to a distance of 20 miles" in the event of a plant mishap. It could not be assumed, as the Rasmussen report contended, that an evacuation could be achieved while only 5% of the population was in automobiles and 90% were indoors.

The report said the AEC study also failed to consider the fact that plutonium, the more lethal fuel which industry hoped soon to use in its reactors, posed a far greater danger than that considered in the commission's own reactor safety study.

AEC member Saul Levine defended the Rasmussen report Nov. 24, saying that the scientists' criticism of the study's form of analysis would have been correct for the 1960s, but "we have advanced that methodology considerably ... our members are in touch with reality." The AEC, he said, "is confident that the techniques used in the Rasmussen report are the best available."

The U.S. Environmental Protection Agency (EPA) Dec. 4 lauded the methods used in the AEC study but said casualties could be "about 10 times higher than those estimated" in the AEC survey.

The EPA statement was contained in an AEC-requested analysis of the commission's 14-volume study completed in the summer. It said the commission's study was "an innovative forward step in risk assessment of nuclear power reactors."

A second report issued by the EPA expressed concern about the possible effects the plutonium-fueled reactors proposed by the AEC would have on the environment. Before this new fuel was used on a full scale, the problem posed by plutonium disposal should be resolved and a new accident survey should be completed, the EPA said.

Few penalties for violations reported. Reporting on a study of AEC records, the New York Times said Aug. 25 that the commission had imposed penalties on only a small fraction of nuclear installations at which violations had been found, despite the fact that many of the violations could have created significant radiation hazards.

According to records for the year ended June 30, the AEC found 3,333 violations in 1,288 of the 3,047 installations inspected. The commission imposed punishment in eight cases: license revocations involving two companies and civil penalties totaling $37,000 against six others.

Under the AEC's three-level classification of the seriousness of violations, 98 were in the top category (violations which had caused or were likely to cause radiation exposures in excess of permissible limits). In category two—defined as violations that, if not corrected, might lead to exposures above permissible limits—there were 2,132 violations. Category three—infractions involving documentation and procedural matters—included 1,103 violations.

The report also noted that a composite ratio of violations and penalties over the past five years was similar to that of the most recent fiscal year: 10,320 inspections; 3,704 installations with one or more violations; and 22 cases involving imposition of penalties.

A-accident bill vetoed. President Ford Oct. 12, 1974 vetoed a bill that would have extended the federal program for reimbursement of loss in the case of a nuclear power plant accident. Congress had adopted the bill by 376–10 House vote Sept. 24 and Senate voice vote Sept. 30.

The President opposed a provision allowing Congress to disapprove the extension if a committee recommended it after evaluation of a study. Ford said that was contrary to constitutional processes for signing legislation into law since it gave a Congressional panel authority to recommend nullification of the measure after signature by the President.

Safety data suppression denied. AEC Chairman Dixy Lee Ray denied Nov. 15 that the AEC had suppressed data on the safety of nuclear plants. She conceded that "while there may be some validity for such accusations in the past, the situation has changed today."

Ray's statement was in apparent response to a New York Times report

Nov. 10 citing AEC documents showing that since 1964 the commission had sought to conceal studies by its own scientists that found nuclear reactors were more dangerous than officially admitted or raised doubts about safety reactor devices.

Ray cited a number of examples to demonstrate the AEC's openness with the public: the release earlier in 1974 of 25,000 pages of documents developed during the deliberations of the Advisory Committee on Reactor Safety. Furthermore, she pointed out that in 1973 the AEC had made public "an uncompleted 1965 study" that sought to update a 1957 report on the consequences of a possible major accident at a large nuclear power plant. Such a mishap could kill up to 45,000 persons, and "the possible size of such a disaster might be equal to that of the state of Pennsylvania," the report said.

The cases cited by Ray were questioned by Daniel Ford, an official of the Union of Concerned Scientists, that had been critical of many AEC policies. Ford said the Advisory Committee released the documents only after being faced with a suit under the Freedom of Information Act, and that many of its pages had been heavily edited. As for the 1973 report, AEC memorandums had shown that the 1965 study had been deliberately withheld from publication by the commission for seven years after its completion and that this study also was finally released under the threat of a Freedom of Information Act suit, Ford said.

The New York Times report said some of the AEC's suppressed documents had been leaked to the Union of Concerned Scientists by commission officials.

Warning on hazards. A group of eight U.S. scientists warned against the possible dangers of nuclear reactors on public health, the environment and national security during a symposium in Washington Nov. 15–16. The experts suggested at the conclusion of a two-day conference, called Critical Mass 74, that leaders of the U.S. Senate and House investigate the hazards posed by the reactors. The meeting, organized by Ralph Nader, drew 650 experts from the U.S., Britain, France and Japan.

A petition drawn up by the eight Americans conceded that scientists had initially welcomed the advent of nuclear power as "a valuable energy source for mankind." But they noted that "this early optimism has been steadily eroded" by the growing realization of the problems posed by relying on commercial nuclear power plants. Among them were the possibility of a catastrophic reactor accident, the chance that some terrorists would steal radioactive materials and build a nuclear bomb and providing safe storage of nuclear wastes for hundreds of years.

Hazards at Oklahoma plant. A report released Jan. 7, 1975 by the AEC upheld charges by the Oil, Chemical and Atomic Workers Union of health hazards and other dangers at the plutonium processing plant of the Kerr-McGee Corp. in Crescent, Okla.

The report, based on the findings of AEC investigators, substantiated 20 of the 39 union claims of danger to the health of the firm's workers and said some X-ray negatives of fuel rods being manufactured for a plutonium-powered reactor had been falsified and that some data concerning the rods had not been used properly.

As a result of the inquiry, the AEC said it would file three charges against Kerr-McGee: failure to report to the AEC a processing equipment failure that had closed the plant for 40 hours, permitting on two occasions an excess amount of plutonium in a specific work area and using a small amount of plutonium in an unauthorized form.

Among the health hazards cited were the sending of a worker into a dangerous area without informing him that a respirator was required to protect him from possible radiation, failure to make certain that the respirators were working and "errors" that resulted in contamination. Since Kerr-McGee started its plutonium operations in 1970, 17 safety lapses in which 73 employes had been contaminated were reported, according to AEC records.

A previous AEC report released Jan. 6 said an employe and critic of Kerr-McGee's safety measures, Karen G. Silkwood, had swallowed microscopic amounts of plutonium seven days before she had been killed in a car accident Nov.

13, 1974. According to the AEC's findings, Miss Silkwood's contamination "probably did not result from an accident or incident within the plant"; two urine samples submitted by Miss Silkwood "contained plutonium which was not present when the urine was excreted" but the subsequent commission inquiry shed no light on the case.

Reactor schedule delay proposed. The Atomic Energy Commission (AEC) proposed Jan. 17 delay in the introduction of commercial breeder reactors from the 1980s to the 1990s because of "major financing difficulties" being encountered by electrical utilities.

The agency pointed to continued development of the liquid metal fast breeder

NUCLEAR POWER UNITS IN OPERATION

(As of Dec. 31, 1974)

Name of System or Sponsor	Name or Location	Name-Plate Rating, Kw	Type of Reactor	Placed in Commercial Operation
Duquesne Light Company	Shippingport No. 1	100,000	Pressurized Water	1957
Commonwealth Edison Company	Dresden No. 1	208,675	Boiling Water	1960
Yankee Atomic Electric Company	Yankee No. 1	185,000	Pressurized Water	1961
Consolidated Edison Company of N.Y., Inc.	Indian Point No. 1	275,000	Pressurized Water	1962
Consumers Power Company	Big Rock Point	75,000	Boiling Water	1962
Pacific Gas & Electric Company	Humboldt Bay	60,000	Boiling Water	1963
Washington Public Power Supply System	Hanford	800,000	Graphite Moderated, Water Cooled	1966
Philadelphia Electric Company	Peach Bottom No. 1	46,000	High Temperature, Helium Cooled and Graphite Moderated	1967
Connecticut Yankee Atomic Power Company	Haddam No. 1	600,300	Pressurized Water	1968
Southern California Edison Company and San Diego Gas & Electric Company	San Onofre	450,000	Pressurized Water	1968
Jersey Central Power & Light Co.	Oyster Creek No. 1	550,000	Boiling Water	1969
Niagara Mohawk Power Corp.	Nine Mile Point No. 1	641,750	Boiling Water	1969
Rochester Gas and Electric Corp.	Ginna No. 1	517,140	Pressurized Water	1969
Northeast Utilities	Millstone Point No. 1	661,500	Boiling Water	1970
Wisconsin Michigan Power Company	Point Beach No. 1	502,841	Pressurized Water	1970
Carolina Power & Light Company	Robinson No. 2	739,328	Pressurized Water	1971
Dairyland Power Cooperative	La Crosse	60,000	Boiling Water	1971
Northern States Power Company	Monticello No. 1	542,730	Boiling Water	1971
Commonwealth Edison Co.	Dresden No. 2	810,000	Boiling Water	1971
Boston Edison Co.	Pilgrim No. 1	655,397	Boiling Water	1972
Commonwealth Edison Co.	Quad Cities No. 1	810,000	Boiling Water	1972
Commonwealth Edison Co.	Quad Cities No. 2	810,000	Boiling Water	1972
Florida Power & Light Co.	Turkey Point No. 3	728,300	Pressurized Water	1972
Maine Yankee Atomic Power Co.	Wiscasset	830,000	Pressurized Water	1972
Vermont Yankee Nuclear Power Corp.	Vernon	537,261	Boiling Water	1972
Virginia Electric and Power Co.	Surry No. 1	822,600	Pressurized Water	1972
Commonwealth Edison Co.	Dresden No. 3	810,000	Boiling Water	1973
Consumers Power Co.	Palisades No. 1	810,000	Pressurized Water	1973
Duke Power Co.	Oconee No. 1	886,667	Pressurized Water	1973
Duke Power Co.	Oconee No. 2	886,669	Pressurized Water	1973
Florida Power & Light Co.	Turkey Point No. 4	728,317	Pressurized Water	1973
Northern States Power Co.	Prairie Island No. 1	559,600	Pressurized Water	1973
Omaha Public Power District	Fort Calhoun No. 1	481,477	Pressurized Water	1973
Tennessee Valley Authority	Browns Ferry No. 1	1,098,420	Boiling Water	1973
Virginia Electric and Power Co.	Surry No. 2	822,600	Pressurized Water	1973
Wisconsin Electric Power Co.	Point Beach No. 2	502,841	Pressurized Water	1973
Arkansas Power & Light Co.	Arkansas Nuclear No. 1	909,748	Pressurized Water	1974
Commonwealth Edison Co.	Zion No. 1	1,085,391	Pressurized Water	1974
Commonwealth Edison Co.	Zion No. 2	1,085,391	Pressurized Water	1974
Consolidated Edison Company of N.Y., Inc.	Indian Point No. 2	1,021,800	Pressurized Water	1974
Duke Power Co.	Oconee No. 3	893,271	Pressurized Water	1974
Iowa Electric Light & Power Co.	Arnold No. 1	565,700	Boiling Water	1974
Metropolitan Edison Co.	Three Mile Island No. 1	837,478	Boiling Water	1974
Nebraska Public Power District	Cooper Nuclear No. 1	800,838	Boiling Water	1974
Northern States Power Co.	Prairie Island No. 2	559,600	Pressurized Water	1974
Philadelphia Electric Co.	Peach Bottom No. 2	1,098,305	Boiling Water	1974
Philadelphia Electric Co.	Peach Bottom No. 3	1,098,305	Boiling Water	1974
Wisconsin Public Service Corp.	Kewaunee No. 1	563,101	Pressurized Water	1974

From "1974 Year-End Summary of the Electric Power Situation in the United States" (Edison Electric Institute)

reactor (LMFBR). The nation's electrical energy needs were growing "to such an extent that existing methods of energy production will be insufficient to meet the requirements," it said. "Additional electrical energy technology options will need to be developed if the nation is to be assured a secure energy supply. The LMFBR is a well-advanced technology which has reached the demonstration plant stage . . . [and] can be developed as a safe, clean, reliable and economic electric power generation system."

The proposal was contained in the AEC's final environmental statement. The agency went out of existence Jan. 19 under legislation replacing it with a new Energy Research and Development Administration and Nuclear Regulatory Commission.

Continued development of nuclear power as a commercial energy source was recommended by 34 scientists in a statement read in Washington Jan. 16 by Dr. Hans A. Bethe of Cornell University. The energy crisis, they said, "is the new and predominant fact of life in industrialized societies" and there was no currently viable alternative to increased use of nuclear power to meet energy needs. "The U.S. choice is not coal or uranium; we need both," the statement said. It also deplored "the fact that the public is given unrealistic assurances that there are easy solutions."

A statement issued the same day by Ralph Nader's Union of Concerned Scientists cautioned that "a wide range of public safety problems must be resolved before nuclear power plant construction proceeds in the U.S."

23 Atomic plants ordered closed. The U.S. Nuclear Regulatory Commission (NRC) Jan. 29 ordered the shutdown of 23 reactors within the next 20 days to check for possible defects in their power plants. The action was prompted by the discovery of five small cracks in the emergency cooling system of an atomic reactor operated by the Commonwealth Edison Co., near Morris, Ill. The leak did not result in the release of any radioactivity to the environment, the NRC said.

All 23 plants were of the boiling-water reactor type.

The NRC announced March 7 that 21 of the plants had passed safety inspection but that another crack had been found at the Morris plant and one other reactor inspection, also at Morris, was deferred.

At a Congressional hearing Feb. 5, NRC officials had said the cracks' cause was unknown but they considered the shutdown of the plants for the inspection "appropriately prudent." Several senators at the hearing questioned the reliability of safety inspections conducted by public utilities operating the plants.

Another witness, Daniel Ford, executive director of the Union of Concerned Scientists, a public interest environmental group, disagreed. He questioned the fitness of the regulatory licensing procedures, in view of the emergency shutdowns, as well as the safety and reliability of reactors themselves.

Documents made public by the Nuclear Regulatory Commission at Ford's request disclosed concern within the agency about the safety issue. The New York Times published March 9 a report on a policy study by the NRC's Edwin G. Triner concluding that utilities owning most of the nuclear reactors were not sufficiently concerned about safety and performance. A N.Y. Times report published April 6, also based on the newly released material, revealed that several NRC scientists and technicians viewed the test for the cracks in the reactors as unreliable.

A different nuclear power reactor safety issue was raised by a fire March 22 in an Athens, Ala. plant operated by the Tennessee Valley Authority. The fire, caused by a candle held by an electrician hunting an air leak, closed down, for an estimated two or three months, the world's largest nuclear generating station. One of the two reactors lost some of its nuclear core cooling water and an emergency cooling system was disabled. Other pumps were operated to stabilize the reactor-core situation and no radiation was said to have escaped.

The Baltimore Gas & Electric Co.'s nuclear power plant in Calvert Cliffs, Md. had announced Jan. 25 it would close Feb. 1 because of unacceptable radiation leaks from the door to the reactor container. The NRC had reported Jan. 24 that the emissions were from three to 10 times above federal standards.

A-plant near populated area barred. The U.S. Seventh Circuit Court of Appeals in Chicago April 1, 1975 barred the building of a 660-megawatt A-power plant near Portage, Ind. on the southern shore of Lake Michigan. The court ordered the site to be filled in by Northern Indiana Public Service Co., which had been granted a license for the plant by the Atomic Energy Commission.

The court said the AEC had violated its own rules in licensing a plant within two miles of a densely populated area, defined as 25,000 or more people. The population of Portage was projected to exceed that by the time the plant was scheduled for operation, and the court noted that a state park abutting the site had up to 87,000 visitors a weekend.

The court was critical of the AEC for "clustering" nuclear plants on the Lake Michigan shore within relatively short distances of Chicago; eight were within 75 miles of downtown Chicago; six more planned, in addition to the Portage site.

The court acted on a suit to halt construction by a unit of the Izaak Walton League and a business civic group.

Plutonium decision delayed. The Nuclear Regulatory Commission (NRC) announced May 8 it would postpone for at least three years a final decision on whether plutonium should be used to help power American nuclear reactors. The predecessor regulatory agency, the Atomic Energy Commission, had recommended in August 1974 that plutonium be used to supplement uranium as a fuel in boiling-water reactors as a way to reduce the cost of producing electric power.

Some experts, such as Dr. Theodore Taylor, a former AEC scientist, contended that widespread use of plutonium could be dangerous because it could be converted without much sophisticated training into homemade weapons.

The theory was put to the test by television producer John Angier, who asked a 20-year old undergraduate student at Massachusetts Institute of Technology to see if he could develop a nuclear bomb in his spare time without talking to experts and using only publicly available reference works. He was given five weeks for the assignment. His plan was submitted for evaluation to a nuclear scientist at the Swedish defense ministry, who concluded, after consulting several colleagues, that there was a "fair chance" the device could explode, with a yield range of from 100,000 to two million pounds of TNT.

The student reported that designing and building the bomb, assuming one had the plutonium, was not much harder "than building a motorcycle." He thought the bomb could be made for $10,000 to $30,000. It would be about the size of a large desk, weigh 550-1,000 pounds and require 11-19 pounds of plutonium.

A copy of his plan was sent to the AEC. The only other copy was destroyed, along with the notes of the student, who had demanded anonymity.

The test gained national prominence by broadcast on public television in March as part of a scientific series called Nova.

A bill to prohibit the NRC from licensing the widespread use of plutonium as a reactor fuel unless expressly authorized by Congress was introduced in the House Feb. 26 by Rep. Les Aspin (D, Wis.), who cited the potential of making a homemade nuclear bomb from plutonium.

The Environmental Protection Agency had recommended further study of the safety and radioactive-waste disposal problems of a plutonium-breeding nuclear reactor April 27. While it was not "necessarily advocating a delay" in development of such a reactor, it said, "sufficient evidence exists to warrant reexamination" of the timing of the breeder program. The suggestion in the report was for a delay of from four to 12 years.

The EPA was critical of some AEC projections—of the probable growth rate of electric power demand, which the EPA found overstated; of the chance of a major accident being one in 10 million, which the EPA called premature; and of the time and effort needed for adequate resolution of environmental and safety problems, which the EPA deemed "highly optimistic."

A study commissioned by the American Physical Society found no reason April 28 for "substantial short-term concern" about the safety of U.S.

nuclear reactors. But it recommended "a continuing effort to improve reactor safety as well as to understand and mitigate the consequences of possible accidents."

The study, by 12 independent physicists, found the safety record so far "excellent," with "no major release of radioactivity." But the group did express concern about the long-term, when an increasing number of reactors would be operating and the likelihood of an accident, although "improbable," became correspondingly greater.

The physicists agreed with the old AEC estimate that the chance of a major accident in an atomic power plant was one in 10 million. But they disagreed with the AEC's estimate that an accident could cause about 310 deaths from cancer. There would be "substantial long-term consequences" from an accident in which radiation was released over a populated area, the panel thought. The AEC estimate was said to be off by as much as a factor of 50 because the cloud of radiation could move 500 miles and endanger people in a 10,000–20,000 square-mile area.

Ford for private A-fuel plants. President Ford asked Congress June 26, 1975 to open the nuclear fuel field to private industry. The government had had exclusive operation in that area since the birth of the atomic bomb.

Under the President's suggestion: Private industry would be permitted to build and operate new uranium enrichment plants, where uranium was processed into fuel for atomic plants producing electrical power. The government would sell its enrichment technology for royalties. The government would step back in on projects that failed because of regulatory or financial problems, the private participants would be paid off and the facility built by the government. Congress was to have veto power over construction plans.

The proposal contained provisions to guarantee the secrecy of the enrichment technology and to subject exports to safeguards and controls. The plants built in the U.S. by private enterprise would be controlled and regulated by the federal government.

Much of the world's enriched uranium currently was produced in three U.S. plants constructed originally for the weapons program. They were operated by private contractors for the Energy Research and Development Administration (ERDA). The plants were operating at capacity and no new orders had been taken since 1974. The Administration proposed to retain government ownership of these three plants.

The President told Congress more nuclear fuel plants were urgently needed "if we are to insure an adequate supply of enriched uranium for the nuclear power needs of the future and if we are to retain our position as a major supplier of enriched uranium to the world."

He said his proposal would result in "enormous savings to the American taxpayer." Other arguments for private entry into the field were that the market for nuclear fuel was "predominantly in the private sector" and the process of uranium enrichment was "clearly industrial in nature."

ERDA had already received various proposals from private industry for building such plants.

Country-by-Country Developments

AUSTRALIA

Uranium mining set. Minerals and Energy Minister Reginald F.X. Connor said Oct. 31, 1974 that Australia would begin mining and processing uranium in the Northern Territory and would lift a 12-year ban on exports to foreign countries.

In a report to Parliament, Connor said initial production would be from the Ranger deposits, to be developed jointly by the government's Atomic Energy Commission and a private firm. A uranium treatment plant was to be established there at a cost of $A100 million. The first shipment of the mineral would go to Japan under contracts already approved, Connor said.

Enrichment advance. Connor told Parliament May 28, 1975 that Australia had made sufficient progress to "go it alone" in commercial development of the gas centrifuge method of uranium enrichment.

Conner said he believed his country was "well on the high road to ranking with the best in the world in this field" and that "the results which we will be able to show in the very near future will be outstanding."

CANADA

New uranium policy announced. Donald S. Macdonald, minister for energy, mines and resources, announced new Canadian export policies Sept. 5, 1974 designed to protect the domestic uranium market. "We're perhaps forfeiting some future export possibilities," he said, "but we're not cutting back on exports." The regulations, effective immediately, established a reserve formula to guarantee supplies for domestic use for the next 30 years with the nation's nuclear reactors operating at 80% capacity.

Canada had six nuclear reactors in operation, four under construction and another 13 planned. According to official figures, domestic needs were expected to rise from the present rate of 500 tons of uranium annually to 6,000 tons by 1990. Canada's known reserves totaled about 400,000 tons, of which some 5,000 tons were being exported annually. International demand was expected to increase 4–6 times within the next decade, Macdonald said.

Features of the new policy:

- Canadian suppliers of uranium would be required to limit export commitments to 10 years, with a five-year extension possible with government approval.
- Foreign purchasers would be forbidden, without the specific authority of the Canadian Atomic Energy Control Board, to re-export uranium bought from Canada.
- Part of the uranium exported would be subject to recall if needed for Canadian domestic use.
- A government-owned stockpile of about 7,500 tons of uranium would be disposed of only on the domestic market, contrasting with a stockpile sale in 1973 to overseas customers for undisclosed sums.
- Uranium would be offered for export in its most advanced possible form (i.e., fabricated nuclear fuel, enriched uranium or, for the U.S. market, electricity) so that Canada could profit from economic advantages of domestic processing.

It was reported July 17 that Canada had assured Japan a 30-year supply of uranium if Tokyo purchased Ottawa's Candu nuclear reactor.

According to a June 16 report, Canada had suspended its nuclear assistance to Nationalist China.

GREAT BRITAIN

U.S.-style A-reactor opposed. A report by a parliamentary committee appointed in November 1973 advised the government Feb. 4, 1974 not to switch to U.S.-designed nuclear power stations for Britain's next generation of reactors. Citing possible safety hazards of the U.S.-style light-water or pressurized water reactors, the document favored reliance on the advanced gas-cooled reactors currently being built in Britain and based on proven British technology. (Britain's Central Electricity Generating Board had recommended in 1973 that the U.S.-designed reactors be built in Britain under license.)

JAPAN

2 A-plants planned. The Electric Power Development council July 4, 1974 proved construction of 22 electric power plants, including two nuclear facilities, that would have a capacity of 5.27 million kilowatts.

The council acted as an estimated 300 persons demonstrated outside the building against construction of the nuclear plants in fear of possible radioactive effects. The nuclear plants were to be built in Tokyo and Kyushu to bring the number of Japan's nuclear power generators to eight.

THE NETHERLANDS

A-stations planned. Economics Minister Ruud Lubbers announced that the Netherlands would build three more nuclear power stations of 1,000 megawatts each by 1985 (reported Sept. 28, 1974). He said the new stations would reduce annual crude oil imports by five million tons, aiding the balance of payments by more than $333 million a year.